PLANT CONSERVATION

PLANT CONSERVATION

A NATURAL HISTORY APPROACH

EDITED BY GARY A. KRUPNICK AND W. JOHN KRESS

With a foreword by Daniel H. Janzen

THE UNIVERSITY OF CHICAGO PRESS · CHICAGO AND LONDON

Gary A. Krupnick is director of the plant conservation unit in the
Department of Botany at the National Museum of Natural History, Smithsonian Institution.
W. John Kress is a research scientist and chairman of the Department of Botany at the
National Museum of Natural History, Smithsonian Institution.

The University of Chicago Press, Chicago 60637
The University of Chicago Press, Ltd., London
© 2005 by The University of Chicago
All rights reserved. Published 2005
Printed in the United States of America

14 13 12 11 10 09 08 07 06 05   1 2 3 4 5

ISBN: 0-226-45512-2 (cloth)
ISBN: 0-226-45513-0 (paper)

No copyright is claimed for the preface, the chapter introductions,
chapters 2.2, 3.1, 4.1, 5.3, 5.5, 5.7, 5.9, 6.4, 6.5, 10.1, 10.2, 11.2, 12.2, 12.3,
and the conclusion.

Library of Congress Cataloging-in-Publication Data

Plant conservation : a natural history approach / Edited by Gary A. Krupnick and W. John Kress ; With
a foreword by Daniel H. Janzen.
    p. cm.
Includes bibliographical references and index.
ISBN 0-226-45512-2 (cloth : alk. paper) — ISBN 0-226-45513-0 (pbk. : alk. paper)
1. Plant conservation. 2. Plant diversity. I. Krupnick, Gary A. II. Kress, W. John.
QK86.P58 2005
639.9'9—dc22

                                                                2005000678

This book is printed on acid-free paper.

CONTENTS

Color plates follow page 158.

FOREWORD

Daniel H. Janzen

HOW TO CONSERVE wild plants? Give the world the power to read them.

If you are illiterate, a library is at best a tempting storehouse of thin sheets of firewood. If you are bioilliterate, that great green mass is simply something in the way, something to be replaced, turned to compost, or fed to the inanimate and animate extensions of the human genome.

Krupnick and Kress have herded a fine batch of botanical cats into a collective review of where the plant priesthood stands on the natural history of plant conservation. I would like to hope that every graduate student and practitioner of wild-plant biology invests a weekend in reading it from cover to cover. But my nonbotanist lifetime of enjoying, using, and relying on many thousands of species of tropical plants leaves me feeling that this book, and all this book's antecedents and parallels, constitutes a massive body of priceless information that realizes only a very small portion of its potential. It is as though we had the Library of Congress and all the world's other massive repositories of the written word forlornly dotted across a firmly illiterate society.

I am standing in a Costa Rican rain forest, magnificent subjects of this book in all directions and underfoot. There are a thousand species of plants within a long stone's throw. Nearly every one of them is a described species with a proper scientific name, a handle that you can plug into Google and come up with something. Nearly all of these species or their near relatives have been studied, sampled, and thought about, and are in "the literature." And I cannot identify a single species.

Oh yes, I can collect a branch, press and dry that branch, take it to INBio or the Missouri Botanical Garden or the U.S. National Herbarium or Kew or the many other dead plant depositories. I can take that branch to a botanical library and slog my way through keys and descriptions with a vocabulary of abaxial, tomentose, spicate, cespitose, and cornifoliar stipules. I can seduce a world-level authority into spending a day in the forest and peering at the branch. I can wait 2.3 years until the branch flowers, put a digital image on the Web, and plead for someone to give me the name—which they will from some far distant country. Now multiply all that by 1,000. Stalemate.

I am a user. I use plants. They mean something to me. They are the food for the caterpillars that I inventory. They are the fruits for the birds that feed my caterpillars to their nestlings. They are the puzzling sources for amazing lead molecules such as the caffeine you had for breakfast. They are the personal bank account and investment portfolio of the ocean of landowners that surrounds the conserved wildlands I and many others struggle to facilitate into survival through nondamaging biodiversity development. We need to be able to read them—now, on the spot, in real time—but we cannot. They are to me as the bright colors on the walls of an art museum are with no guide, no brochure, no labels. And you cannot see the morphine, caffeine, turpentine, lignins, and genes that change from book to book. It's all green. Small surprise that their destiny is that of firewood and biomass shoveled into the craw of the great consumer.

So we have on the one hand a great, burgeoning and decaying body of academic—with some practical flavors—knowledge and experience about wild plants, encapsulated in the world's herbaria, botanical gardens, libraries, URLs, and plant biologists' brains. We have on the other hand, fortunately, some still substantial amounts of wild plant diversity, especially, but certainly not exclusively, in the tropics. And standing between the two, 99.999% bioilliterate, we have 5 billion people. And when any one of those people wants to do something with one of those plant species, now for real purposes—with the potential of leading that person or some greater part of society to want to see that plant species survive—that person quite simply cannot now, in real time, get the name for that plant. The name, the unique identifier. Why the name? Because without a name it is just another green blob. Without a name there is no way into the collective knowledge of specific plants. Try using Google without a name. Type in green and brown, 3 m tall, sawtoothed leaf margins, growing on wet soil, at 10 degrees north, 85 degrees west, and see what you get. Or try to look that up in *Index Kewensis*.

It is time for a zoological observation. It is fascinating to now know, through the agility and miracle of sequencing, that hippos are the ancestors of whales. But what I want to know is how to tell apart all the species of whales when I am standing in a supermarket looking at a whale steak, or standing on a hot tropical beach wondering whether to call the vultures or the Smithsonian to the newly stranded whale. Sequencing has another function besides revealing phylogeny.

Imagine what it would do to any and all aspects of human interactions with wild plants if you—any of you—could walk up to any plant anywhere—seedling, sapling, 40-m tree, grass, root, pressed leaf, or fallen log—and know in a few seconds its scientific name. I need not describe how today's technology would then let you use that name to get into the warehouse of collective botanical knowledge. That capacity would transform far more than the science of plant biology, the conservation of plants, and the superficial ways we currently make use of the incredible diversity of form, physiology, genetics, and chemistry of plants. It would be to plants what the printing press was to stories, education, science, law, medicine, and communication.

Science fiction? Yes, it has been for many decades if not centuries. Step out of the space ship and point the gadget, and it says—friendly, unfriendly, poisonous, curative, edible, spit it out. But no more. Today we have the capacity to essentially barcode animals with a 650-base-pair sequence of the *CO1* mitochondrial gene, build libraries of those sequences, and miniaturize the sequencing process into a cheap, fast, hand-held gadget. One for the belt or backpack of anyone on the planet. Thirty million sequences in a chip the size of your thumbnail. Thirty million pieces of collateral information in another chip. And the cell-phone uplink for a dialogue with Google and its equivalents. The emergence of this process is now in motion, in its nascent germination through initiatives bubbling up in the free-standing museums and pilot projects spearheaded by Paul Hebert at the University of Guelph (e.g., Godfray 2002; Hebert et al. 2003a; Hebert et al. 2003b; Stoeckle 2003; Janzen 2004; and see Barrett et al. 2004) and the marine biodiversity inventory initiative (see Stoeckle et al. 2004).

Unfortunately, the *CO1* mitochondrial gene sequence that works so well with animals does not appear to be a good species-level unique identifier for plants. Additionally, all those neat secondary compounds long evolutionarily designed to gum up the inner workings of herbivores—be they fungi, bacteria, or animals—are quite unpleasant for sequencing engineering. However, my plea is single-minded. It should take only a minute fraction of the energy that has been invested in plant phylogenies and within-population variation to locate the "right" gene sequence to serve as a species-level unique identifier for plants, so that the engineering process can get on with creating a hand-held barcorder for plants as one and the same as the hand-held barcorder for animals and fungi. Perhaps it is already known—the usable *CO1* gene sequence for animals was not a new discovery, to say the least, but rather the novelty was Hebert's decision to use it as a species-level identifier.

Yes, we can still fill tomes arguing whether *Sweitenia macrophylla* is one or ten species, or whether the rain forest population of *Cordia alliodora* is the same species as the dry forest population of *Cordia alliodora*. But with a barcorder, I—and the farmer, forester, conservation biologist, ecologist, taxonomist, biodiversity prospector, ecotourist—can know that this sapling of *S. macrophylla* is not one of its ten sympatric look-alikes, and that it is indeed *C. alliodora* rather than any one of the other 15 species of sympatric *Cordia*. And since I probably know where I am, my barcorder does not have to tell the *Cordia* in front of me from the 100+ species growing in other countries. Now multiply all that by the 13,000 species of plants in Costa Rica. Or the 400,000 in the world.

I—a hard-core field biologist and biodiversity developer dependent on being able to identify wild living plants on the spot—have always viewed, and used, herbaria and their constructors as incredible libraries of plant biodiversity. But the need for a barcorder serendipitously multiplies their value enormously. What if you could snip a square centimeter of paper out of each book in the world's libraries and instantly have a hand-held index to all books (and if properly linked, the readable contents of those books)? In my not-so-humble opinion, the vastly greatest contribution to plant conservation—and all the rest of wild plant science—that could be made by the botanical community at this time would be to figure out the right gene (or two, if absolutely necessary), determine the methodological tricks to get its sequence quickly and cheaply out of most herbarium specimens, and join that proactively to the above-mentioned initiative to create and arm the barcorder. It is a no-brainer that every subsequent sequence from a plant at the time of collection, and from all the users who are motivated enough to allow the georeferenced sequences they obtain with their barcorder to be included, will rapidly expand the geographic coverage and functionality of such a plant sequence library.

Lean back and imagine. You step into that great green living mass, on your way home from school, as you file your chainsaw, as you taste your samples. A chip of tissue into the gadget in your hand tells you its name, now. You ask first those plants that to you have some salient trait. Soon many are old friends with names, the basis for communication. And when/if your chip of collaterals is large, or you have Google or whatever access through your cell phone, you can let your fingers do the walking.

What is missing from this scenario? The hardest part. Think about bioprospecting for rain forest drugs to support conservation. If you could put a penny tax on every cup of coffee—we make 3 trillion a year—you could pay all conservation and taxonomic costs for all of the world forever. Caffeine is *the* number one rain forest drug. The trick is not finding *Coffea arabica* or the technology of the cup of coffee, or even its medical ramifications. The trick is getting the penny tax and having it go to conservation and the taxonomy behind it. At this late stage, impossible.

Yes, there will be huge gains in bioliteracy and all that entails from creating the hand-held barcorder. But once again we are on the brink of producing something extremely useful to global society, like a cup of coffee, while leaving the profession—botanists, other biologists, taxonomists, conservationists, and all the other pro bono users—without a budget to gather and maintain the information the user links to once the plant is identified, and to massage that information into more digestible formats. Somehow the barcorder *must* have built into it a system whereby every time you ask it for a name, a penny drops into a bucket, the bucket that supports conservation and taxonomy. Many millions of user events, many millions of pennies. More users, more pennies. More bioliteracy, more pennies. We buy a postage stamp without a blink, we pay for our phone calls with only the mildest of grumping, we do not begrudge our highway taxes. The emergence of the barcorder demands a three-member team—the science of plant sequencing, the microengineering of miniaturization of a currently laboratory process, and the emergence of an entrepreneurial force aimed at delivery of a self-supporting social good.

Do we want to conserve wild plant diversity? Yes. The answer does not lie in better keys, more keys, more images on the web, more web sites, more species pages, more descriptions, more phylogenies, more specimens, more maps, more anythings. Those are necessary collaterals, but not sufficient. The answer lies in a process that will for the first time connect the collective species-level biodiversity knowledge of the world to any and all users, on the spot, now. Fast, cheap, and on-site single (or very few) gene sequencing—developed for the purpose of identification—has the potential to deliver the species-specific linkage between the species and its human-known collaterals. There is a huge opportunity for the botanical community to thrust itself into a position of friendly social global prominence—just as have education, agriculture, medicine, and communication. Now is the time.

LITERATURE CITED

Barrett, T., Meyer, P., Stoeckle, M., and Yung, J. 2004. Taxonomy, DNA, and the Barcode of Life. Program for the Human Environment, Rockefeller University. http://phe.rockefeller .edu/BarcodeConference/index.html.

Godfray, H. C. J. 2002. Challenges for taxonomy. Nature 417:17–19.

Hebert, P. D. N., Cywinska, A., Ball, S. L., and deWaard, J. R. 2003a. Biological identifications through DNA barcodes. Proceedings of the Royal Society B 270:313–322.

Hebert, P. D. N., Ratsingham, S., and deWaard, J. R. 2003b. Barcoding animal life: cytochrome C oxidase subunit 1 divergences among closely related species. Proceedings of the Royal Society B 270, supplement O3BL0066.1–4.

Janzen, D. H. 2004. Now is the time. In Godfray, C., Knapp, S., and Gauld, I., eds., Taxonomy for the 21st Century. Philosophical Transactions of the Royal Society of London, B 359:731–732.

Stoeckle, M. 2003. Taxonomy, DNA, and the bar code of life. BioScience 53:796–797.

Stoeckle, M., Bucklin, A., Knowlton, N., and Hebert, P. 2004. Census of Marine Life DNA Barcoding Protocol. Program for the Human Environment, Rockefeller University. http://www.coreocean.org/Dev2Go.web?id=255158.

PREFACE

AT FIRST GLANCE, natural history museums appear to hold a collection of ancient things—fossils of dinosaurs that lived millions of years ago, skeletons of animals from faraway continents, rocks and minerals from beneath the earth's surface. To the untrained eye, these objects tell stories of how things existed on the planet centuries ago. But can they tell us anything about today's earth or what the future may hold for us?

Natural history museums, herbaria, and botanical gardens across the world house within their walls an amazing collection of botanical specimens—a representation of the earth's plant species. These botanical collections, both living and preserved, have been developed over several centuries and continue to grow today. They contain millions of specimens from all parts of the world that encompass a diversity of organisms (e.g., fungi, algae, mosses, and vascular plants). These collections serve as an encyclopedia of the earth's flora and provide valuable information regarding the diversity of species, where they grow, the habitats they come from, and their native communities. Many specimens also contain information about taxa that have vanished from environments and ecosystems previously undisturbed by human activities. These plant samples in some cases represent the last record of the species and provide us with the only information available about the plants in question.

The Smithsonian Institution is one establishment that holds such an important collection. The U.S. National Herbarium at the Smithsonian was officially founded in 1848. The herbarium contains some 4.7 million specimens dating back two centuries, making this collection among the ten largest in the world and representing about 8% of the plant-collection resources of the country. The herbarium has important historical collections, including specimens from the U.S. Exploring Expedition (1838–1842), the U.S. North Pacific Exploring Expedition (1853–1856), the LaPlata Expedition (1853–1856), the Mexican Boundary Survey (1854–1855), the California Geological Survey (1860–1867), the International Boundary Commission, United States and Mexico (1892–1894), and the Colombia *Cinchona* missions (1940–1945), as well as numerous recent expeditions to countries around the world.

The curators at the Smithsonian have contributed significantly to the field of plant taxonomy—the study and identification of plant life at all levels of biological organization (i.e., from ecosystems down to the unicellular organisms) and in various geographic and environmental regions. They have identified new plant species, described and revised systematic relationships among species, and provided data about plant communities across the continents and in the seas.

It is now universally accepted that many of the habitats previously explored by botanists are gone or are under immediate threat of destruction. Governments generate policies to save threatened and endangered species, but historically plants, lichens, and algae have been ignored in favor of the "charismatic megafauna"—the engaging panda, the majestic bald eagle. But what about the moss growing in the understory of a rain forest, the lichen clinging to a boulder near the coast, or the annual alpine wildflower seen during a brief window of time each year? How do we best determine conservation plans for these species as well?

In order to answer this question, herbaria and botanical gardens have entered a new phase in their careers. In addition to collecting and describing the world's flora, the information stored in herbaria and the research that takes place in their laboratories have become invaluable tools to the conservation community. Natural history has always been the foundation of conservation biology, and increasingly conservation programs are turning to natural history institutions to obtain knowledge of endangered and threatened species. In the global world of conservation, to save endangered species we must know about their basic biology. At an even greater scale, to save biodiversity, we must first know what it is and where it lives. This task of discovering, identifying, and describing plants is the central work of botanists at museums and botanical gardens.

A recent example illustrates how natural history has contributed to the field of conservation biology. Conservation International's Biodiversity Hotspots project (Myers et al. 2000) and World Wildlife Fund's Terrestrial Ecoregions of the World project (Olson and Dinerstein 1998) were both based on science generated at natural history museums. Both programs used data on species richness and endemism (knowledge found within the collections and among the curators) to determine areas of conservation priority (chapter 11 provides more detail).

The mission of the Smithsonian's U.S. National Herbarium is to discover the diversity of plant life in marine and terrestrial environments, to describe this diversity, to interpret the evolutionary origin of this diversity, and to explain the processes responsible for this diversity. In addition, the Smithsonian seeks to understand how humans are affected by and have altered plant diversity on the planet. *Plant Conservation: A Natural History Approach* provides a unique view of conservation biology from the perspective of Smithsonian botanists and their colleagues at other natural history institutions (plate P.1).

The book is divided into four parts. The first part (chapters 1–3) provides a framework for understanding plant diversity. A broad historical perspective traces the early evolution and diversification of plants. When did the major lineages evolve and how many have persisted to this day? Estimates are provided on how many plant species have existed in the past, how many perished during global mass extinctions, and how many are present today. Background rates of extinction are examined and contrasted with current rates of human-induced plant extinction. Several models are presented that describe the current distribution of plant diversity at multiple scales (e.g., latitudinal, continental, and oceanic).

The evolutionary and taxonomic threats and consequences of habitat alteration and plant extinction are explored in the second part (chapters 4–5). Areas of prime botanical importance—regions rich in plant diversity with high rates of endemism—are explored. Some of these sites are major wilderness areas that are still largely intact, whereas others are sites faced with constant pressures of heavy exploitation and fragmentation. At the taxonomic level, plant families may respond to threats differently. We explore the threats and responses for an array of plant families, from the microscopic marine dinoflagellates growing in tropical waters to grass species adapted to the arctic tundra.

The third part (chapters 6–9) reviews the varied causes of today's plant extinctions. Habitat fragmentation, degradation, and destruction occur on both terrestrial and marine landscapes. The alteration of these habitats has a direct effect on the native flora and an indirect effect on community interactions. Plants introduced from foreign lands have taken over habitats and are pushing the native flora aside. Plants face an additional indirect threat as human activities have changed the climate patterns of the planet. Furthermore, as plant populations are reduced in size, plant species face the threats of genetic loss and inbreeding depression.

With the natural history of plant biodiversity as the foundation, the fourth part (chapters 10–14) reviews the assessments, management strategies, and conservation actions undertaken to safeguard the remaining species on the planet. The role that museums and botanical gardens have in assessing plant biodiversity is explored in detail, from identifying priority habitats for protection to assessing the conservation status of species at the genetic, taxon, and community level. Ex situ and in situ conservation programs are evaluated. As an example of work in progress, a sustainable management model is proposed and applied to a major plant community. International and local laws, treaties, and coalitions are examined to address society's responsibility in taking the necessary steps to protect global plant biodiversity from further decline. Finally, the role of grassroots conservation organizations that are in the forefront of local activism is discussed from the perspective of a longtime participant.

Plant Conservation: A Natural History Approach provides a unique perspective on biodiversity and the current threats to its future from the viewpoint of botanists who have traversed the globe discovering and documenting the world's plant species. The information contained here is not an exhaustive treatment of the science of plant conservation. Rather, a diverse sampling of the conservation activities of botanists at museums and botanical gardens is provided as an introduction to the varied contributions to the field that natural history scientists are making. This volume, therefore, is a vital complement to ecological and management perspectives on plant conservation that can be found in other books. The chapters that follow emphasize the role that museums and botanical gardens will play in the present and future conservation of the earth's species and habitats.

We would like to acknowledge the assistance of those who have taken the time to review these chapters: Brian Boom, New York Botanical Garden; Daniel Faith, the Australian Museum; Peggy Fong, University of California, Los Angeles; Michael Foster, Moss Landing Marine Laboratories; Michael Grayum, Missouri Botanical Garden; Leo Hickey, Yale University; William Laurance, Smithsonian Tropical Research Institute; Thomas Lovejoy, the Heinz Center; Herb Meyer, National Park Service; Larry Morse, NatureServe; Nigel Pitman, Duke University; Tom Ranker, University of Colorado; Sarah Reichard, University of Washington; Jan Salick, Missouri Botanical Garden; George Schatz, Missouri Botanical Garden; Paul Wolf, Utah State University; George Yatskievych, Missouri Botanical Garden; and two anonymous reviewers. This volume is dedicated to botanists around the world who have devoted their careers to understanding and conserving plants.

LITERATURE CITED

Myers, N., Mittermeier, R. A., Mittermeier, C. G., da Fonseca, G. A. B., and Kent, J. 2000. Biodiversity hotspots for conservation priorities. Nature 403:853–858.

Olson, D., and Dinerstein, E. 1998. The Global 200: a representation approach to conserving the earth's most biologically valuable ecosystems. Conservation Biology 12:502–515.

PART I

PLANT DIVERSITY
PAST AND PRESENT

CHAPTER 1

Plant diversity has evolved and is evolving on the earth today according to the processes of natural selection. All basic "body plans" of vascular plants were in existence by the Early Carboniferous, approximately 340 million years ago, and each major radiation occupied a particular habitat type, for example, wetland versus terra firma, pristine versus disturbed environments. Today the great species diversity we see on the planet is dominated by just a few of these basic plant types. This chapter sets the stage to understanding the origin and radiation of plant diversity on the earth before the significant influence of humans. The implication is that plants may become more diverse in the number of species level over time, but less diverse in overall form. The history of life tells us that plants that are dominant and diverse today will certainly be inconspicuous or even extinct in the distant future.

EVOLUTION OF LAND PLANT DIVERSITY:
MAJOR INNOVATIONS AND LINEAGES
THROUGH TIME

William A. DiMichele and Richard M. Bateman

PLANT DIVERSITY VIEWED through the lens of deep time takes on a considerably different aspect than when examined in the present. Although the fossil record captures only fragments of the terrestrial world of the past 450 million years, it does indicate clearly that the world of today is simply a passing phase, the latest permutation in a string of spasmodic changes in the ecological organization of the terrestrial biosphere. The pace of change in species diversity and that of ecosystem structure and composition have followed broadly parallel paths, unquestionably related but not always changing in unison. Extending back to the Silurian and possibly the Ordovician, the landscapes of the emergent surface experienced vast changes. Accompanying these changes were increases in species diversity within an ever-narrowing phylogenetic spectrum of the total evolutionary tree of vascular plants. Compared with the Carboniferous, today's world has vastly more species but fewer major clades encompassing the ecologically dominant elements. Flowering plants hold sway in most nonboreal biomes (large-scale, climatically limited, ecological units). The totality and reliability of this pattern suggest that future events will see a continued narrowing of the phylogenetic spectrum from which dominant groups are drawn.

THE EARLIEST RADIATIONS

The lineages of plants we recognize today trace their roots to the Late Silurian and Early Devonian. During that time, early embryophytes (bryophytes and vascular land plants) appeared, and the ancestral groups of subsequent radiations began to establish themselves in terrestrial ecosystems. The principal vascular-plant groups in this early monophyletic radiation encompassed three distinct body plans: rhyniophyte, lycophyte, and euphyllophyte (Kenrick and Crane 1997). The rhynio-phytes were likely the basal-most group from which the others were descended (Banks 1975; Raubeson and Jansen 1992), and hence were markedly paraphyletic. Rhyniophyte morphology was extremely simple (plate 1.1A), consisting of rhi-zomes bearing small upright dichotomizing axes with some branch tips terminated in sporangia; the presence of roots is equivocal, and leaves were absent. The lyco-phytes included two lineages, the Zosterophyllopsida (plate 1.1C) and Lycopsida; all lycopsids probably are descendants of a single zosterophyll ancestor. Similar in many aspects of morphology, the lycopsids survive until today and bear the leaves and roots that are so conspicuously lacking in the zosterophylls (Gensel 1992; Hue-ber 1992). The euphyllophytes were a complex group that encompassed enormous structural diversity. Early euphyllophytes included what have been called trimero-phytes, a paraphyletic group possessing anisotomous branching, terminal sporan-gia, isospores, complex stelar morphology, and often enations of various sorts, but lacking true leaves (plate 1.1B). From this plexus the remainder of the vascular plants arose during the Devonian (Gensel and Andrews 1984; Kenrick and Crane 1997).

The phylogenetic distinctiveness of the basal lineages of vascular plants has certainly been clarified by subsequent evolutionary divergence among their de-scendants. Had human taxonomists existed during the Late Silurian and Early De-vonian, they would have faced a significantly greater challenge than we do in rec-ognizing those aspects of body plan that differentiate these early plant lineages, lacking knowledge of subsequent evolutionary divergence among the plant line-ages. Nonetheless, the often-asserted extinction of a spectrum of intermediate forms formerly linking these groups is not supported by the fossil record. No doubt there is a considerable body of plants of which we have no record and no knowl-edge, but the vast majority of these would likely have been minor variants of exist-ing body plans. Missing "intermediates," however, are merely an unsubstantiated expectation derived from uncritical acceptance of evolution as a gradual unfolding of morphological form rather than the discontinuous process apparent in the vast wealth of fossil data (Gould 2002; Bateman and DiMichele 2003).

The species diversity encapsulated by this early radiation has proven difficult to assess. The most comprehensive compilations (Niklas et al. 1980; Knoll et al. 1984) suggest a sigmoidal increase in number of taxa, reaching a plateau in the Middle

Devonian. The reality of this pattern was challenged by Raymond and Metz (1995), who re-evaluated the effects of sampling intensity and taphonomy on this estimate. They concluded that, although diversity increased through time, the shape of the pattern of increase could not be reconstructed reliably due to the spatio-temporal unevenness of sampling.

Biogeography of the early radiation is similarly subject to interpretation as a consequence of sampling inequalities. Raymond (1987) and Edwards (1990) analyzed the available data and concluded that three major phytogeographic regions existed: the northern latitudes, the southern latitudes, and the equatorial tropics, the latter divisible into subprovinces. Kenrick and Crane (1997) concluded, however, that the data supporting even this basic tripartite pattern are weak and should be approached with caution. Thus, at this time, the loci of origin of vascular plants per se and of the major sublineages, and the pattern of their global spread, cannot be identified with confidence.

ACCRUAL OF MORPHOLOGICAL COMPLEXITY

Throughout the Devonian, plants accrued morphological complexity, which can be benchmarked by the first appearances of particular structural features. Chaloner and Sheerin (1979) made one of the earliest attempts to measure this phenotypic diversity by compiling the first occurrence of features such as laminate leaves, stomata, and sporangial characteristics, but without regard to the evolutionary lineages in which the attributes appeared, making this a crude but nonetheless interesting measure of the slow accretion of morphological innovations. Knoll et al. (1984) examined the patterns with greater consideration of phylogenetic relationships, developing a measure of morphological advancement and plotting the average scores for intervals of the Late Silurian and Devonian in both the euphyllophyte and lycophyte lineages. They recovered a sigmoidal pattern of increasing complexity in each group, reaching a plateau in the Middle through Late Devonian before beginning to rise again in the Carboniferous.

DiMichele and Bateman (1996) ascribed this rapid rise to the attainment of a threshold level of morphological-developmental complexity. Recognizing that morphology is an expression of degrees of developmental regulation (Stein 1993; Cronk et al. 2002), they speculated that certain levels of canalized (inherently constrained) morphology and developmental control were needed before major changes in body plan could be attained. Prior to reaching this threshold, any single lineage would have given rise only to minor phenotypic innovations as structure and its regulatory control accrued. In other words, there may have been an interval when differences between ancestral and descendant species were relatively minor; this is not to say that evolution was "gradual," characterized by insensibly graded steps. We would argue that the size of the discrete steps was, perforce, smaller than later only due to the relative simplicity of body plans. Once at the threshold, major changes in body plan became possible between ancestor and descendant due to the greater opportunities for heterotopy (change in the relative position of expression of canalized morphological modules) and heterochrony (change in the relative timing of expression of canalized morphological modules). This is especially the case in organisms with minimal developmental interdependencies among organ systems ("epigenetics": morphology created by developmental interaction, not specifically coded by the genome). The subsequent progressive increase in developmental constraints limiting contingency (the "epigenetic ratchet" of Levinton 1988) may have played a major role in later limitations on the origin of novelties. The diversification of lineages that accompanied this global increase in morphological complexity was considered to be a "novelty radiation" (Erwin 1992) by Bateman et al. (1998) because major increases in complexity were accompanied by relatively low rates of appearance of new species. This occurred in an ecologically undersaturated terrestrial landscape in which competition was minimal, permitting an unusually high probability for survival of derived forms (DiMichele et al. 1992; Bateman et al. 1998).

APPEARANCE OF MODERN BODY PLANS

Modern vascular-plant body plans began to make their appearance in the Middle Devonian, and all were present by the Early Carboniferous (DiMichele and Bateman 1996). The most difficult part of this pattern to explain is the similar timing of the radiation in the two major branches of vascular-plant diversity: the lycopsids and the euphyllophytes. Long separated evolutionarily, these major clades essentially simultaneously gave rise to a variety of distinctive body plans. In the lycopsids (Lycopsida) there appeared the lycopodioids (Lycopodiales), selaginelloids (Selaginellales; plate 1.1D), and isoetoids (Isoetales; plate 1.1E). From the euphyllophyte lineage arose the sphenopsids (Sphenopsida; plate 1.1F), ferns (Filicopsida; plate 1.1G), progymnosperms (Progymnospermopsida), seed plants (Spermatopsida; plate 1.1H), and several enigmatic pteridophytic groups (e.g., Cladoxylales, Iridopteridales). The parallel patterns of these phenotypic radiations suggests that morphological evolution may have had a "clocklike" component as complexity accrued within such simple morphological backgrounds, the threshold being reached approximately contemporaneously in lycopsids and euphyllophytes (Bateman 1999).

The origin of vascular-plant body plans is the terrestrial equivalent of the metazoan radiation in the Cambrian. The Late Silurian–Early Devonian origination of the euphyllophyte and lycophyte clades, although profound, was not as important a step in restricting the pathways of evolutionary change as was the later diversification, reflecting what Gould (1991) termed contingency. The body plans that appear during this radiation differ radically from one another in design and mode of reproduction, and represent the fundamental architectural groups of vascular plants extant from that time onward. Subsequent evolution was nested within these body plans; as noted by Bateman et al. (1998), each subsequent radiation generated more species but fewer major morphological divergences than the last, the average degree of phenotypic divergence between putative ancestor and descendant decreasing over time.

ECOLOGICAL SORTING

The Devonian-Carboniferous radiation was accompanied by clade-level ecological partitioning in which each of the major architectural groups was centered in a different part of the terrestrial landscape or had a distinct ecological role (DiMichele et al. 2001b). The principal division was between seed plants and isoetalean lycopsids. Seed plants apparently originated in wetland habitats in the Late Devonian but invaded and then radiated in terra firma environments during the Carboniferous (Gillespie et al. 1981; DiMichele and Bateman 1996). The terra firma landscape, vast in extent and environmentally heterogeneous, was an unexploited resource space permissive of radiations and survival of highly derived forms. Consequently, numerous variations on the seed-plant body plan evolved and survived, recognized taxonomically as orders, including the Carboniferous Lyginopteridales, Medullosales, and Callistophytales; the Permian Coniferales, Cycadales, and Peltaspermales; and the Mesozoic Caytoniales, Bennettitales, Corystospermales, Gnetales, and "Angiospermales." The Isoetales, particularly the rhizomorphic lycopsids, also originated in wetlands (Scheckler 1986) but radiated there and became the dominant elements in wetlands during the Carboniferous, extending into the Permian in eastern parts of the Old World tropics. Although wetlands were extensive during much of the Paleozoic, the physical heterogeneity of these habitats was (and is) considerably less than that of terra firma, resulting in fewer evolutionary opportunities, which is reflected in much lower diversity in isoetalean body plans when compared to seed plants. The main dichotomy in isoetaleans sensu lato separates the paraphyletic "lepidodendrids" from the isoetaleans sensu strictu (Bateman et al. 1992). The lepidodendrids had complex aerial branching systems, elaborated "stigmarian" root systems, and often huge size. The isoetaleans sensu strictu were much smaller forms with cormose bases and mostly unbranched shoot systems. Forms with cormose bases first appear in the Late Devonian (Chitaley and McGregor 1988), followed by the lepidodendrids in the Early Carboniferous.

Of the other major clades, the sphenopsids were primarily plants of physically stressful, aggradational habitats, such as streamsides or lake margins. As the only group to evolve tree size while retaining a rhizomatous habit, the arboreous sphenopsids demonstrated remarkable ability to recover from burial by sediment (Gastaldo 1992). This is a subset of the environment of exceptionally narrow breadth, providing little opportunity for the survival of new forms in the face of competition from incumbents. As a consequence, sphenopsids always have been of low diversity, both in species numbers and in variations on their body plan. The ferns appear monophyletic on the basis of molecular evidence (Pryer et al. 2001), but less clearly so on morphological evidence (Rothwell 1996). Early ferns radiated in environments of disturbance as opportunists (Scott and Galtier 1985). Although later becoming dominant elements in some kinds of habitats, the opportunistic strategy persisted throughout the evolutionary history of ferns.

This pattern of ecological partitioning characterized the Carboniferous into the Early Permian, a period of more than 50 million years. It began to break down at higher latitudes in the early Late Carboniferous (Meyen 1982; Knoll 1984). In the tropics the pattern of disassembly was complex. Vegetation typical of the Late Carboniferous persisted in tropical rain forest areas of what is now China until the Late Permian. Elsewhere in the tropics, climatic changes disrupted the dominance-diversity patterns, leading first to ecological reorganization within the wetland species pool (DiMichele and Phillips 1996), and then to major changes in floras, plants from xeric paratropical areas replacing the ancestral wetland vegetation (Mapes and Gastaldo 1987; Broutin et al. 1990; DiMichele and Aronson 1992).

EVOLUTIONARY DYNAMICS: THE SPATIAL COMPONENT OF EVOLUTIONARY INNOVATION

Data from the fossil and modern records suggest that speciation occurs everywhere on the landscape where organisms exist. Areas of high diversity have been considered "hot spots," where the establishment and longevity (but perhaps not rate of origin) of new species are elevated; such areas may be found both in continental and island settings. "Background" speciation, however, seems to occur widely in space and approximately uniformly across the landscape, setting aside major environmental perturbations of heterogeneous effect such as glaciations.

The terrestrial fossil record also indicates a bias in the landscape position of major phenotypic divergence. All speciation does not entail the same ancestor-descendant distances. The vast majority of descendants differ from their immediate ancestor in minor ways. A very few differ in more significant ways, including fundamental differences in body plan and/or reproductive morphology. Such forms will be expected to possess developmental irregularities or structural abnormalities that initially may compromise fitness (Valentine 1980; Bateman and DiMichele 1994, 2002), requiring establishment in environments of little or no competition for resources.

For example, conifers first appear as fragments of charcoal in depositional basins that formed proximal to contemporaneous rising upland regions (Lyons and Darrah 1989). Typical macrofossil remains occur only later in lowland basins in association with other xeromorphic plants and with physical indicators of dry climates (Broutin et al. 1990; DiMichele and Aronson 1992). Similarly, evidence of many plants that came to dominate Mesozoic lowlands appears first as fragmentary remains in dry-climate deposits of late Paleozoic basinal lowlands, intercalated within more typical wetland deposits. Included are cycads (Mamay 1976), osmundaceous ferns (Miller 1971), and various conifer groups (Schweitzer 1986; DiMichele et al. 2001a). The angiosperms also evidently evolved in remote areas and moved into lowlands along river margins as weeds (Hickey and Doyle 1977;

Crane 1989). These data suggest that major clades originated in environmentally marginal areas, especially those where moisture limitation was significant. Such areas had the lowest levels of resource competition. Through time, such loci became ever more remote from the basinal lowlands as more proximal areas filled with species, increasing resource competition and the exclusionary effects of incumbency (occupation of a particular niche at a particular location, often termed "incumbent advantage": reviewed by DiMichele and Bateman 1996).

ECOLOGICAL REPLACEMENT DYNAMICS

At some point in history, lineages that had evolved in remote, environmentally marginal areas moved into basinal lowlands, and thus into the preservational window of the fossil record. Most such movements appear to have been examples of environmental tracking rather than competitive displacement of lineages already established in the lowlands. The Paleozoic tropical wetland flora, for example, remained compositionally distinct from the xeric flora. Where fossil deposits of these two floras are found intercalated in lowland basins, they have different environmental signatures and have few species in common (DiMichele and Aronson 1992). It appears that the xeric flora was moving into the basins during times of climatic drying. The ultimate replacement of the wetland flora by the xeric flora occurred as the Paleozoic tropics became severely dry during the Permian (Ziegler 1990), eliminating the wetland flora. When wet conditions returned, the elements of the prior flora were gone, resulting in the evolution of new wetland lineages from the surrounding xeric lineages. Thus, the evolution of major innovations in marginal habitats, becoming progressively more remote from ancient lowland regions, was complemented by the movement of these lineages back into the lowlands following major climatic changes and consequent ecological disruption.

EXPANSION OF THE ANGIOSPERMS

The angiosperms are the most species-rich group of vascular plants. As a consequence, they merit special attention in any consideration of diversity. The modern diversity of this group is accompanied by ecological dominance of most of the world's terrestrial ecosystems. The complexity and character of those ecosystems is a reflection of the great diversity of form that has evolved within the flowering plants.

Evidence from the fossil record strongly suggests that angiosperms originated in the tropics during the Early Cretaceous. These early angiosperms were likely small, opportunistic, dominantly woody plants that exploited disturbed habitats (Hickey and Doyle 1977). Multigene molecular phylogenies based entirely on extant species (e.g., Chaw et al. 2000) have recently undermined the paleobotanically favored "anthophyte hypothesis" (e.g., Doyle and Donoghue 1986; Donoghue and Doyle 2000). This theory interpolated gymnosperms that possess some angiosperm-like reproductive features (the extant Gnetales, together with the extinct Bennettitales and Pentoxylales) between modern angiosperms and the paraphyletic pteridosperms of the Late Paleozoic and Mesozoic. The recent molecular phylogenies once more allow the possibility of origination of the angiosperm clade from one of the more derived pteridosperms (cf. Harris 1964; Long 1977). This lower phylogenetic placement should not be viewed as indicating that the modern angiosperms are relatively primitive, as they have accumulated many molecular and morphological apomorphies over the long period since they diverged from their gymnospermous ancestor.

The most significant characteristic of the flowering plants, relative to other seed plants, is their species diversity. The angiosperms elaborated the basic body plan into a wide range of phenotypic variation, encompassing a broader spectrum of differences than in the rest of the seed-plant clade. They have undergone an almost unbroken increase in species numbers throughout the late Mesozoic and Cenozoic (Lidgard and Crane 1990). The increase in species numbers was accompanied, initially, by the origin of many of the extant arborescent families by the end of the Late Cretaceous, a novelty radiation in its combination of high rates of innovation within a background of initially relatively low numbers of species; this mimics, but encompasses less morphological disparity than, other radiations at higher taxonomic rank that took place in the Devonian and Carboniferous. In the course of this dynamic, the angiosperms may have evolved greater genetic novelty than ever before, though perhaps largely through more diverse combinations of genes than those characterizing other, earlier groups. At the same time, the probability of successful establishment of profound phenotypic novelties, such as those that characterized the Middle and Late Devonian architectural novelty radiation, has never been lower (Bateman and DiMichele 2002, 2003).

Just as the angiosperms expanded at the expense of other Cretaceous plant clades, narrowing the phylogenetic spectrum of dominance, so too the early angiosperm clades have been replaced by the expansion of later-appearing clades within the angiosperms. This is particularly true of grasses and composites, which have become important to dominant elements in many ecosystems. We speculate that there may be an underlying fractal process in operation within the terrestrial vascular plants (Bateman and DiMichele 2003). The changes in dominant angiosperm groups through time may be driven by the same underlying process that controlled the larger-scale replacement of lower vascular plants by seed plants during the late Paleozoic, and the replacement of early seed-plant clades by newer seed-plant groups in the Mesozoic.

CONCLUSIONS

The history of plant diversity, resolved at the level of major evolutionary lineages, is a record of ever-shrinking phylogenetic dispersion of ecological dominance. In the Paleozoic, the land was divided among four (arguably, more than four) classes of plants and many orders, the latter especially among the seed plants. By latest Paleozoic and early Mesozoic, most of the lower vascular-plant clades had either gone extinct (particularly those involved in the initial terrestrial radiation) or at least had retreated into the ecological shadows, largely the victims of climatic changes (Knoll 1984). A plethora of seed plants replaced these pteridophytes, bringing a great diversity of form, but primarily experimenting with only one broad reproductive theme, the seed habit.

With the rise of flowering plants, species-level diversity sky-rocketed, and with it a wide range of subtle variations on both vegetative and reproductive themes. Angiosperms explored more aspects of architecture than all other groups of seed plants combined. In a Linnean sense, however, they are an ordinal-level group (a fact masked by neobotanically focused classifications that routinely inflate the ranks of taxonomic groups within the angiosperms). In the Cenozoic, the same process began to occur within the angiosperms; the most recently originating groups, such as grasses and composites, are often dominantly herbaceous and began to expand at the expense of older lineages.

In this sense, the history of diversity reveals an ever-attenuating spectrum of dominant major clades. The pattern appears to be broadly "fractal," beginning with all vascular plants, replaced by seed plants and a variety of ferns, then angiosperms and polypodiaceous ferns, and most recently subgroups within the angiosperms. Extrapolation from the paleobotanical record suggests that this taxonomic attenuation is likely to continue, even if it is associated with further gentle increases in overall species diversity in those lineages that rise to dominance. Its natural projection, however, does not take into account the very recent depredations of human civilization.

ACKNOWLEDGMENTS

We thank Mary Parrish for preparing the illustration.

LITERATURE CITED

Banks, H. P. 1975. Reclassification of Psilophyta. Taxon 24:401–413.

Bateman, R. M. 1999. Integrating molecular and morphological evidence of evolutionary radiations. Pp. 422–471 in Hollingsworth, P. M., Bateman, R. M., and Gornall, R. J., eds., Molecular Systematics and Plant Evolution. Taylor and Francis, London.

Bateman, R. M., Crane, P. R., DiMichele, W. A., Kenrick, P. R., Rowe, N. P., Speck, T., and Stein, W. E. 1998. Early evolution of land plants: phylogeny, physiology, and ecology of the primary terrestrial radiation. Annual Review of Ecology and Systematics 29:263–292.

Bateman, R. M., and DiMichele, W. A. 1994. Saltational evolution of form in vascular plants: a neo-Goldschmidtian synthesis. Pp. 63–102 in Ingram, D. S., and Hudson, A., eds., Shape and Form in Plants and Fungi. Linnean Society Symposium 16. Academic Press, New York.

———. 2002. Generating and filtering major phenotypic novelties: neo-Goldschmidtian saltation revisited. Pp. 109–159 in Cronk, Q. C. B., Bateman, R. M., and Hawkins, J. A., eds., Developmental Genetics and Plant Evolution. Taylor and Francis, London.

———. 2003. Genesis of phenotypic and genotypic diversity in land plants: the present as the key to the past. Systematics and Biodiversity 1:13–28.

Bateman, R. M., DiMichele, W. A., and Willard, D. A. 1992. Experimental cladistic analysis of anatomically-preserved arborescent lycopsids from the Carboniferous of Euramerica: an essay on paleobotanical phylogentics. Annals of the Missouri Botanical Garden 79:500–559.

Broutin, J., Doubinger, J., Farjanel, G., Freytet, F., Kerp, H., Langiaux, J., Lebreton, M.-L., Sebban, S., and Satta, S. 1990. Le renouvellement des flores au passage Carbonifère Permien: approaches stratigraphique, biologique, sédimentologique. Académie des Sciences Paris, Comptes Rendues 311:1563–1569.

Chaloner, W. G., and Sheerin, A. 1979. Devonian macrofloras. Special Papers in Palaeontology 23:145–161.

Chaw, S.-M., Parkinson, C. L., Cheng, Y., Vincent, T. M., and Palmer, J. D. 2000. Seed-plant phylogeny inferred from all three plant genomes: monophyly of extant gymnosperms and origins of Gnetales from conifers. Proceedings of the National Academy of Sciences USA 97:4086–4091.

Chitaley, S., and McGregor, D. C. 1988. *Bisporangiostrobus harrisii* gen. et sp. nov.: an eligulate lycopsid cone with *Duosporites* megaspores and *Geminospora* microspores from the Upper Devonian of Pennsylvania, U.S.A. Palaeontographica 210B:127–149.

Crane, P. R. 1989. Patterns of evolution and extinction in vascular plants. Pp. 153–187 in Allen, K. C., and Briggs, D. E. G., eds., Evolution and the Fossil Record. Belhaven, Chichester, United Kingdom.

Cronk, Q. C. B., Bateman, R. M., and Hawkins, J. A., eds. 2002. Developmental Genetics and Plant Evolution. Taylor and Francis, London.

DiMichele, W. A., and Aronson, R. B. 1992. The Pennsylvanian-Permian vegetational transition: a terrestrial analogue to the onshore-offshore hypothesis. Evolution 46:807–824.

DiMichele, W. A., and Bateman, R. M. 1996. Plant paleoecology and evolutionary inference: two examples from the Paleozoic. Review of Palaeobotany and Palynology 90:223–247.

DiMichele, W. A., Hook, R. W., Beerbower, R., Boy, J., Gastaldo, R. A., Hotton, N. III, Phillips, T. L., Scheckler, S. E., Shear, W. A., and Sues, H.-D. 1992. Paleozoic terrestrial ecosystems. Pp. 204–325 in Behrensmeyer, A. K., Damuth, J. D., DiMichele, W. A., Sues, H.-D., and Wing, S. L., eds., Terrestrial Ecosystems through Time. University of Chicago Press, Chicago.

DiMichele, W. A., Mamay, S. H., Chaney, D. S., Hook, R. W., and Nelson, W. J. 2001a. An Early Permian flora with Late Permian and Mesozoic affinities from north-central Texas. Journal of Paleontology 75:449–460.

DiMichele, W. A., and Phillips, T. L. 1996. Climate change, plant extinctions, and vegetational recovery during the Middle-Late Pennsylvanian transition: the case of tropical peat-forming environments in North America. Geological Society Special Publication 102:210–221.

DiMichele, W. A., Stein, W. E., and Bateman, R. M. 2001b. Ecological sorting during the Paleozoic radiation of vascular plant classes. Pp. 285–335 in Allmon, W. D., and Bottjer, D. J., eds., Evolutionary Paleoecology. Columbia University Press, New York.

Donoghue, M. J., and Doyle, J. A. 2000. Demise of the anthophyte hypothesis? Current Biology 10:R106–R109.

Doyle, J. A., and Donoghue, M. D. 1986. Seed plant phylogeny and the origin of the angiosperms: an experimental cladistic approach. Botanical Review 52:321–431.

Edwards, D. 1990. Constraints on Silurian and Early Devonian phytogeographic analysis based on megafossils. Pp. 233–242 in McKerrow, W. S., and Scotese, C. R., eds., Palaeozoic Palaeogeography and Biogeography. Geological Society, London.

Erwin, D. 1992. A preliminary classification of evolutionary radiations. Historical Biology 6:133–147.

Gastaldo, R. A. 1992. Regenerative growth in fossil horsetails following burial by alluvium. Historical Biology 6:203–219.

Gensel, P. G. 1992. Phylogenetic relationships of the zosterophylls and lycopsids: evidence from morphology, paleoecology, and cladistic methods of inference. Annals of the Missouri Botanical Garden 79:450–473.

Gensel, P. G., and Andrews, H. N. 1984. Plant Life in the Devonian. Praeger Press, New York.

Gillespie, W. H., Rothwell, G. W., and Scheckler, S. E. 1981. The earliest seeds. Nature 293: 462–464.

Gould, S. J. 1991. The disparity of the Burgess Shale arthropod fauna and the limits of cladistic analysis: why we must strive to quantify morphospace. Paleobiology 17:411–423.

———. 2002. The Structure of Evolutionary Theory. Belknap Press, Cambridge, Massachusetts.

Harris, T. M. 1964. The Yorkshire Jurassic Flora. II. Caytoniales, Cycadales, and Pteridosperms. British Museum (Natural History), London.

Hickey, L. J., and Doyle, J. A. 1977. Early Cretaceous fossil evidence for early angiosperm evolution. Botanical Review 43:3–104.

Hueber, F. M. 1992. Thoughts on the early lycopsids and zosterophylls. Annals of the Missouri Botanical Garden 79:474–499.

Kenrick, P., and Crane, P. R. 1997. The Origin and Early Diversification of Land Plants: A Cladistic Study. Smithsonian Series in Comparative Evolutionary Biology. Smithsonian Institution Press, Washington, DC.

Knoll, A. H. 1984. Patterns of extinction in the fossil record of vascular plants. Pp. 21–68 in Nitecki, M., ed., Extinctions. University of Chicago Press, Chicago.

Knoll, A. H., Niklas, K. J., Gensel, P. G., and Tiffney, B. H. 1984. Character diversification and patterns of evolution in early vascular plants. Paleobiology 10:34–47.

Levinton, J. 1988. Genetics, Paleontology, and Macroevolution. Cambridge University Press, Cambridge.

Lidgard, S., and Crane, P. R. 1990. Angiosperm diversification and Cretaceous floristic trends: a comparison of palynofloras and leaf macrofloras. Paleobiology 16:77–93.

Long, A. G. 1977. Lower Carboniferous pteridosperm cupules and the origin of the angiosperms. Transactions of the Royal Society of Edinburgh 70B:13–35.

Lyons, P. C., and Darrah, W. C. 1989. Earliest conifers in North America: upland and/or paleoecological indicators? Palaios 4:480–486.

Mamay, S. H. 1976. Paleozoic origin of the cycads. United States Geological Survey Professional Paper 934:1–48.

Mapes, G., and Gastaldo, R. A. 1987. Late Paleozoic non-peat accumulating floras. University of Tennessee Studies in Geology 2:115–127.

Meyen, S. V. 1982. The Carboniferous and Permian floras of Angaraland: a synthesis. International Society of Applied Biology, Biological Memoirs 7:1–109.

Miller, C. N. 1971. Evolution of the fern family Osmundaceae based on anatomical studies. University of Michigan, Contributions from the Museum of Paleontology 23:105–169.

Niklas, K. J., Tiffney, B. H., and Knoll, A. H. 1980. Apparent changes in the diversity of fossil plants. Evolutionary Biology 12:1–89.

Pryer, K. M., Schneider, H., Smith, A. R., Cranfill, R., Wolf, P. G., Hunt, J. S., and Sipes, S. D. 2001. Horsetails and ferns are a monophyletic group and the closest living relatives of seed plants. Nature 409:618–622.

Raubeson, L. A., and Jansen, R. R. 1992. Chloroplast DNA evidence on the ancient evolutionary split in vascular land plants. Science 255:1697–1699.

Raymond, A. 1987. Paleogeographic distribution of Early Devonian plant traits. Palaios 2:113–132.

Raymond, A., and Metz, C. 1995. Laurussian land-plant diversity during the Silurian and Devonian: mass extinction, sampling bias, or both? Paleobiology 21:74–91.

Rothwell, G. W. 1996. Phylogenetic relationships of ferns: a paleobotanical perspective. Pp. 395–404 in Camus, J. M., Gibby, M., and Johns, R. J., eds., Pteridology in Perspective. Royal Botanic Gardens, Kew, London.

Scheckler, S. E. 1986. Geology, floristics, and paleoecology of Late Devonian coal swamps from Appalachian Laurentia (USA). Annales Societé Geologique Belgique 109:209–222.

Schweitzer, H. J. 1986. The land flora of the English and German Zechstein sequences. Pp. 31–54 in Harwood, G. M., and Smith, D. B., eds., The English Zechstein and Related Topics. Geological Society Special Publication 22. Blackwell, London.

Scott, A. C., and Galtier, J. 1985. Distribution and ecology of early ferns. Proceedings of the Royal Society of Edinburgh 86B:141–149.

Stein, W. E. 1993. Modeling the evolution of stelar architecture in vascular plants. International Journal of Plant Science 154:229–263.

Valentine, J. W. 1980. Determinants of diversity in higher taxonomic categories. Paleobiology 6:444–450.

Ziegler, A. M. 1990. Phytogeographic patterns and continental configurations during the Permian Period. Pp. 363–379 in McKerrow, W. C., and Scotese, C. R., eds., Palaeozoic Palaeogeography and Biogeography. Geological Society, London.

CHAPTER 2

DIVERSITY AND DISTRIBUTION
OF PLANTS

Plant diversity in both terrestrial and marine environments is not evenly dis-
tributed around the globe. For example, tropical rain forests and coral reefs have
a far more diverse flora than desert, tundra, arctic intertidal, and mangrove
communities. Diversity is measured in a number of ways according to the scien-
tific or conservation question being addressed. The overall number of plant
species in existence today, the uneven distribution of this diversity, and the
mechanisms that regulate this distribution are topics that scientists have been
investigating for centuries. Assessing, managing, and protecting the earth's
species can best be accomplished with a clear understanding of how many spe-
cies are present, where they are found, and what factors control or limit their
distribution.

2.1 TERRESTRIAL PLANT DIVERSITY

Jens Mutke, Gerold Kier, Gary A. Krupnick, and Wilhelm Barthlott

KNOWLEDGE ABOUT the diversity and distribution of land plants is highly
uneven. Vascular plants, including ferns, gymnosperms, and flowering plants, are
relatively well documented, thanks to their ubiquitous presence in our lives and our
environment, and their direct importance for humanity. Not only are they at the
base of the terrestrial food chain, but plants are also one of the most economically
important major groups of organisms. At the beginning of the 20th century Holden
and Wycoff (1911–1914) published their *Bibliography Relating to the Floras,* listing
some 7,750 entries. Much progress has been made since then (Frodin 2001b).

This situation for vascular plants is in contrast with the other groups of plants—
the algae, liverworts, and mosses. Though a world checklist of mosses with nearly
12,800 accepted names has already been published (Crosby et al. 2000), the spatial
patterns of bryophyte diversity (including mosses, liverworts, and hornworts) are
still insufficiently known. Regional checklists such as the LATMOSS catalogue of
Neotropical mosses (Delgadillo et al. 1995) and the list of O'Shea (1997) for sub-
Saharan Africa allow some comparisons between the floras with regard to their
composition and diversity. The total number of moss species on a regional basis
might give a first impression about the magnitude of diversity, such as 2,800 docu-

mented species in tropical Africa (O'Shea 1997), 2,500 species in China (Redfearn 1994), some 1,900 species in Brazil (Yano 1981), 1,320 species in North America (Vitt and Buck 1992), and around 1,400 species in Europe (J.-P. Frahm, personal communication). The overall known distribution of bryophyte species, however, still reflects a pattern of research intensity rather than a pattern of genuine diversity. At a finer resolution, the unrealistically low figure of only eight documented species in Benin (O'Shea 1997), or five species known from the Distrito Federal of Brazil (Yano 1981), reveals the lack of adequate information on mosses (Mutke and Barthlott, in press).

For this reason the review of spatial patterns of land-plant diversity presented in this section refers exclusively to vascular plants (algae diversity will be covered in section 2.2). Though the knowledge of regional floras is still uneven for vascular plants (Jäger 1976; Frodin 2001a, 2001b), they are the largest systematic group of organisms where such analyses are available (Malyshev 1975; Williams et al. 1994; Barthlott et al. 1996; Barthlott et al. 1999a; Groombridge and Jenkins 2002; Mutke and Barthlott, in press). The first world map of species richness of vascular plants, giving total numbers of species for large landscapes without referring to a standardized area size, was published by Wulff (1935). Much earlier, Alexander von Humboldt (1806) discussed some of the most prominent patterns of plant diversity, such as the gradient of increasing species richness with decreasing latitude (Hawkins 2001).

Species-richness maps are a useful tool in analyzing patterns and trends in the distribution of biodiversity. There are two main approaches for the production of richness maps. The taxon-based approach (Barthlott et al. 1999b) charts the distributional ranges of every single taxon known (species, infraspecific taxa, genera, etc., depending on the level of the analysis). The total richness map is then produced, using an overlay of the various individual species maps. This approach allows for multiple analyses at various levels, such as a comparison of the diversity patterns of various subgroups. However, a vast amount of data, resources, and time is needed to undertake such an approach, especially considering that distribution data are not yet available for every known species. Thus, an alternative technique is the inventory-based approach, which uses summary diversity data of entire geographic areas, such as species richness of the 50 states of the United States. These figures are readily available from several thousands of floras on a global scale (Frodin 2001b) and have often been presented in tables and have been charted on maps without reference to the surface area of the geographic units (Wulff 1935; WRI 1997; Groombridge 1992). To allow comparison of these figures for the individual geographic areas, however, it is necessary to convert the species-richness data to a standard area using species-area relationship models (Arrhenius 1920, 1921; Malyshev 1975, 1991; Rosenzweig 1995). For the map in plate 2.1, species-richness figures for some 1,400 geographic areas around the world were plotted on the map, and species richness was interpolated in areas without primary data using information on climate, vegetation, and topography. This approach is comparable to the production of a human population-density map. To analyze the "habitats" of every single citizen of North America separately, for example, would be quite difficult for a geographer. The logical method is to use data on the number of inhabitants in the various cities, counties, and so on, and to plot these figures on a map.

THE UNEVEN DISTRIBUTION OF PLANT DIVERSITY

The world map in plate 2.1 shows that plant diversity is highly unevenly distributed on earth. The small South American country of Ecuador, for example, with its ca. 16,000 species (Jørgensen and León-Yánez 1999) on a surface area comparable to Nevada, harbors some 40% more species than Europe (according to data in Tutin et al. 1964–1980). An area the size of a soccer stadium is needed in the lowland rain forest of Ecuador to find some 1,000 species of trees, shrubs, herbs, lianas, and epiphytes (Valencia et al. 1994). South Africa, including the highly distinct Cape Province with many species occurring nowhere else, is home to 30% more species than the entire United States, though covering an area smaller than Alaska (Groombridge 1992; Davis et al. 1997).

Global species-richness maxima are found in highly structured—geodiverse (see below)—tropical and subtropical areas of the Chocó–Costa Rica center, the tropical eastern Andes, eastern Brazil (Mata Atlantica), the eastern Himalaya–Yunnan center, northern Borneo, and New Guinea. In addition, some Mediterranean-type climate areas are regional maxima of plant diversity, such as the Cape Province in South Africa and southwestern Australia (plate 2.2).

Moist tropical broadleaf forests harbor the highest number of species, with up to 10,000 species per 10,000 km^2 in the La Amistad Biosphere Reserve in Costa Rica and Panama (according to Davis et al. 1997) or ca. 5,000 species on only 1,200 km^2 of land on Mount Kinabalu, Borneo (after Beaman, in press). As stated above, however, these absolute maxima are linked to areas with a high heterogeneity of the landscape, including many different types of forests. Lowland rain forests, often bearing an exceptional alpha diversity on small sample plots, normally show somewhat lower diversity at landscape levels (e.g., Valencia et al. 1994). Nevertheless, more than 10,000 plant species in the Chiribiquete-Araracuara-Cahuinarí region of Amazonian Colombia or some 6,000 species in the Gran Sumaco and upper Napo River region of eastern Ecuador (Davis et al. 1997) still make tropical rain forests the most species-rich terrestrial biome on a global scale. The next most species-rich biomes are tropical and subtropical coniferous forests and Mediterranean forests, woodlands, and scrub. Lowest mean species richness per biome can be found in tundra and taiga regions. Absolute minima of species diversity at a regional scale can be found in hyperarid regions, such as the Sahara and Atacama deserts, as well as in Arctic and Antarctic environments. There are several deserts, however, such as the southern African Succulent Karoo and Namaqualand and the deserts of southwestern North America, that harbor relatively high species diversity. The Namaqualand harbors the same number, ca. 2,750 plant species (Cowling et al. 1999) as North Carolina, on less than half the area.

Family richness of seed plants was studied by Williams et al. (1994) on a global scale using a 611,000 km^2 grid. The World Conservation Monitoring Centre did the same exercise for nonaquatic plant families at a higher spatial resolution (Groom-bridge and Jenkins 2002). Williams et al. (1994) found more than 200 families co-occurring on the Malaysian peninsula in contrast to about 100 to 130 families in different areas of Europe or North America. The second richest area for plant family diversity in addition to the whole Indo-Malayan region is Central America with more than 180 families per grid cell. In North America, the humid southeast—a Pleistocene refuge area (Delcourt and Delcourt 1993)—harbors the greatest family richness (Thorne 1993; Mutke and Barthlott 2000).

At the genus level, the Neotropics show the highest diversity, with more than 2,700 genera of vascular plants co-occurring in the tropical Andes (N. Brummit, personal communication). South and Southeast Asia, with more than 2,500 genera in India and a similar number in China, ranks second. This shift in diversity of the floras at different taxonomic levels is even more dramatic when comparing species numbers. The Neotropics with their estimated 90,000 species (Gentry 1982) harbor a higher diversity at this level than Africa and Southeast Asia together (Davis et al. 1997).

Highest proportions of endemic species—taxa that are restricted to a defined area—can often be found on islands, such as New Guinea and Madagascar (Myers 1988; Davis et al. 1994) with some 90 and 80%, respectively, of the flora occurring nowhere else (plate 2.2). Mainland areas with topographically or climatically iso-lated—"insular"—positions harbor high numbers of endemic species as well. The most prominent example is the Cape region in South Africa, with some 5,800 of the 8,600 native vascular-plant species being endemic (Cowling and Richardson 1995). Almost half of the native flora of the California floristic province is endemic (Thorne 1993), whereas no species are endemic to Delaware, Iowa, or Kansas (Mutke 2002, based on an earlier version of USDA 2004).

Several approaches for setting priorities in nature conservation combine these measures of species richness and endemism with data on threats to the area by human activities (e.g., Myers 1988, 1990; Olson and Dinerstein 1998; Mittermeier et al. 1999; Myers et al. 2000; Olson et al. 2001). Some of these methodologies are discussed in chapter 10.

DIVERSITY PATTERNS OF DIFFERENT SUBGROUPS OF VASCULAR PLANTS

Overall diversity of vascular plants shows a high correlation with environmental parameters (fig. 2.1). Barthlott et al. (1999b) found comparable r^2 for the correla-tion of vertebrate and insect diversity with overall plant diversity when comparing species richness at country level. It is quite clear, however, that these patterns are the result of the diversities of different subgroups of organisms with distinct ecology and evolutionary history. To get a better understanding of the overall patterns it is important to examine these subgroups and to compare diversity patterns of taxa with different evolutionary history and age. To demonstrate, one might compare the diversity patterns of the ca. 820 species of gymnosperms, an evolutionarily

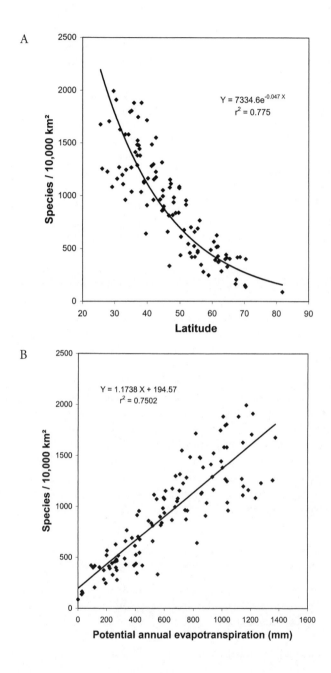

A

$Y = 7334.6e^{-0.047\ X}$
$r^2 = 0.775$

B

$Y = 1.1738\ X + 194.57$
$r^2 = 0.7502$

Figure 2.1 Plant species richness of North American ecoregions in relationship to (A) latitude and (B) potential annual evapotranspiration (Mutke 2002, computed on the basis of species-richness figures in Ricketts et al. 1999).

ancient lineage, with overall vascular-plant diversity. Main centers in Yunnan/ Sichuan (up to 59 species per 10,000 km²), northern Borneo (≤40 species per 10,000 km²), and New Guinea (≤37 species per 10,000 km²) are the same for gymnosperms as for the vascular plants in general (Mutke and Barthlott, in press). In contrast to these Southeast Asian centers, low gymnosperm diversity exists in other parts of the tropics and subtropics, such as Brazil, the Congo basin, central Australia, and parts of India. The exceptions to this trend are Mexico and California, where high gymnosperm diversity can be found (≤ 41 species per 10,000 km²). The absolute maximum of gymnosperm diversity at the family level is located at the Chinese-Vietnamese border, with nine families per 10,000 km². Even within the gymnosperms, the major subgroups show strongly diverging patterns. The conifers with ca. 630 species show highest diversity in the Northern Hemisphere, with the main centers as described above (compare Farjon 1998). The cycads have a mainly tropical to subtropical distribution with centers in Mexico, South Africa, and eastern Australia (Jones 2002). The Gnetales include three genera in three families with totally independent distribution patterns (*Gnetum* in the humid tropics, *Ephedra* in temperate and subtropical dry areas, and *Welwitschia* in the Namib Desert) (Kubitzki 1990).

The diversity patterns of the almost 10,000 species of ferns are not as well-documented as those of the gymnosperms. While many new species are waiting to be discovered, especially in the tropical humid montane forests (compare Kessler 2001), the well-documented floras of North America and Europe show a less pronounced latitudinal gradient of fern species richness compared to the overall vascular-plant diversity. Whereas the total numbers of fern species rise, for example from 58 species in Finland to 107 in Spain, the proportion of ferns in the vascular-plant flora of Finland is 5% in contrast to less than 2% in some countries of southern Europe (according to *Flora Europaea*, Tutin et al. 1964–1980). On a global scale, the highest proportions of fern species in relation to the overall vascular-plant diversity are found on oceanic islands such as Puerto Rico (15% according to Groombridge 1992) or Juan Fernandez (24%, Marticorena 1990) and in humid montane forests of the tropics, with often greater than 10% of the species (Kessler et al. 2001).

EVALUATION AND MEASUREMENT OF BIODIVERSITY

Numerous criteria exist for assigning a value to biodiversity. For most of them several indexes can be used for their quantitative measurement. Listed below are some of the most frequently used criteria.

1. Taxon richness: The number of species present in a given area is the most common measure of biodiversity. This is mainly due to the fact that species are the unit of biodiversity that can be defined most easily; hence, data on species richness are more abundant than others. Measuring the richness of higher taxa such as genera or families, however, can reveal the fundamental differences among the species of a flora or fauna and thus provides another invaluable indicator of

diversity. For other purposes it might be appropriate to look at the infraspecific level. Spatial patterns of diversity can vary considerably with taxonomic level, as shown by Quian (1999), Mutke and Barthlott (2000), and Mutke (2002).

2. Abundance structure: High species richness in an area may be impressive, but a different value may be placed on it if it were known that only a few species were dominant while most others were represented by a very small number of individuals. Hence, the distribution of individuals or biomass among the species within a community (evenness) is another important criterion for the evaluation of its diversity. These measures can also be combined with taxon richness (e.g., Shannon-Wiener index; for a discussion of several diversity indexes see Magurran 1988).

3. Phylogenetic and character diversity: A community can be regarded as being more diverse the more features (genetic, phenetic, or functional) its species show (Humphries et al. 1995; Webb 2000). Feature diversity can be approximated by phylogenetic diversity and can be applied as an indicator of potential use or conservation value (Williams 1993; Faith 1994).

4. Range sizes and degree of endemism: A given area is usually regarded as being more valuable when the percentage of endemics is high or when the species hosted by an area have on average small range sizes (mean or median range-size rarity; Williams 1999). Degree of endemism can be combined with species richness to represent endemism richness (Kier and Barthlott 2001). There is often a high correlation between degree of endemism and species richness, but patterns can also differ. For example, a comparison between plates 2.1 and 2.2 reveals that the number of vascular-plant species per 10,000 km^2 is higher in the Congo basin than in west Madagascar, whereas the latter has a higher level of endemism richness.

5. Intactness: The degree to which an area is affected by habitat destruction or by nonnative invasive species usually influences the indicators mentioned above but can also be regarded as a criterion of biodiversity value on its own.

6. Relevance for ecosystem function: The role of species with regard to ecosystem processes such as the flux of matter (carbon dioxide, nitrogen, water, etc.) often influences the value assigned to them.

7. Current and potential economic value: This criterion emerges from the anthropocentric point of view, but in practical terms it is one of the most important. The value that biodiversity has on its own should be an argument strong enough for its protection, but, in reality, the direct or indirect economic effects usually turn out to be among the strongest causes for action.

LATITUDINAL AND CLIMATIC GRADIENTS

Within the last few decades, the scientific literature has featured a discussion about the trends of biological diversity along various environmental gradients (e.g., Pianka 1966; Begon et al. 1990; Currie 1991; Rohde 1992; O'Brien 1993; Rosenzweig 1995; Brown 1995; Gaston 2000; Willis and Whittaker 2002). Apart from species-area relationships (see below), the most prominent patterns in this context are the latitudinal and altitudinal gradients of species richness (Humboldt 1806; Hawkins 2001; Rohde 1992; Rosenzweig 1995; Gaston 1996; Rahbek 1995; Willig et al. 2003). As shown in figure 2.1A, species richness in many groups of organisms increases with decreasing latitude, though opposite patterns for single groups can be found as well (e.g., Janzen 1981; Williams 1993). Even though this trend is among the best documented in ecology, the discussion on the causes and mechanisms behind latitudinal gradients is still lively. Rohde (1992) names more than 25 different mechanisms that have been suggested for generating latitudinal gradients. The most popular explanations are geographic area, environmental stability, historical influences, climatic factors, energy availability, and habitat heterogeneity (Kerr and Packer 1997).

One remarkably simple hypothesis explaining why the tropics has the greatest species richness is that it has a larger total area size than other ecoclimatic zones (Terborgh 1973; Rosenzweig 1995). However, there is no systematic decline in area size of ecoclimatic zones with increasing latitude. Some high-latitude zones, such as boreal forests or tundra, tend to be large. As a sole model the geographic-area hypothesis seems to be insufficient. The few statistical tests of the hypothesis show no clear result (Blackburn and Gaston 1997; Hawkins and Porter 2001).

Historical events, such as glaciations during the ice ages, have important influences on the floras of the respective areas. The well-known impoverishment of the central European flora compared to Pleistocene refugia in China and the southeastern United States serves as a prominent example (Boufford and Spongberg 1983). For tropical regions it is postulated that the large humid forest areas of today were restricted to smaller patches surrounded by dryer climate during the ice ages (Haffer 1982, 1997; White 1993; Haffer and Prance 2001). Today centers of species richness and restricted-range species can be found in these areas (for further discussion, see chapter 10). However, Nelson et al. (1990) found a close relationship of the proposed refuge areas in the Amazon basin to sampling density. In this context, the climatic stability of an area is increasingly discussed as one parameter for the evolution of diverse floras (Fjeldsa and Lovett 1997; Cowling and Procheş, in press). Jansson (2003) shows a high correlation of centers of range-restricted species for several taxa of animals and plants with areas of low levels of past climatic changes.

Species richness has also been shown to be closely associated with measures of available environmental energy (e.g., potential evapotranspiration, PET) in several analyses studying various groups of organisms (Currie and Paquin 1987; O'Brien 1993, 1998; Mutke 2000, 2002; Mutke et al. 2002; Francis and Currie 2003; compare as well fig. 2.1B). The species-energy theory (Wright 1983) suggests that the energy supply supporting a community limits the capacity of that community to contain species. The relationship is scale-dependent. At a relatively local scale there is a tendency for a hump-shaped unimodal relationship (Rosenzweig 1995), whereas at larger geographic scales species richness shows a positive monotonic relationship to energy availability (Gaston 2000). It is argued that greater energy availability enables an area to support a greater biomass. By enabling more individuals to coexist, more species have enough individuals to maintain viable populations. Criticizing the theory, McGlone (1996) argues that the evolutionary process of speciation, together with extinction, mainly shapes the diversity of a given area on the geographic scale, which is far too slow to generate the patterns of diversity on the basis of the current climate. A strong correlation of regional climate with diversity arises on the basis that present environmental conditions and spatial variation are a proxy for past situations. Cowling and Lombard (2002) and Cowling and Proches (in press) explain the regional differences of plant species in the South African Cape region mainly by differences in the speciation/extinction histories instead of contemporary climate or environmental heterogeneity. Nevertheless, Tuomisto and Ruokolainen (1997) and Kerr and Currie (1999) argue that in most analyses parameters of the contemporary environment show a more consistent relation to species diversity than does evolutionary history and that the influence of the latter is difficult to test. Hawkins and Porter (2003) found far stronger statistical relationships of North American mammal and bird diversity to current climate than to historical factors. The long-lasting discussions on this issue may be due to the facts that (1) the influences are difficult to partition, (2) some of the hypotheses are difficult to test, and (3) different groups of organisms in different biogeographic regions were studied at different spatial scales, resulting indeed in differences of the importance of the explanatory variables.

Finally, so-called null models show the possibility of generating a gradient in species richness solely by a random association between the geographic midpoint and the size of the geographic ranges of the species (Colwell and Hurtt 1994; Colwell and Lees 2000; Jetz and Rahbek 2001). This middomain effect is often also cited as an explanation for the midaltitude species-richness peaks, which are found, for example, for bryophytes and various epiphytic taxa in the tropical Andes (Gradstein and Pócs 1989; Ibisch et al. 1996; Kessler 2000). It has been shown, however, that these altitudinal patterns may vary greatly among different groups and may be explained by different reactions to geologic processes in the past and to climatic factors such as the levels of condensation and resulting higher precipitations at midaltitudes (Gentry 1982; Lauer 1989; Ibisch et al. 1996; Kessler 2000; Mutke 2002; Braun et al. 2002). Ibisch et al. (1996) show as well that the shape of the species-richness curve along the altitudinal gradient varies greatly with the resolution of the altitudinal zones used. It is probably a great simplification, however, to expect one or a few mechanisms to explain a pattern that is built on a wide range of different life histories and ecological adaptations of a large number of species.

DIVERSITY AT DIFFERENT SCALES: THE HETEROGENEITY OF THE LANDSCAPE

When studying biodiversity at different scales, the relative importance of different parameters and processes changes (e.g., see Rahbek and Graves 2001; Willis and Whittaker 2002). Nevertheless, local species sets are assembled from a regional pool of species. Thus, regional processes influencing this pool are important drivers structuring local communities as well. These include the effects of geophysical properties and history of the region and large-scale ecological and evolutionary processes, such as speciation, regional extinctions, or species migrations (Huston 1999). Correspondingly, in many systems it is found that local species richness is lower but proportional to regional richness (Caley and Schluter 1997; Gaston

2000). The relationship between species richness and area is not linear, however, nor is it uniform in different taxonomic groups, in different habitats, or at different scales (Connor and McCoy 1979; Rosenzweig 1995; Crawley and Harral 2001). For example, Lennon et al. (2001) found no statistically significant relationship between richness patterns of British birds at resolutions of 10 × 10 km and 90 × 90 km. Most analyses found that the number of species increases at a declining rate with increasing area size (Connor and McCoy 1979; Rosenzweig 1995; Plotkin et al. 2000). The most commonly used equation to describe this relation is the power model by Arrhenius (1920, 1921): $S = cA^z$, where S is the species number on an area of size A, and c the species richness on an area of size 1. This function describes the linear regression fitted to a log (species) versus log (area) plot with the parameter z being the slope of the regression line. Though the parameter z commonly shows values about 0.25 (Rosenzweig 1995), it varies considerably among different systems studied. For vascular plants, we found values for this parameter ranging from 0.11 in hot deserts and 0.13 in the Arctic tundra to 0.33 in Neotropical moist forests (G. Kier, J. Mutke, and W. Barthlott, unpublished data). Biologically, this parameter can be interpreted as an index of the spatial ecological heterogeneity or the floristic heterogeneity (Malyshev 1991, 1993).

Whittaker (1972, 1977) differentiated between alpha, beta, and gamma diversity to analyze biodiversity at different scales. Alpha diversity addresses the species richness at the local scale or the within-habitat diversity. Beta diversity, also called between-habitat diversity, describes the species turnover comparing different sites. Gamma diversity is the term for species richness at larger areas comprising different habitats. The relationship among these three types of diversity is described by the equation gamma = alpha × beta (Whittaker 1977). The magnitude of beta diversity varies among ecosystems. For example, there is lower species turnover in boreal forests compared to tropical moist forests (Condit et al. 2002). However, the patterns of beta diversity within the Amazon lowland rain forests, for example, are still less understood compared to alpha diversity in these ecosystems (Tuomisto et al. 1995; Pitman et al. 1999; Condit et al. 2002; Tuomisto et al. 2003). To a large degree, beta diversity is influenced by the diversity of the abiotic environment, the geodiversity (Faith and Walker 1996; Barthlott et al. 1996; Barthlott et al. 2000; Jedicke 2001; Braun et al. 2002). This is especially so in mountain systems with a high diversity of elevational zones, slopes, aspects, steep climatic gradients, and mosaics of habitat patches that permit a high diversity of different vegetation types and ecological communities. Barthlott et al. (2000) showed that all species-richness maxima for vascular plants in South America are found in areas with high geodiversity values. Thus, especially for floristically poorly studied areas, indicators are derived from known environmental parameters or vegetation maps generated with the use of remote sensing to describe the heterogeneity of the landscape (Wickham et al. 1995; Rahbek and Graves 2001; Kerr et al. 2001; Braun et al. 2002; Mutke 2002). Most of these studies found that heterogeneity of the environment is one of the important parameters in multivariate models predicting species diversity on the basis of environmental data, though the relation is not always this clear in univariate analyses (Mutke 2002). Analyzing mammal diversity in North America, Kerr and Packer (1997) found that habitat heterogeneity is a good predictor for species richness in "high-energy regions" with potential evapotranspiration (PET) greater than 1,000 mm yr^{-1}, whereas this relation was weak in "low-energy regions" at higher latitudes where PET is a stronger predictor. Deutschewitz et al. (2003) and Kühn et al. (2004) showed close correlation of the diversity of native and introduced species with measures of geodiversity in central Europe. Even the fact that towns are relatively species-rich today might partly be explained by the fact that humans tend to settle in geodiverse areas (Kühn et al. 2004).

THE ROAD AHEAD

As discussed in the first paragraphs of this chapter, the full story of plant diversity is far from complete. Every year more than 2,000 new species of vascular plants are described. There has been no year since the work of Linnaeus in the 18th century without the publication of a new genus of vascular plants (analyses based on the *Index Kewensis*, Royal Botanic Gardens Kew 1993). Thomas (1999) estimates that some 25% of the Neotropical plant species are still awaiting discovery. Half of all vascular-plant genera, however, were described by the mid-19th century. Even at the species level half of all names were published by the start of the 20th century (analyses based on the *Index Kewensis*, Royal Botanic Gardens Kew 1993). It is estimated that some 85% of the expected total vascular-plant diversity has been described (Heywood 1995). There are few things yet to be discovered that will represent completely new lineages of vascular plants. This is in contrast to the situation found in many groups of animals, such as the insects, where new orders have recently been discovered (Klass et al. 2002). Regardless, knowledge of regional floras is highly uneven. Frodin (2001a) has discussed the "areas that most need floras." It is very promising to see new initiatives for the discovery of plant biodiversity, especially those that analyze the data gathered during the last two centuries with the help of modern computerized information systems (e.g., Chatelain et al. 2001; Müller et al. 2003; Linder et al., in press).

2.2 MARINE PLANT DIVERSITY

Walter H. Adey

PLANTS IN THE SEA play an extremely important role in the functioning of the earth's biosphere. Seawater covers nearly three-quarters of the earth's surface. Even though water blocks solar radiation to varying degrees, roughly 40% of the photosynthesis occurring on earth takes place in the oceans. Equally important, much of the carbon that is taken up by marine plants is eventually sequestered long-term in the deep ocean; oceanic photosynthesis thereby plays a key role in structuring geologic-scale climatic patterns (Saltzman 2002).

The plants that dominate the land play a very minor role in the sea. Gymnosperms, ferns, and mosses are absent. Lichens play only a subsidiary role in the intertidal and shallow subtidal, the boundary zone between marine and terrestrial realms. Only two families of flowering plants, the Potamogetonaceae and the Hydrocharitaceae, both monocot families and both with freshwater species, have given rise to marine submerged aquatic vegetation (SAV). These two plant lines invaded the ocean from terrestrial environments in the later Tertiary, probably through estuarine systems. Marine SAVs or "seagrasses," with only 12 genera and 48 species, occur along the coasts of much of the world ocean, except for the strictly Arctic and Antarctic zones. Only a few species are pantropic (*Halophila decipiens*) or pantemperate (*Zostera marina*). Most species occur throughout their respective biogeographic regions, and a few regions are unusually rich in endemic species (South Australia, Japan). Almost all marine SAVs occur on weakly disturbed muddy to sandy bottom. A few may grow over rock or coral from their sediment base. One genus of several species in temperate waters (*Phyllospadix*) is more adapted to rock substrate (Phillips and Menez 1988).

Seagrasses can be highly productive, locally extensive, and of high standing crop. They are subject to significant human degradation of habitat and present serious conservation issues. However, they represent only a tiny fraction of marine plant biodiversity. Algae, belonging to a number of phyla or divisions, make up the vast majority of the ocean's plant species, and it is to those groups that the following analysis is directed.

Algae are highly polyphyletic. Depending upon the author or the approach, at least seven different lineages have arisen (Andersen 1992), and at least three of those lineages existed in the ocean during the Precambrian, long before life moved onto the land. In general, the basic or genomic diversity of marine algae is much higher than in land plants. The view of Norton et al. (1996) that "surely vastly more of the world's current genetic diversity is lodged in different phyla rather than in a plethora of closely related species" is brought sharply into focus by Sepkoski's note (1995) that twice as many phyla reside in the sea as on land. There are likely over a million species of algae, as compared to somewhat over 300,000 species of land plants (e.g., Groombridge 1992). In addition marine algae represent a greater array of basic diversity than the land plants. The dominating marine benthic algae of the Rhodophyta and Phaeophyta are virtually absent from fresh water, where the majority of algae are evolutionary offshoots of the single division of green algae (division/phylum Chlorophyta).

Plants in the open ocean, or the pelagic environment, are represented by algal phytoplankton, mostly unicells or sometimes filaments, with dominant members derived from several algal divisions, the Bacillariophyceae (diatoms), Dinophyceae (dinoflagellates), Coccolithophoraceae, and Cryptophyceae being especially important (Andersen 1992). Longhurst (1998) provides a modern analysis of the distribution of plankton, including phytoplankton throughout the world ocean. The emphasis in this section is on benthic algae, although the uncertainty regarding the number of species (one to two orders of magnitude) applies just as well to phytoplankton as it does to the benthic algae discussed below. While coastal algal biodiversity is clearly more subject to human perturbations, it should not be assumed that the effects of human activity in the open ocean are negligible. That roughly half of anthropogenic carbon is thought to be lost to an oceanic sink (initially through phytoplankton) should dispel that notion. Also, ultraviolet radiation increase due to stratospheric ozone breakdown is likely to affect the vertical distribution of phytoplankton.

In a recent review of global algal biodiversity, Norton et al. (1996) concluded that the current approximation of 30,000 to 40,000 species vastly understates the true value. Recent work with the electron microscope and molecular taxonomy has shown that some algal groups will eventually lose species to synonymy (Andersen 1992). However, such groups tend to be confined to microalgae that can be rather plastic relative to season, reproductive stage, and specific environment; "few taxonomists doubt that [those lost to synonymy] will be far outstripped by those awaiting discovery" (Norton et al. 1996).

It is puzzling that scientists have not devoted more time to understanding the biodiversity represented by the morphologically and physiologically diverse algae, since algae perform roughly 40% of the photosynthesis on earth. There is, however, a vast array of benthic algae in the sea that are small-coin to tree size (the seaweeds); these plants are extraordinarily diverse in color, structure, and reproductive process. Seaweeds, clothed and bathed in microscopic algal epiphytes and epibenthic plankton, highly dominate on hard (rocky or coral) bottoms.

BIOGEOGRAPHY AND THERMOGEOGRAPHY

Species of marine benthic organisms are not evenly or randomly distributed along coastlines. They are grouped into several dozen regions or provinces of vastly different dimensions. For more than a century the study of these regions has been carried out within the disciplines of biogeography and zoogeography (e.g., Ekman 1953; Briggs 1974). The same basic patterns exist among seaweeds (e.g., van den Hoek 1984; Alvarez et al. 1988), although Bolton (1994) and Luning (1990) have suggested that significant anomalies exist for the seaweeds relative to the distribution of animal biodiversity.

Adey and Steneck (2001) have demonstrated that water climates (summer and winter temperature regimes) are not evenly or randomly distributed along the world's coastlines, and that there is a correlation between the distinctive areas of these climatic regimes and biogeographic groupings. Those authors have shown that most of the biogeographic regions can be physically modeled by graphically examining the Pleistocene continuity of the climatic units. Summer and winter climatic regimes that continue to occupy large, contiguous coastal regions through-

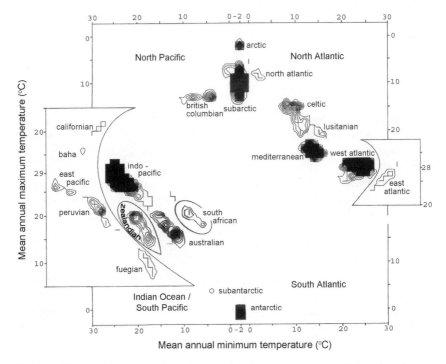

Figure 2.2 Climatically characterized areas of rocky coast that are contiguous and continuous throughout the 3–5 million years of Pleistocene climate cyclicity. The contoured or z-dimension represents the amount of coastal area continuously available. See Adey and Steneck 2001 for derivation.

out the 3–5 million years of Pleistocene glacial-interglacial shifts have produced species concentrations that can be characterized as biogeographic regions. These theoretical regions, based on summer and winter temperatures, coastal area, isolation, and contiguous continuity over Pleistocene time, are called thermogeographic regions (plate 2.3, fig. 2.2). Since the majority of living algal species have evolved over Pleistocene time, it is likely that the biogeographic regions for most species will be the same as the thermogeographic regions. Adey and Steneck (2001) have examined that relationship for coralline algae in the North Atlantic and have shown a striking correlation.

The distribution and abundance of generic and family-level taxa can be examined by looking at Tertiary climate distribution along coastlines. Deep-ocean drilling and core analysis of microfossils are providing increasingly long-range climate information that will allow extension of the thermogeographic model to higher taxa. The climate of today's coastal geographic unit is less than 5,000 to 10,000 years old; thus, most biogeographic groupings attached to a specific segment of coastline have existed for a much shorter time than most species have required for their evolution. Coastal biotas have been quite mobile during the Pleistocene. For example, the Atlantic boreal (Celtic) today (and in many interglacials) is centered in the northern British Isles. During the glacials, however, it was centered farther south in the western Mediterranean and the adjacent eastern Atlantic. In terrestrial biogeography, the most recent of these shifts are not only well recognized but also mapped, using fossil pollen, especially from ancient lake-bed sediments (Wilson et al. 2000). Lacking fossil pollen, and in many cases for rocky shores even the clear evidence of greatly changed ecology, benthic biogeographers have failed to fully recognize this very significant situation.

The net result of climatically driven biogeographic cyclicity is that biogeographic regions are "focused" to relatively narrow zones, with sometimes large transition areas, occupied by species from adjacent core regions. The transition zones are thus hot spots of biodiversity that constantly shift their character with time. Perhaps in some cases, the species in these transition zones are ecologically and reproductively supported by their "home" core regions over thousands of years. Also, such hot spots are not likely stable, but constantly shifting with time as new species arrive and older species become locally extinct.

THERMOGEOGRAPHIC REGIONS AND SEAWEED BIODIVERSITY

Because the described biogeographic model is based on relative coastal area (table 2.1) and a theoretical basis for a species-area curve exists ($S = cA^z$; see section 2.1), it is possible to predict the expected biodiversity of poorly known regions from extensively studied regions. Several reasonably well described seaweed floras apply more or less closely to regions that are well-defined in the theoretical model (e.g., Abbott and Hollenberg 1976; Adams 1994). These floras are almost certainly incomplete. However, they provide a reasonable framework within which to estimate global species diversity.

Table 2.1 lists the 16 well-defined thermogeographic regions for which Adey and Steneck (2001) were able to provide relative areas in square nautical miles. Arctic and Antarctic regions are omitted because the areas present in the glacial state (CLIMAP Project Members 1976) are too small to provide an adequate multiplier for current area. The concept that net Pleistocene polar coastal area is small, because it virtually disappears during glacials, applies well to the Arctic, where few species have evolved and there are even fewer distinct genera (e.g., Luning 1990). The Antarctic is similarly depauperate in species; however, because (unlike the Arctic) it is so remote and has been glaciated for a much larger period of time, it is rich in endemics. The Arctic is very rich in Subarctic species because the latter region is large, has great Pleistocene continuity, and surrounds the Arctic. Since the numbers used in table 2.1 are based on a uniform coastal width, they are a relative rather than an absolute measure.

Table 2.1 lists the expected regional red, brown, and green seaweed biodiversities based on the direct, log/log relationship for large regions (Hubbell 2001) and extrapolated from the floras of New Zealand (Adams 1994) and California (Abbot and

Table 2.1 Estimated global seaweed biodiversity

Thermogeographic regions[a]	Area (n.m.²)[b]	Seaweed biodiversity based on	
		New Zealand	California
Indo-Pacific	175,020	88,908	287,000
West Atlantic	6,168	3,135	10,109
Subarctic	5,016	2,550	8,220
Mediterranean	3,354	1,705	5,497
British Columbian	840	427	1,377
Celtic (Atlantic boreal)	810	412	1,328
Australian (combined)	726	369	1,190
Zealandian	714	363[c]	1,170
Peruvian	522	265	856
Fuegian	300	153	492
Lusitanian	300	153	492
South African	270	137	443
East Atlantic	210	107	344
East Pacific	210	107	344
Californian	180	92	295[c]
Baja California & Gulf	84	43	138
Totals (rounded)		99,000	320,000

Sources: Estimates based on the thermogeographic regions of Adey and Steneck 2001 and the floras of New Zealand (Adams 1994) and California (Abbott and Hollenberg 1976).

[a]Arctic and Antarctic are not provided a quantitative area for reasons discussed by Adey and Steneck (2001). Those regions are likely quite distinct but very impoverished.

[b]In effect, a ratio rather than real area; see Adey and Steneck 2001. n.m., nautical miles.

[c]Number has been determined from floras by separating out cosmopolitan species (those ranging beyond adjacent biogeographic regions) and then evaluating from distribution and type data the likely region in which each species dominates. Note: these values are best referred to the concept of "home region"—the region in which a species reaches its maximum abundance. These are not "endemic" species—see Adey and Steneck (2001). These are also minimum values in that they do not include cosmopolitan species, and especially for the New Zealand region represent incomplete floras. They also do not include cyanobacteria and diatoms, and the latter omission is likely significant.

Hollenberg 1976). These two floras cover coastal territory that is relatively isolated from adjacent regions and are similar in geographic coverage to the equivalent region of the thermogeographic model. Also there is enough ecological and distributional data to allow tentative regional placement for most overlapping species. The California flora has been studied extensively by many workers over more than a century, compared to the New Zealand flora, which is well-studied but has significant geographic gaps in coverage. For the three major algal divisions, the numbers provided represent a likely minimum (based on New Zealand) to maximum (based on California) range of biodiversities. These values do not include benthic cyanophytes and diatoms. This may not represent a serious omission for temperate, Arctic, and Subarctic regions; however, it is likely a problem for tropical regions (see chapter 6).

It has been known for over a century that most larger and higher taxa show a latitudinal diversity gradient (fewer species in colder waters). In recent years, this gradient has been correlated with solar energy input and quantified based on mean sea-surface temperatures (Roy et al. 1998; Buzas and Culver 1999) for the very well known mollusks and foraminifera. While a typical gradient from Arctic to tropics is on the order of 4–5 to 1, the ranges from temperate regions (indexes of Californian and New Zealandian) to tropics or Arctic is less. In a general way, it is expected that Subarctic species diversities are half as large as the numbers presented based on area; tropical diversities may be twice as high. Rare species show stronger gradients than abundant species (i.e., more proportionally rare species in the tropics), and a more detailed analysis of available data might take that relationship into account.

GLOBAL SEAWEED BIODIVERSITY

As predicted by the earlier-cited literature, the species estimates derived from the sum of the marine biogeographic regions (even when the Arctic, Antarctic, and several less significant regions and cosmopolitan species are excluded) are an order of magnitude higher than the published listings of seaweeds for these major groups. Many of the smaller regions are more or less similar in terms of already known species and expected species counts. However, the model suggests that the tropical Indo-Pacific alone would be expected to have over 80% of the total species biodiversity, far more than the number already identified for the entire world ocean. As noted, likely latitudinal gradients would increase the numbers above the estimates derived in this paper.

Bolton (1994) and Norton et al. (1996) have suggested that the apparent low level of seaweed biodiversity in the Indo-Pacific results from limitation by space-competing corals and intensive grazing. Typical levels of grazing in reef environments increase primary productivity, however, and would tend to also increase algal diversity based on classical grazer-algae interactions (Adey 1998). On the other hand, red, brown, and green algal turfs dominate coral-reef carbonate surfaces (Price and Scott 1992; Adey 1998). This ubiquitous coral-reef subcommunity provides a large part of the photosynthesis of these highly productive ecosystems and the rapid carbon removal that supports the intensive calcification needed for reef maintenance. Yet the biodiversity of algal turf species, and especially the generally omitted cyanophytes and diatoms, over the immense Indo-Pacific (and to a lesser extent the Caribbean), is vastly under-studied. In providing a species biodiversity estimate of over 3.5 million biological species for pantropic coral reefs, using a microcosm analog, the Small and Adey (1998) data also show an estimate of pantropic reef algae at 875,000 species. Of that number, roughly 43% (373,000 species) would be red, green, and brown seaweeds. The remainder are benthic diatoms, cyanophytes, and to a lesser degree dinoflagellates, the diatoms being particularly important. If algal diversity for the tropics in table 2.1 is doubled for latitudinal gradient, the resulting number of 600,000 species is close to that derived in Small and Adey (1998).

The thermogeographic-predicted seaweed diversity for the Mediterranean, at over 1,700 to 5,000 species, would seem high as compared to existing floras (e.g., see Norton et al. 1996, approximately 830 species). However, the eastern warm region shows relatively few species in the Norton et al. diagram, and it is suspected that field studies are as yet inadequate, with the same caveat (under-studied algal turfs) as for the equatorial tropics. While the late Miocene refilling of the Mediterranean from the adjacent Atlantic would not have allowed an influx of tropical species (van den Hoek and Breeman 1989), it would seem likely that the over 5 million years since that event has been fully adequate to develop a rich warm temperate-tropical species complex. While the generic framework would perhaps be skewed as a result of the refilling, it is not likely that species number would have been significantly affected.

The most anomalous thermogeographic-predicted algal biodiversity figure is that for the Pacific Subarctic. At 2,100 to 11,000 species, this is far above the level presently described, and even if the number is halved for latitudinal gradient, the number is still very large. Yet this is a very large area (Sea of Okhotsk, Kamchatka, and the western and northern Bering Sea), and it has been areally and climatically stable through the Pleistocene (i.e., the biota need only be shifted a short distance south to find an equally large area during glacial maxima). Most of the Arctic Ocean and Atlantic Canada (ca. 200 seaweed species) are likely centered on the Pacific Subarctic region. However, it seems probable that the Okhotsk/Kamchatka region is greatly under-studied with regard to its algal flora—proportionally the most under-studied region on the planet.

Issues of endemicity and presence-absence data relate to biogeographic regions (Adey and Steneck 2001). For the most part, endemic species are rare and provide poor definition for regional delimitation; endemic genera are far more valuable, as they can indicate relict Tertiary regions. The use of presence-absence produces a search for rarity and is heavily dependent on variation in field effort. The key character applicable to determining species distribution in relation to biogeographic regions is abundance (cover, standing crop, etc.). As Adey and Steneck (2001) demonstrated, for species of Corallinales (where cover was the abundance measure used), individual species can reach a very high relative percentage of cover within their "home regions." However, they also extend through transition areas and into adjacent biogeographic regions in small percentages, often in marginal microclimates. Presence-absence as applied to coralline species would have failed to delimit biogeographic regions and additionally would provide biodiversity lists that have to be "sifted" for redundant species from flora to flora. While it is likely that some species are eurythermic and cosmopolitan, ranging widely over several regions, most are restricted in significant abundance to a "home region." This is markedly so for species of the Corallinales in the Northern Hemisphere but, according to studies in progress for the Atlantic Subarctic (Adey and Hayek, in progress), applies also, though to a lesser degree, to the primary intertidal seaweeds, where widespread eu-rythermic character would be expected.

North and south of the tropics, biogeographic regions that were areally contin-uous during the Pleistocene often develop large transitional zones where many spe-cies of adjacent regions occur. In some cases, for example western Europe, where distinct biogeographic regions lie geographically close and connected, zones of high species overlap are to be expected. This is true in the tropical east Atlantic region, where both Lusitanian and perhaps even Mediterranean species are to be found. This situation is likely responsible for the large reported species number (approx. 650 species; Prud'homme van Reine and van den Hoek 1988) as compared to thermogeographic-predicted numbers (120 to 340 spp., table 2.1). Whether such diversity hot spots are likely subjects for conservation, when for many of the spe-cies the "home regions" are elsewhere, is a matter for debate. Such a transition region is very likely to be unstable with time, constantly changing in species com-position as each interglacial proceeds and colder region species are lost while warmer-region species continue to expand range.

CONCLUSION

There are many more species of marine algae in the ecosystems of the oceans than have been described by scientists. Roughly 20,000 have been described (about 50% of the total given by Groombridge 1992 for all algal species). That same global bio-diversity analysis suggested that 100,000 species (using the same 50%) would be a good working figure for what actually exists, though noting that a number of 10 million species has been proposed (mostly diatoms). Based on the analysis of a quantitative biogeographic model, two extensive algal floras, and a microcosm-analog modeling of coral reefs, however, the information provided in this chapter suggests that 300,000 to 1,000,000 species exist for the marine benthic realm alone. Some biogeographic regions are reasonably well known (Atlantic boreal [Celtic], California), others are vastly understudied due to lack of attention (the Pacific Sub-arctic) or the enormity of the task (the tropical Indo-Pacific).

It is extremely difficult to frame questions of potential loss of seaweed species due to direct degradation of local environment, movement of invasive species, or change in the flora during global warming without a baseline understanding of seaweed dis-tribution. Prior to the development of the theoretical thermogeographic model by Adey and Steneck (2001), no universally accepted pattern for examining seaweed distribution and biogeography existed. Many potentially valuable floras, difficult in their own right to produce, overlapped biogeographic regions, and because they were based on presence-absence, usually without significant regard to abundance or seasonality, provided only a very limited view of large-scale biodiversity. In short, an enormous research effort has been proceeding without a working hypothesis or framework, and the results ask more questions than they answer. It is virtually im-possible to evaluate seaweed biodiversity from the table provided by Norton et al. (1996) because of the extensive overlap of species from adjacent regions. Even by consulting such floras in detail, it is quite difficult to evaluate the distribution and abundance of many species, some of which have been found once or twice and many others of which have been collected from very few locations.

Most researchers have failed to apply even the most basic understanding of Pleistocene climate fluctuations. The view has been that of a preexisting, stationary global picture of the distribution of seaweed floras. Except for the core tropical Indo-Pacific and tropical western Atlantic regions, the reality has been extremely fluid. Seaweed biogeography is sometimes framed in the context of Tertiary climate or geography (e.g., Luning 1990), but this framework is unlikely to have direct species significance, since seaweed species evolution likely happened mostly in the Pleistocene.

The dynamic yet well-patterned biogeography of many regions has not been appreciated by the workers of the 20th century, and many species-distribution characterizations, especially with limited field information, have been stated in error. While we can begin to frame conservation questions of seaweed (and general algal) floras, only intensive further field research framed by working hypotheses can lead us to meaningful answers. Equally critical, as long as seaweed distribution is primarily framed qualitatively as presence-absence, rarity continues to be equated with abundance. Until sufficient fieldwork produces even quasi-quantitative data, the basic conservation questions, except for the occasional explosive invasive, will largely go unanswered.

LITERATURE CITED

Abbott, I., and Hollenberg, G. 1976. Marine Algae of California. Stanford University Press, Stanford, California.

Adams, N. 1994. Seaweeds of New Zealand. Canterbury University Press, Christchurch.

Adey, W. 1998. Coral reefs: algal structured and mediated ecosystems in shallow, turbulent, alkaline waters. Journal of Phycology 34:393–406.

Adey, W., and Hayek, L. C. In progress. The biogeographic structure of the northwestern Atlantic rocky shore intertidal.

Adey, W., and Steneck, R. 2001. Thermogeography over time creates biogeographic regions: a temperature/space/time-integrated model and an abundance-weighted test for benthic marine algae. Journal of Phycology 37:677–698.

Alvarez, M., Gallardo, T., Ribera, M., and Gómez Garreta, A. 1988. A reassessment of northern Atlantic seaweed biogeography. Phycologia 27:221–233.

Andersen, R. A. 1992. Diversity of eukaryotic algae. Biodiversity and Conservation 1:267–292.

Arrhenius, O. 1920. Distribution of the species over the area. Meddelanden Från K. Vetenskapsakademiens Nobelinstitut 4:1–6.

———. 1921. Species and area. Journal of Ecology 9:95–99.

Barthlott, W., Biedinger, N., Braun, G., Feig, F., Kier, G., and Mutke, J. 1999a. Terminological and methodological aspects of the mapping and analysis of global biodiversity. Acta Botanica Fennica 162:103–110.

Barthlott, W., Kier, G., and Mutke, J. 1999b. Biodiversity: The Uneven Distribution of a Treasure. NNA Reports 12/1999 (special issue 2): 18–28.

Barthlott, W., Lauer, W., and Placke, A. 1996. Global distribution of species diversity in vascular plants: towards a world map of phytodiversity. Erdkunde 50:317–328.

Barthlott, W., Mutke, J., Braun, G., and Kier, G. 2000. Die ungleiche globale Verteilung pflanzlicher Artenvielfalt: Ursachen und Konsequenzen. Berichte der Reinhold Tüxen-Gesellschaft 12:67–84.

Beaman, J. H. In press. Mount Kinabalu: hotspot of plant diversity in Borneo. In Friis, I., and Balslev, H., eds., Plant Diversity and Complexity Patterns: Local, Regional, and Global Dimensions. Royal Danish Academy of Sciences and Letters, Copenhagen.

Begon, M. E., Harper, J. L., and Townsend, C. R. 1990. Ecology: individuals, populations, and communities. 2nd edition. Blackwell Scientific, Boston.

Blackburn, T. M., and Gaston, K. 1997. The relationship between geographic area and the latitudinal gradient in species richness in New World birds. Evolutionary Ecology 11:195–204.

Bolton, J. J. 1994. Global seaweed diversity: patterns and anomalies. Botanica Marina 37:241–245.

Boufford, D. E., and Spongberg, S. A. 1983. Eastern Asian–eastern North American phytogeographical relationships: a history from the time of Linnaeus to the twentieth century. Annals of the Missouri Botanical Garden 70:423–439.

Braun, G., Mutke, J., Reder, A., and Barthlott, W. 2002. Biotope patterns, phytodiversity, and forestline in the Andes, based on GIS and remote sensing data. Pp. 75–89 in Körner, C., and Spehn, E. M., eds., Mountain Biodiversity: A Global Assessment. Parthenon Publishing, London.

Briggs, J. 1974. Marine Zoogeography. McGraw-Hill, New York.

Brown, J. H. 1995. Macroecology. University of Chicago Press, Chicago.

Buzas, M., and Culver, S. 1999. Understanding regional species diversity through the log series distribution of occurrences. Diversity and Distributions 8:187–195.

Caley, M. J., and Schluter, D. 1997. The relationship between local and regional diversity. Ecology 78:70–80.

Chatelain, C., Gautier, L., and Spichiger, R. 2001. Application du Sig Ivoire á la distribution potentielle des espéces en fonction des facteurs écologiques. Systematics and Geography of Plants 71:321–326.

CLIMAP Project Members. 1976. The surface of ice-age earth. Science 191:1131–1137.

Colwell, R. K., and Hurtt, G. C. 1994. Nonbiological gradients in species richness and a spurious Rapoport effect. American Naturalist 144:570–595.

Colwell, R. K., and Lees, D. C. 2000. The mid-domain effect: geometric constraints on the geography of species richness. Trends in Ecology and Evolution 15:70–76.

Condit, R., Pitman, N., Leigh, E. G. Jr., Chave, J., Terborgh, J., Foster, R. B., Núñez V., P., Aguilar, S., Valencia, R., Villa, G., Muller-Landau, H. C., Losos, E., and Hubbell, S. P. 2002. Beta-diversity in tropical forest trees. Science 295:666–669.

Connor, E. F., and McCoy, E. D. 1979. The statistics and biology of the species-area relationship. American Naturalist 113:791–833.

Cowling, R. M., Esler, K. J., and Rundel, P. W. 1999. Namaqualand, South Africa: an overview of a unique winter-rainfall desert ecosystem. Plant Ecology 142:3–21.

Cowling, R. M., and Lombard, A. T. 2002. Heterogeneity, speciation/extinction history, and climate: explaining regional plant diversity patterns in the Cape Floristic Region. Diversity and Distributions 8:163–179.

Cowling, R. M., and Proches, Ş. In press. Patterns and evolution of plant diversity in the Cape Floristic Region. In Friis, I., and Balslev, H., eds., Plant Diversity and Complexity Patterns: Local, Regional, and Global Dimensions. Royal Danish Academy of Sciences and Letters, Copenhagen.

Cowling, R. M., and Richardson, D. 1995. Fynbos: South Africa's Unique Floral Kingdom. Fernwood Press, Vlaeberg, South Africa.

Crawley, M. J., and Harral, J. E. 2001. Scale dependence in plant biodiversity. Science 291: 864–868.

Crosby, M. R., Magill, R. E., Allen, B., and He, S. 2000. A Checklist of the Mosses. Missouri Botanical Garden, St. Louis.

Currie, D. J. 1991. Energy and large-scale patterns of animal- and plant-species richness. American Naturalist 137:27–49.

Currie, D. J., and Paquin, V. 1987. Large-scale biogeographical patterns of species richness of trees. Nature 329:326–327.

Davis, S. D., Heywood, V. H., and Hamilton, A. C. 1994. Centres of Plant Diversity: A Guide and Strategy for Their Conservation. Vol. 1, Europe, Africa, and the Middle East. IUCN Publications Unit, Cambridge.

Davis, S. D., Heywood, V. H., Herrera-MacBryde, O., Villa-Lobos, J. L. and Hamilton, A. C. 1997. Centres of Plant Diversity. A Guide and Strategy for Their Conservation. Vol. 3, The Americas. IUCN Publications Unit, Cambridge.

Delcourt, P. A., and Delcourt, H. R. 1993. Paleoclimates, paleovegetation, and paleofloras during the Late Quaternary. Pp. 71–94 in Flora of North America Editorial Committee, ed., Flora of North America North of Mexico. Oxford University Press, New York.

Delgadillo Moya, C., Bello, B., and Cardenas Soriano, A. 1995. LATMOSS: a catalogue of Neotropical mosses. Monographs in Systematic Botany from the Missouri Botanical Garden 56.

Deutschewitz, K., Lausch, A., Kühn, I., and Klotz, S. 2003. Native and alien plant species richness in relation to spatial heterogeneity on a regional scale in Germany. Global Ecology and Biogeography 12:299–311.

Ekman, S. 1953. Zoogeography of the Sea. Sidgewick and Jackson, London.

Faith, D. P. 1994. Phylogenetic diversity: a general framework for the prediction of feature diversity. Pp. 251–268 in Forey, P. L., Humphries, C. J., and Vane-Wright, R. I., eds., Systematics and Conservation Evaluation. Clarendon Press, Oxford.

Faith, D. P., and Walker, P. A. 1996. Environmental diversity: on the best-possible use of surrogate data for assessing the relative biodiversity of sets of areas. Biodiversity and Conservation 5:399–415.

Farjon, A. 1998. World Checklist and Bibliography of Conifers. Royal Botanical Gardens, Kew, Richmond, United Kingdom.

Fjeldsa, J., and Lovett, J. C. 1997. Geographical patterns of old and young species in African forest biota: the significance of specific montane areas as evolutionary centres. Biodiversity and Conservation 6:325–346.

Francis, A. P., and Currie, D. J. 2003. A globally consistent richness-climate relationship for angiosperms. American Naturalist 161:523–536.

Frodin, D. G. 2001a. Floras in retrospective and for the future. Plant Talk 25:36–39.

———. 2001b. Guide to Standard Floras of the World. Cambridge University Press, Cambridge.

Gaston, K. J. 1996. Biodiversity-latitudinal gradients. Progress in Physical Geography 20: 466–476.

———. 2000. Global patterns in biodiversity. Nature 404:220–227.

Gentry, A. H. 1982. Neotropical floristic diversity: phytogeographical connections between Central and South America, Pleistocene climatic fluctuations, or an accident of the Andean orogeny? Annals of the Missouri Botanical Garden 69:557–593.

Gradstein, S. R., and Pócs, T. 1989. The biogeography of tropical rainforest bryophytes. Pp. 311–325 in Lieth, H., and Werger, M. J. A., eds., Tropical Rain Forest Ecosystems. Ecosystems of the World, vol. 14A. Elsevier, Amsterdam.

Groombridge, B., ed. 1992. Global Biodiversity: Status of the Earth's Living Resources. Compiled by World Conservation Monitoring Centre. Chapman and Hall, London.

Groombridge, B., and Jenkins, M. D. 2002. World Atlas of Biodiversity: Earth's Living Resources in the 21st Century. University of California Press, Berkeley.

Haffer, J. 1982. General aspects of the refuge theory. Pp. 6–24 in Prance, G. T., ed., Biological Diversification in the Tropics. Columbia University Press, New York.

———. 1997. Alternative models of vertebrate speciation in Amazonia: an overview. Biodiversity and Conservation 6:451–476.

Haffer, J., and Prance, G. T. 2001. Climatic forcing of evolution in Amazonia during the Cenozoic: on the refuge theory of biotic differentiation. Amazoniana 16:579–607.

Hawkins, B. A. 2001. Ecology's oldest pattern? Trends in Ecology and Evolution 16:470.

Hawkins, B. A., and Porter, E. E. 2001. Area and latitudinal diversity gradient for terrestrial birds. Ecology Letters 4:595–601.

———. 2003. Relative influences of current and historical factors on mammal and bird diversity patterns in deglaciated North America. Global Ecology and Biogeography Letters 12:475–481.

Heywood, V. H., ed. 1995. Global Biodiversity Assessment. Cambridge University Press: Cambridge.

Holden, W., and Wycoff, E. 1911–1914. Bibliography relating to the floras. Bibliographic Contribution of the Lloyd Library 1:1–513.

Hubbell, S. 2001. The Unified Neutral Theory of Biodiversity and Biogeography. Princeton University Press, Princeton, New Jersey.

Humboldt, A. von. 1806. Ideen zu einer Physiognomie der Gehölze. Cotta, Tübingen.

Humphries, C. J., Williams, P. H., and Vane-Wright, R. I. 1995. Measuring biodiversity value for conservation. Annual Review of Ecology and Systematics 26:93–111.

Huston, M. A. 1999. Local processes and regional patterns: appropriate scales for understanding variation in the diversity of plants and animals. Oikos 86:393–401.

Ibisch, P. L., Boegner, A., Nieder, J., and Barthlott, W. 1996. How diverse are Neotropical epiphytes? An analysis based on the "Catalogue of Flowering Plants and Gymnosperms of Peru." Ecotropica 2:13–28.

Jäger, E. J. 1976. Areal- und Florenkunde (floristische Geobotanik). Progress in Botany 38: 314–330.

Jansson, R. 2003. Global patterns in endemism explained by past climatic change. Proceedings of the Royal Society B 270:583–590.

Janzen, D. H. 1981. The peak in North American ichneumonid species richness lies between 38° and 42°N. Ecology 62:532–537.

Jedicke, E. 2001. Biodiversität, Geodiversität, Ökodiversität: Kriterien zur Analyse der Landschaftsstruktur: ein konzeptioneller Diskussionsbeitrag. Naturschutz und Landschaftsplanung 33:59–68.

Jetz, W., and Rahbek, C. 2001. Geometric constraints explain much of the species richness pattern in African birds. Proceedings of the National Academy of Science USA 98:5661–5666.

Jones, D. L. 2002. Cycads of the World. 2nd edition. Smithsonian Institution Press, Washington, DC.

Jørgensen, P. M., and León-Yánez, S. 1999. Catalogue of the Vascular Plants of Ecuador. Missouri Botanical Garden Press, St. Louis.

Kerr, J. T., and Currie, D. J. 1999. The relative importance of evolutionary and environmental controls on broad-scale patterns of species richness in North America. Ecoscience 6: 329–337.

Kerr, J. T., and Packer, L. 1997. Habitat heterogeneity as a determinant of mammal species richness in high-energy regions. Nature 385:253–254.

Kerr, J. T., Southwood, T. R. E., and Cihlar, J. 2001. Remotely sensed habitat diversity predicts butterfly species richness and community similarity in Canada. Proceedings of the National Academy of Science USA 98:11365–11370.

Kessler, M. 2000. Elevational gradients in species richness and endemism of selected plant groups in the central Bolivian Andes. Plant Ecology 149:181–193.

———. 2001. Pteridophyte species richness in Andean forests in Bolivia. Biodiversity and Conservation 10:1473–1495.

Kessler, M., Parris, B. S., and Kessler, E. 2001. A comparison of the tropical montane pteridophyte floras of Mount Kinabalu, Borneo, and Parque Nacional Carrasco, Bolivia. Journal of Biogeography 28:611–622.

Kier, G., and Barthlott, W. 2001. Measuring and mapping endemism and species richness: a new methodological approach and its application on the flora of Africa. Biodiversity and Conservation 10:1513–1529.

Klass, K. D., Zompro, O., Kristensen, N. P., and Adis, J. 2002. Mantophasmatodea: a new insect order with extant members in the Afrotropics. Science 296:1456–1459.

Kubitzki, K. 1990. Gnetatae. Pp. 378–391 in Kramer, K. U., and Green, P. S., eds., The Families and Genera of Vascular Plants, vol. 1. Springer-Verlag, Berlin.

Kühn, I., Brandl, R., and Klotz, S. 2004. The flora of German cities is naturally species rich. Evolutionary Ecology Research 6:749–764.

Lauer, W. 1989. Climate and weather. Pp. 19–41 in Lieth, H., and Werger, M. J. A., eds., Tropical Rain Forest Ecosystems. Ecosystems of the World, vol. 14B. Elsevier, Amsterdam.

Lennon, J. J., Koleff, P., Greenwood, J. J. D., and Gaston, K. J. 2001. The geographical structure of British bird distributions: diversity, spatial turnover, and scale. Journal of Animal Ecology 70:966–979.

Linder, H. P., Lovett, J. C., Mutke, J., Barthlott, W., Jürgens, N., Rebelo, T. and Küper, W. In press. A numerical re-evaluation of the sub-Saharan phytochoria. In Friis, I., and Balslev, H., eds., Plant Diversity and Complexity Patterns: Local, Regional, and Global Dimensions. Royal Danish Academy of Sciences and Letters, Copenhagen.

Longhurst, A. 1998. Ecological Geography of the Sea. Academic Press, San Diego.

Luning, K. 1990. Seaweeds and Their Environment, Biogeography, and Ecophysiology. John Wiley, New York.

Magurran, A. E. 1988. Ecological Diversity and Its Measurement. Chapman and Hall, London.

Malyshev, L. I. 1975. The quantitative analysis of flora: spatial diversity, level of specific richness, and representativity of sampling areas. In Russian. Botanicheskii Zhurnal 60:1537–1550.

———. 1991. Some quantitative approaches to problems of comparative floristics. Pp. 15–34 in Nimis, P. L., and Crovello, T. J., eds., Quantitative Approaches to Phytogeography. Kluwer Academic, Dordrecht.

———. 1993. Ecological background of floristic diversity in northern Asia. Fragmenta Floristica et Geobotanica, supplement 2:331–342.

Marticorena, C. 1990. Contribución a la estadística de la flora vascular de Chile. Gayana Botanica 47:85–113.

McGlone, M. S. 1996. When history matters: scale, time, climate, and tree diversity. Global Ecology and Biogeography Letters 5:309–314.

Mittermeier, R. A., Myers, N., Robles Gil, P., and Mittermeier, C. G. 1999. Hotspots: Earth's Biologically Richest and Most Endangered Terrestrial Ecoregions. Cemex, Mexico City.

Müller R., Nowicki, C., Barthlott, W., and Ibisch, P. L. 2003. Biodiversity and endemism mapping as a tool for regional conservation planning: case study of the Pleurothallidinae (Orchidaceae) of the Andean rain forests in Bolivia. Biodiversity and Conservation 12: 2005–2024.

Mutke, J. 2000. Methodische Aspekte der räumlichen Modellierung biologischer Vielfalt: das Beispiel der Gefäßpflanzenflora Nordamerikas. IANUS Working Paper 5/2000:15–29.

———. 2002. Räumliche Muster Biologischer Vielfalt: die Gefäßpflanzenflora Amerikas im globalen Kontext. Ph.D. diss., Botanical Institute, University of Bonn.

Mutke, J., and Barthlott, W. 2000. Some aspects of North American phytodiversity and its biogeographic relationships. Pp. 435–447 in Breckle, S.-W., Schweizer, B., and Arndt, U., eds., Results of Worldwide Ecological Studies. Verlag Günter Heimbach, Stuttgart.

———. In press. Patterns of vascular plant diversity at continental to global scales. In Friis, I., and Balslev, H., eds., Plant Diversity and Complexity Patterns: Local, Regional, and Global Dimensions. Royal Danish Academy of Sciences and Letters, Copenhagen.

Mutke, J., Kier, G., Braun, G., Schulz, C., and Barthlott, W. 2002. Patterns of African vascular plant diversity: a GIS based analysis. Systematics and Geography of Plants 71:1125–1136.

Myers, N. 1988. Threatened biotas: "hot spots" in tropical forests. Environmentalist 8:187–208.

———. 1990. The biodiversity challenge: expanded hot-spots analysis. Environmentalist 10:243–256.

Myers, N., Mittermeier, R. A., Mittermeier, C. G., da Fonseca, G. A. B., and Kent, J. 2000. Biodiversity hotspots for conservation priorities. Nature 403:853–858.

Nelson, B. W., Ferreira, C. A. C., da Silva, M. F., and Kawasaki, M. L. 1990. Endemism centers, refugia, and botanical collection density in Brazilian Amazonia. Nature 345:714–716.

Norton, T. A., Melkanian, M., and Andersen, R. 1996. Algal biodiversity. Phycologia 35:308–326.

O'Brien, E. M. 1993. Climatic gradients in woody plant species richness: towards an explanation based on an analysis of southern Africa's woody flora. Journal of Biogeography 20: 181–198.

———. 1998. Water-energy dynamics, climate, and prediction of woody plant species richness: an interim general model. Journal of Biogeography 25:379–398.

Olson, D. M., and Dinerstein, E. 1998. The Global 200: a representative approach to conserving the earth's most biologically valuable ecoregions. Conservation Biology 12:502–515.

Olson, D. M., Dinerstein, E., Wikramanayake, E. D., Burgess, N. D., Powell, G. V. N., Underwood, E. C., D'Amico, J. A., Itoua, I., Strand, H. E., Morrison, J. C., Loucks, C. J., Allnutt, T. F., Ricketts, T. H., Kura, Y., Lamoreux, J. F., Wettengel, W. W., Hedao, P., and Kassem, K. R. 2001. Terrestrial ecoregions of the world: a new map of life on earth. Bio-Science 51:933–937.

O'Shea, B. J. 1997. The mosses of sub-Saharan Africa. Part 2, Endemism and biodiversity. Tropical Bryology 13:75–85.

Phillips, R., and Menez, E. 1988. Seagrasses. Smithsonian Contributions to the Marine Sciences 34:1–104.

Pianka, E. R. 1966. Latitudinal gradients in species diversity: a review of concepts. American Naturalist 100:33–46.

Pitman, N. C. A., Terborgh, J., Silman, M. R., and Núñez V., P. 1999. Tree species distribution in an upper Amazonian forest. Ecology 80:2651–2661.

Plotkin, J. B., Potts, M., Yu, D., Bunyavejchewin, S., Condit, R., Foster, R., Hubbell, S., LaFrankie, J., Manokaran, N., Seng, L., Sukumar, R., Nowak, M. A., and Ashton, P. S. 2000. Predicting species diversity in tropical forests. Proceedings of the National Academy of Science USA 97:10850–10854.

Price, I., and Scott, F. 1992. The Turf Algal Flora of the Great Barrier Reef. Part 1, Rhodophyta. James Cook University, Townsville, Australia.

Prud'homme van Reine, W., and van den Hoek, C. 1988. Biogeography of Cape Verdean seaweeds. Courier Forschungsinstitut Senckenberg 105:35–49.

Quian, H. 1999. Spatial pattern of vascular plant diversity in northern America north of Mexico and its floristic relationship with Eurasia. Annals of Botany 83:271–283.

Rahbek, C. 1995. The elevational gradient of species richness: a uniform pattern. Ecography 18:200–205.

Rahbek, C., and Graves, G. R. 2001. Multiscale assessment of patterns of avian species richness. Proceedings of the National Academy of Science USA 98:4534–4539.

Redfearn, P. L. Jr. 1994. List of Mosses of China. Ozarks Regional Herbarium, Springfield, Missouri.

Ricketts, T. H., Dinerstein, E., Olson, D. M., Loucks, C. J., Eichbaum, W., DellaSala, D., Kavanagh, K., Hedao, P., Hurley, P. T., Carney, K. M., Abell, R., and Walters, S. 1999. Terrestrial Ecoregions of North America: A Conservation Assessment. Island Press, Washington, DC.

Rohde, K. 1992. Latitudinal gradients in species diversity: the search for the primary cause. Oikos 65:514–527.

Rosenzweig, M. L. 1995. Species Diversity in Space and Time. Cambridge University Press, Cambridge.

Roy, K., Jablonski, D., Valentine, J., and Rosenburg, G. 1998. Marine latitudinal diversity gradients: tests of causal hypotheses. Ecology 95:3699–3702.

Royal Botanic Gardens Kew. 1993. Index Kewensis on Compact Disc. Oxford University Press, Oxford.

Saltzman, B. 2002. Dynamic Paleoclimatology. Academic Press, San Diego.

Sepkoski, I. 1995. Large scale history of biodiversity. Pp. 202–212 in Heywood, V. H., ed., Global Biodiversity Assessment, UNEP. Cambridge University Press, New York.

Small, A., and Adey, W. 1998. Are current estimates of coral reef biodiversity too low? The view through the window of a microcosm. Atoll Research Bulletin 458:1–20.

Terborgh, J. 1973. On the notion of favorableness in plant ecology. American Naturalist 107: 481–501.

Thomas, W. W. 1999. Conservation and monographic research on the flora of tropical America. Biodiversity and Conservation 8:1007–1015.

Thorne, R. F. 1993. Phytogeography. Pp. 132–153 in Flora of North America Editorial Committee, ed., Flora of North America North of Mexico, vol. 1. Oxford University Press, Oxford.

Tuomisto, H., and Ruokolainen, K. 1997. The role of ecological knowledge in explaining biogeography and biodiversity in Amazonia. Biodiversity and Conservation 6:347–357.

Tuomisto, H., Ruokolainen, K., Kalliola, R., Linna, A., Danjoy, W., and Rodriguez, Z. 1995. Dissecting Amazonian biodiversity. Science 269:63–66.

Tuomisto, H., Ruokolainen, K., and Yli-Halla, M. 2003. Dispersal, environment, and floristic variation of western Amazonian forests. Science 299:241–244.

Tutin, T. G., Heywood, V. H., Burges, N. A., Valentin, D. H., Walters, S. M., Webb, D. A., Ball, P. W., and Moore, D. M. 1964–1980. Flora Europaea. Cambridge University Press, Cambridge.

USDA (U.S. Department of Agriculture), National Resources Conservation Service. 2004. The PLANTS Database, Version 3.5. National Plant Data Center, Baton Rouge. http://plants.usda.gov.

Valencia, R., Balslev, H., and Paz y Mino, C. G. 1994. High tree alpha-diversity in Amazonian Ecuador. Biodiversity and Conservation 3:21–28.

van den Hoek, C. 1984. World-wide latitudinal and longitudinal seaweed distribution patterns and their possible causes as illustrated by the distribution of Rhodophyta genera. Helgoländer Meeresuntersuchungen 38:227–257.

van den Hoek, C., and Breeman, A. M. 1989. Seaweed biogeography of the North Atlantic: where are we now? Pp. 55–86 in Garbury, D., and South, G., eds., Evolutionary Biogeography of the Marine Algae of the North Atlantic. NATO ASI Series G, Ecological Sciences, vol. 22. Springer-Verlag, Berlin.

Vitt, D. H., and Buck, W. R. 1992. Key to the moss genera of North America north of Mexico. Contributions from the University of Michigan Herbarium 18:43–71.

Webb, C. O. 2000. Exploring the phylogenetic structure of ecological communities: an example for rain forest trees. American Naturalist 156:145–155.

White, F. 1983. The vegetation of Africa: a descriptive memoir to accompany the UNESCO/ AETFAT/UNSO vegetation map of Africa. Natural Resources Research 20. UNESCO, Paris.

———. 1993. Refuge theory, ice-age aridity, and the history of tropical biotas: an essay in plant geography. Fragmenta Floristica et Geobotanica, supplement 2:385–409.

Whittaker, R. H. 1972. Evolution and measurement of species diversity. Taxon 21:213–251.

———. 1977. Evolution of species diversity in land communities. Evolutionary Biology 10: 1–67.

Wickham, J. D., Wade, T. G., Jones, K. B., Riitters, K. H., and O'Neill, R. V. 1995. Diversity of ecological communities of the United States. Vegetatio 119:91–100.

Williams, P. H. 1993. Measuring more of biodiversity for choosing conservation areas, using taxonomic relatedness. Pp. 194–227 in Moon, T.-Y., ed., International Symposium on Biodiversity and Conservation. Korea University, Seoul.

———. 1999. WORLDMAP Public Demo, Version v4.19.24. Privately distributed, London.

Williams, P. H., Humphries, C. J., and Gaston, K. J. 1994. Centres of seed-plant diversity: the family way. Proceedings of the Royal Society B 256:67–70.

Willig, M. R., Kaufman, D. M., and Stevens, R. D. 2003. Latitudinal gradients of biodiversity: patterns, processes, scale, and synthesis. Annual Review of Ecology, Evolution, and Systematics 34:273–309.

Willis, K., and Whittaker, R. J. 2002. Species diversity: scale matters. Science 295:1245–1248.

Wilson, R. C. L., Drury, S. A., and Chapman, J. L. 2000. The Great Ice Age, Climate Change, and Life. Routledge, London.

WRI (World Resources Institute). 1997. World Resources, 1996–1997. Oxford University Press, Oxford.

Wright, D. H. 1983. Species-energy theory: an extension of species-area theory. Oikos 41: 496–506.

Wulff, E. W. 1935. Versuch einer Einteilung der Vegetation der Erde in pflanzengeographische Gebiete auf Grund der Artenzahl. Repertorium Specierum Novarum Regni Vegetabilis 12:57–83.

Yano, O. 1981. A checklist of Brazilian mosses. Journal of the Hattori Botanical Laboratory 50:279–456.

CHAPTER 3

PLANT EXTINCTIONS

Extinction is a natural phenomenon. Most of the plant species that have been present on the earth over the last 400 million years are now extinct. The pace of extinction has also varied over time, with periodic mass extinctions. In general, as species diversity and speciation rates increase, the background extinction rate has grown as well. During the last several hundred years, extinction events have become more common as human populations have exponentially grown and affected natural environments. Assessing contemporary plant extinctions is problematic given the fundamental paradox of documenting populations that are no longer present. Nonetheless, based on an examination of better-known floras, worldwide plant extinctions probably have at least doubled or tripled since the 17th century. This chapter begins with a discussion of extinctions over paleontological time and then addresses our understanding of contemporary extinctions.

3.1 A PALEONTOLOGICAL PERSPECTIVE ON PLANT EXTINCTIONS

Scott Wing

THE FOSSIL RECORD of land plants extends back over 400 million years of earth history marked by many perturbations—impacts of extraterrestrial objects, large-scale volcanic eruptions, rifting continents, and changes in climate and the composition of the atmosphere. Large environmental changes have the potential to cause extinctions of plants, and indeed over geologic time extinctions are commonplace. The vast majority of plant species that have ever existed are now extinct, as are most genera, many families, and higher taxa. This fossil record of extinction has the potential to provide an understanding about current and future extinctions caused by human activities or by anthropogenic climate change. Ideally, it is necessary to understand how the magnitude and rate of past environmental changes are related to the rate of extinction. In order to use the fossil record in this way, several questions arise: What are normal or background rates of plant extinction? How much do they accelerate during perturbations? What types of environmental change are most likely to cause extinctions—for example, what is the relative importance of climate change, massive disturbance, competition, predation, and loss of co-evolved animals (pollinators or dispersers)? What kinds of plants (functional type, habitat preference, reproductive biology, etc.) are most vulnerable to extinction? Knowing rates of extinction from the fossil record would also provide a baseline against which to compare current extinctions.

The first part of this section reviews where, when, and how fossils are preserved, because the ability to quantify and characterize extinctions in the fossil record is strongly influenced by these biases. The next part offers a synopsis of three major plant mass extinctions as points of comparison with the present. The last part discusses a small number of generalities that seem to emerge from considering the fossil record of plant extinctions. It will become clear that quantitative answers to the questions posed above are difficult to obtain with current data and methods of analysis, and therefore a baseline is not available that would explain how unusual the current anthropogenic extinctions are in the context of natural events. More work on the fossil record of plant extinction is required to develop a geo-historical context for the effect humans have on plants.

CHARACTERISTICS OF THE PLANT FOSSIL RECORD

Fragmentation and Taxonomic Resolution

Plants typically fall apart or partially decay before being preserved, so most species of fossil plants (plate 3.1) have been described from only one or two organs, and even these may be only partially preserved. A small percentage of fossil plant species have been described from multiple organs (Manchester 1986), but even when the type specimen has two or more attached organs (e.g., leaves and fruits), other occurrences of the same species are commonly based on a single type of organ. Because of the limited number of characters available for most fossil species, many no doubt include multiple biological species that could be recognized if more characters were available. The relatively small number of characters of many fossil plant species also can make it difficult to place them in a higher taxon, or to assess their growth form, life history, or reproductive biology.

Taxonomic information and stratigraphic density commonly have an inverse relation in the plant fossil record. The most complex and systematically informative plant organs are preserved less commonly and may be too rare for statistical treatment and too dispersed stratigraphically to resolve short or closely spaced events. Simpler plant organs, like pollen and spores, or fragments, such as dispersed cuticles, have an abundant fossil record that permits statistically significant samples in close stratigraphic position (Upchurch 1989). In general, though, these fossils cannot be identified to species, and they may be diagnostic only to genus or family of plants (Upchurch 1989). Although it is possible to calculate rates and amounts of extinction over time intervals that are comparable to historic or future anthropogenic extinctions (hundreds or thousands of years), it can usually only be done with fossils that provide less taxonomic information.

Which Plants Get Fossilized?

Plant fossils are surprisingly common in a range of terrestrial and nearshore marine depositional environments, but most species of plants probably do not leave a fossil record at all. Most plant fossils are preserved in wet environments where sediment was deposited frequently—river channels and floodplains, peat mires, lakes, deltas, lagoons, and shallow marine settings (Gastaldo 1988; Behrensmeyer et al. 1992; Behrensmeyer et al. 2000). Arid environments have low potential to preserve plants. Although upland areas may preserve plants, over very long intervals of geologic time, erosion will probably destroy the fossils. This is why deposits of, for example, crater lakes are more abundant in the Cenozoic than in the Paleozoic. The fossil record of plants is mostly the fossil record of lowland, wetland environments, and this is increasingly true of older time periods.

Even within wet, lowland habitats many species are unlikely to become fossilized. Herbaceous plants are small and do not regularly abscise leaves; therefore they produce fewer fossilizable units (Scheihing 1980). Plants with thin, easily decomposed leaves are also underrepresented in protofossil assemblages (Spicer 1981). Some of the most diverse living herbaceous families (e.g., Asteraceae, Orchidaceae) have very poor fossil records (Collinson et al. 1993).

Spatio-temporal Scale and Paleontological Uncertainty

The deposition of sediment in terrestrial environments tends to be localized and intermittent, creating a plant fossil record that is discontinuous in both space and time. In large, rapidly subsiding sedimentary basins, beds preserving plant megafossils may be closely spaced stratigraphically, permitting paleobotanists to quantify rates and amounts of extinction over geologically short periods of time, and even to compare some of the biological traits of survivors and victims. One strategy for documenting plant extinctions is to focus on these relatively complete sedimentary sequences (DiMichele and Phillips 1996; Johnson 2002).

Although stratigraphically detailed studies within a basin provide the best information about a small region, they are subject to problems as well. Extinction rates calculated from local records will overestimate true extinction intensity if some of the recorded disappearances are only local extirpations. (This overestimation may be offset by low taxonomic resolution in palynological data, because higher taxa may appear to persist even if many of their component species go extinct. This effect depends on how species rich the genera and families are, and how distinctive their pollen and spores, or palynomorphs, are.) Local records also are strongly influenced by the environment of deposition and preservational processes. Shifts in sedimentary processes may lead to changes in the preservation and deposition of plant remains that mimic floral change and extinction. Conversely, some local environments that preserve plant fossils (e.g., peat swamps) may support only a restricted subset of the regional species that can persist on the unusual substrate. The flora in such specific habitats may not change even when regional floral composition is shifted by climate or other factors. This could lead to underestimation of true regional floral change.

In order to gain the best estimate of extinction, paleontological studies must have a geographic as well as a temporal dimension; however, the older the time period being considered, the more difficult it becomes to correlate records in different sedimentary basins with high temporal resolution. For example, a Triassic pond deposit 220 million years old might contain many bedding planes that preserve fossil leaves. The leaf fossils on a given bedding plane typically accumulated during the same flood, and might have all grown in the same season or year. The multiple layers of leaves thus represent a series of discrete samples of local vegetation ("snapshot" samples), each separated from the next by a season during which sediment without leaves was deposited. Each bedding-plane sample is comparable to a census of the local vegetation, and a series of such samples can yield a "movie" of local vegetational change over a period of decades or centuries that passed hundreds of millions of years ago. Even though there might be many such pond deposits in our Triassic example (and therefore many local movies), it would be difficult to correlate these to one another with millennial resolution because they are far beyond the range of radiocarbon dating (greater than 40,000 years old).

For the more distant past it is seldom possible to correlate records of floral change at the same level of temporal resolution that can be achieved within local sections. In order to study regional or global patterns of extinction or change in the distant past, samples are grouped into periods that are hundreds of thousands or millions of years long, but that are recognizable regionally or globally. This creates a trade-off between temporal resolution and geographic scope—regional or global relevance is achieved only by losing the ability to describe short-lived events. In physics the uncertainty principle states that is not possible to simultaneously know the velocity and mass of a particle; an analogous "paleontological uncertainty principle" makes it hard to simultaneously know the precise timing and the geographic extent of a past event.

In order to calculate extinction rates in global studies, floras generally are assigned to time bins that are millions of years long, and therefore rates of extinction are averaged over very long time periods. This is particularly problematic if the time periods are not of equal length, because rates of extinction can depend strongly on the intervals over which they are calculated. High rates of extinction are seldom maintained for long periods, so with intervals of longer duration, longer periods of background extinction will be averaged with short periods of high extinction. As a result, extinction rates might appear to be lower if longer time intervals are used (Foote 1994).

Another potential confusion introduced by global data is that opposing trends in separate regions may offset or disguise one another (Rees 2002). If high extinction rates on one continent are combined with increasing diversity on another continent at the same time, globally averaged data would show little change.

A variety of extinction metrics have been developed by paleontologists to address problems like unequal interval length and changing levels of standing diversity (Foote 1994; Regan et al. 2001). Perhaps the most widely accepted metric for extinctions is the per-taxon rate of extinction, which is calculated by dividing the number of extinctions observed by the standing diversity of the higher taxon being studied (the number of species available to go extinct) and the number of years in the time interval over which extinctions were observed. Paleontological data commonly yield rates of tenths to tens of species per species million years (where "species million year" is the proportion of the extant species that go extinct per million years; Valentine et al. 1991; Regan et al. 2001). It should be noted, though, that dividing by the period of observation does not necessarily make rates calculated over different time periods comparable because dividing the number of extinctions by the number of years in an interval implicitly assumes that the rate of extinction was uniform during the interval. If extinctions occurred in pulses, the actual rate could have been much higher than the calculated rate, and the likelihood that pulses will be averaged with periods of low extinction increases for longer time intervals. This means that rates of extinction calculated over longer time intervals will generally be slower than those calculated over short intervals.

BACKGROUND EXTINCTION

During the vast majority of geologic time intervals, the number of species of land plants has increased, and this is also seen within each of the constituent higher taxa, such as angiosperms, gymnosperms, and pteridophytes (Niklas et al. 1980, 1985). In other words, most of the time speciation outstrips extinction among land plants. This has led to the large increase in the diversity of land plants over geologic time. The overall increase in plant diversity is in spite of rising extinction rates among plants through geologic time; among the earliest land plants extinction rates were 0.076 species per species million years, roughly one-third of the rate (0.198 species per species million years) calculated for Cretaceous and Cenozoic flowering plants (Valentine et al. 1991).

Although overall extinction rates have risen through geologic time, this appears to result largely from differences among clades: lycophytes (0.06 species per species million years) < sphenophytes (0.10) < Filicales (0.13) < conifers (0.25) < dicots (0.26) (Valentine et al. 1991). Because clades with higher extinction rates have been most diverse later in geologic time, overall extinction rate has risen. In contrast, within several higher taxa (e.g., Filicales, pteridosperms, cycadophytes, ginkgos, and conifers) extinction rates declined through geologic time, suggesting that later-appearing species within each lineage are more resistant to extinction.

The overall increase that plants show in vulnerability to extinction over geologic time contrasts strongly with the record of marine invertebrate animals, in which extinction rates decline over geologic time (Valentine et al. 1991). Valentine et al. suggested two possible explanations for the difference in extinction patterns between plants and animals. One explanation invokes a feature of the terrestrial environment, in that land plants have continually evolved to occupy more stressful (and undersaturated) habitats over geologic time. (This is in contrast to marine animals, which appeared in stressful habitats and came to occupy more stable ones.) The newly appearing higher taxa of plants that invaded unoccupied colder and drier habitats were successful because of their high rates of speciation rather than their low rates of extinction (even though, once established, each taxon had declining extinction rates). The other explanation invokes endogenous factors. Valentine et al. suggested that the more open developmental system of plants permitted the evolution of new higher taxa, and that these new higher taxa were characterized by higher rates of both extinction and speciation than earlier-appearing taxa. The reason for the differences between extinction rates in marine animals and land plants remain unresolved; however, it appears that the successive diversification of lycophytes, ferns, lower seed plants, and angiosperms is controlled by speciation rate rather than by extinction rate.

MASS EXTINCTION

Large-scale compilations of the plant fossil record do not show dramatic declines in diversity or mass extinctions (fig. 3.1; Niklas et al. 1985). Knoll (1984) and Traverse (1988) identified several aspects of plant biology that might influence patterns of extinction on geologic time scales. First, many plants are quite resistant to physical destruction, at least compared with animals. They can regrow or resprout after damage, and they produce seeds or spores that can survive through unfavorable periods. Second, plants are sessile, so that they have no means as individuals to escape rapid environmental change. Third, almost all plants (excepting a small proportion of parasites and saprophytes) make their own food from sunlight, water, and atmospheric carbon dioxide. This similarity in requirements might cause competitive exclusion to be an important factor in plant extinction.

Knoll (1984) and Traverse (1988) inferred that gradual extinction from competition or climate change would be more common among plants than mass extinction from environmental perturbations. Boulter et al. (1988) reviewed palynological data sets from across the record of land plants, and concluded that most plant extinction was gradual rather than sudden. Over the last 15 years, however, detailed stratigraphic studies of plant fossils have documented mass extinctions among plants at three different time periods in the past.

The Westphalian-Stephanian Extinction (~306 Ma)
The Carboniferous and Early Permian were the last time in earth history (prior to the last few million years) that major continental glaciers extended into middle latitudes. The Carboniferous glaciation had global effects on climate and sea level. The cold polar regions of the time probably restricted the wet tropical belt to a narrow range of latitudes that included most of what is now eastern and southern North America and western Europe—a region called tropical Euramerica.

Tropical Euramerica had vast coastal plains where the ocean advanced and retreated with the waning and waxing of glaciers in Gondwana. In the wet tropical climate extensive swamps formed on the coastal plains. Although this is the most ancient known plant extinction, it is relatively well understood. Many of the fossils that document plant life in the Carboniferous come from masses of petrified peat

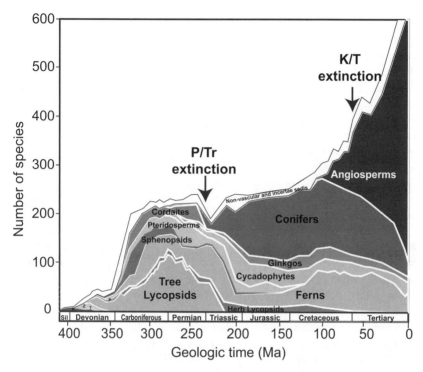

Figure 3.1 Diversity of plant species through time, based on the compilation of Niklas et al. (1985). Note the small diversity declines at the Permian-Triassic and Cretaceous-Tertiary boundaries, which are the two largest mass extinction intervals for marine and terrestrial animals. P, progymnosperms; Z, zosterophylls; T, trimerophytes.

(coal balls) that preserve plants in cellular detail, so that taxonomic resolution is high, and understanding of the biology of these plants is quite good.

Single coal beds of the Westphalian Stage of the Late Carboniferous typically contain some 40–50 species of plants (DiMichele and Phillips 1996). Peat swamp vegetation was dominated by lycopsid trees. Slightly drier areas were inhabited by a variety of small ferns and tree ferns of the family Marattiaceae. Fossils of these plants are associated with fossil charcoal, indicating wildfires. Other plants common in lowland swamps were seed ferns like *Medullosa*, and conifer relatives called cordaites. Tree-sized sphenopsids were also moderately abundant, though probably more common in muddy swamps than in peat swamps proper.

The extinction at the Westphalian-Stephanian boundary (about 306 million years ago [Ma]) is thought to have occurred over a period of 100,000 years or less. At this time about 67% of the species inhabiting the peat swamps and 50% of the species found in muddy swamps were eliminated (DiMichele and Phillips 1996). The palynoflora shows approximately 44% extinction across the same interval. The extinction did not hit all forms equally. Of the tree species in the peat swamps almost 87% did not survive to the Stephanian. Among smaller plants (ground cover, vines, and shrubs) only about 33% of species went extinct. The lycopsids were hit hard not only in terms of species, but also in terms of abundance. Coal balls from the late Westphalian are 60–70% lycopsid by volume, but by the early Stephanian they were less than 10% (DiMichele and Phillips 1996).

The first fossil assemblages deposited following the extinction show fluctuating composition, but within tens to hundreds of thousands of years, tree ferns appear to have dominated all swamp habitats. In general the postextinction peat swamps were more homogeneous habitats, with less area of the standing water that had favored lycopsids. The dominance of ferns in the aftermath of the Westphalian-Stephanian extinction is thought to reflect the success of plants with high dispersal ability, rapid growth, and relatively broad environmental tolerances. In other words, the Westphalian-Stephanian extinction brought about a period of dominance by weeds.

Although the extinction was geologically rapid, it was not global. In parts of China, lycopsid-dominated peat swamp floras persisted through the rest of the Late Carboniferous and into the Early Permian. Most authors attribute the extinction to the drying of the Euramerican tropics as the continents moved northward and glaciation declined (Kerp 2000).

The Permian-Triassic Extinction (251 Ma)

In the southern supercontinent of Gondwana, floral composition has long been known to change dramatically near the Permian-Triassic boundary. Late Permian Gondwanan floras are dominated in both number of species and number of specimens by the seed fern *Glossopteris*. At least some species of *Glossopteris* were large trees with dense, conifer-like wood, and simple elliptical to obovate leaves. Early Triassic floras in the same region are dominated by a different, distantly related seed fern, *Dicroidium*, which had more highly divided leaves. Many *Dicroidium* species appear to have been shrubby plants. Recent studies of floral change and extinction near the Permian-Triassic boundary in Australia have cited up to 97% extinction of plants (based on leaf fossils) in eastern Australian sequences (Retallack 1995). Extinction levels in the microflora appear to be much lower, about 19%, perhaps partly because of reworking of Permian pollen into lowermost Triassic sediments (Retallack 1995), but also because of the lower taxonomic resolution of microfloras.

Although the extinction of the *Glossopteris* flora has long been taken as a marker of the Permian-Triassic boundary, only more recently has it become clear that there was also a major extinction in the northern supercontinent of Laurasia. A global survey of microfloras shows that in many different types of local environments (shallow marine, various types of terrestrial setting) and on all continents there was a dramatic increase in the abundance of fungal spores during the last million years of the Permian (Visscher et al. 1996; Twitchett et al. 2001). This global increase in fungal spores is a unique event in the history of life, and probably records a period during which there was massive die-off of terrestrial plants followed by fungal decay. The end of the Permian in the Euramerican region is now also associated with the disappearance of many types of coniferous pollen at almost precisely the same time as the extinctions in the marine realm (Looy et al. 1999; Twitchett et al. 2001). The cause of the die-off and extinction is uncertain, but it has been related to changes in the atmosphere resulting from the eruption of the vast Siberian flood basalts (Visscher et al. 1996).

In spite of the high levels of extinction in many local sections, a global analysis of plant megafossils shows that the severity of extinction varied geographically (Rees 2002). Globally about 60% of plant genera were lost from the last stage of the Permian to the first stage of the Triassic, but the extinction also coincides with a steep decline in the number of localities from which fossil plants have been recovered (Rees 2002). Lower plant diversity in the Early Triassic is at least in part a consequence of fewer samples, but the magnitude of the sampling effect is difficult to estimate, leaving the severity of the global extinction in doubt.

The collapse of terrestrial productivity in the Late Permian apparently persisted through the first 5–6 million years of the Triassic. Worldwide there are no coals reported from the Early Triassic—the only period of geologic time after the Devonian for which this is true (Retallack et al. 1996). Retallack et al. attributed this gap in coal deposition to highly reduced terrestrial productivity rather than to low preservation of peat. In tropical areas of what is now Europe, palynofloras show that the dominant conifers disappeared at the end of the Permian; Early Triassic palynofloras were composed largely of lycopsids and moss spores (Looy et al. 1999). The moss/lycopsid flora persisted for about 5 million years before conifer pollen once again became abundant. It is still not known if the length of the recovery interval is related to the severity of the extinction and the slow rediversification of life, or to continuing environmental stress. The underlying causes of the Permian-Triassic extinctions (in both marine and terrestrial systems) are also poorly understood, but could include rapid climate change resulting from increases in greenhouse gas concentrations, extraterrestrial impacts, and oceanic anoxia (Erwin et al. 2002).

The Cretaceous-Tertiary Extinction (65.5 Ma)
The mass extinction at the end of the Cretaceous, 65 million years ago, is associated with the impact of a large extraterrestrial object, or bolide (Alvarez et al. 1980). The bolide hit the earth at the northern tip of the Yucatán Peninsula, leaving behind a ~150-km-diameter crater that is somewhat asymmetrical, implying an oblique, northwest-directed impact (Schultz and D'Hondt 1996).

Immediate results of the impact were a high-speed, high-temperature shockwave that moved downrange across North America (Schultz and D'Hondt 1996), a gigantic earthquake that triggered massive slumping and debris flows on continental shelves (Norris and Firth 2002), and the ejection of tens of thousands of cubic kilometers of target rock into the atmosphere (Claeys et al. 2002). This debris was heated frictionally while reentering the earth's atmosphere and started wildfires globally (Wolbach et al. 1988).

Longer-term physical, chemical, and biological effects of the bolide impact are less well understood. Although Alvarez et al. (1980) initially proposed that the K-T extinctions were caused by cooling and darkness induced by a global dust cloud of impact debris, more recent estimates suggest that the quantity of fine dust would not have been sufficient to diminish light levels below those required for photosynthesis (Pope 2002). It has also been proposed that carbon dioxide and water vapor created in the impact might have generated a global greenhouse effect for hundreds or thousands of years after the event (Wolfe 1990). Because the target rock had large amounts of sulfur and carbonate minerals, it is also possible that sulfur dioxide and water generated by the impact might have resulted in acid rain and in global cooling (Pope 2002).

High-resolution studies of the K-T boundary in North Dakota reveal roughly 80% extinction of plant species (Johnson et al. 1989; Johnson 1992; Johnson and Hickey 1990; Johnson 2002). These levels of extinction have a high reliability because they are based on large, closely spaced samples that come from matched depositional environments. High levels of extinction have also been recorded in megafloras from the Raton basin of northern New Mexico and southern Colorado—about 1,000 km closer to the Chicxulub impact structure (Wolfe and Upchurch 1986). Palynomorph extinction across the western United States was closer to 30% (Nichols 2002; Hotton 2002).

Studies of palynomorphs from the K-T boundary interval have revealed disturbance as well as extinction. Palynofloras from the first few centimeters of rock overlying the impact ejecta layer commonly are dominated by one or a few species, usually ferns (Tschudy et al. 1984; Nichols 2002; Nichols and Johnson 2002; Sweet 2001; Vajda et al. 2001). This fern-rich assemblage (called the "fern spike") has been interpreted as an early successional flora that existed during the first phase of recovery from impact-generated disruption (Orth et al. 1981).

Although regional studies document high species extinction at the K-T boundary, particularly among angiosperms, it is difficult to determine if higher taxa went extinct. In the 30 million years preceding the K-T boundary, flowering plants had become by far the most diverse group of land plants (Niklas et al. 1985; Lidgard and Crane 1990; Lupia et al. 1999), and the fossil records of many extant orders begin in the Late Cretaceous (Collinson et al. 1993; Magallon et al. 1999; Wing 2000). Although no higher taxa of angiosperms are known to have been extinguished at the K-T boundary, this could well be an artifact of paleobotanical nomenclature, specifically the reluctance to name extinct families and orders of plants based on fossil material.

Generalizations about Mass Extinctions

Several generalizations apply to all three extinctions discussed above. First, higher taxa appear to survive even when species-level extinction is extreme (>50–80%). This suggests that the loss of species in these mass extinctions is distributed across higher taxa, rather than being concentrated within a single lineage to such an extent that the whole group is lost. Second, although extinctions are not so selective that they eliminate higher taxa, they do appear to hit vegetational dominants (large tree species) hardest. Third, the "recovery" vegetation after the mass extinction seems to be dominated by small, opportunistic plants, ferns in the early Stephanian and early Paleocene, and lycopsids and mosses in the Early Triassic. Fern dominance, especially, may represent the success of weedy and easily dispersed species in post-catastrophe vegetation.

DISCUSSION AND CONCLUSIONS

The quality and type of paleontological data on plant extinctions is highly variable in time and space, and the importance of extinction seems to vary according to the scale of study. Analyses at large spatial and temporal scales (continental to global using time intervals of millions of years) suggest extinction is a relatively minor factor in plant evolution. Rates of extinction are low (<0.3 species per species million years), decreases in total diversity are few and small (<10%), and the replacement of one major group by another appears to be gradual and unrelated to mass extinction. In contrast, smaller-scale studies (several to many stratigraphic sections in a local area using time intervals of thousands or tens of thousands of years) reveal that some mass extinctions have removed a majority of plant species over periods of tens to hundreds of thousands of years, and have particularly affected vegetational dominants.

Reconciling the divergent views of plant extinction obtained from large- and small-scale studies is an issue of interest for plant conservation as well as paleobiology. The higher magnitudes of plant extinction measured in small-scale studies is consistent with the idea that most extinction occurs in short bursts, and that large-scale studies underestimate extinction levels because they average short intervals of high extinction with long intervals of low extinction. The small-scale studies also are closer in temporal scale to ecological observations of recent extinctions, and therefore are probably more relevant to conservation biology. If the rates of extinction measured in large-scale studies are artificially low, then intervals of high extinction may be more common in the history of plants on land than has previously been appreciated.

In conclusion, it is important to emphasize that rates of extinction calculated from the fossil record are not strictly comparable with those calculated from historical data, for a variety of reasons. The fossil record of plant extinctions is based on dispersed fossil organs that may not represent biological species, and it is composed almost exclusively of common species whereas historical plant extinctions are mostly of endemic species. Furthermore, rates of extinction calculated over different time intervals are not strictly comparable.

In spite of difficulties in observing and interpreting the plant fossil record, highly resolved studies of well-preserved, fossiliferous intervals hold the promise of revealing the information needed by conservation biology: What are typical rates of plant extinction measured at an ecologically relevant time scale? What types of environmental change are most likely to cause extinctions? What kinds of plants are most vulnerable to extinction? Although currently there are not enough highly resolved studies to give a robust answer to these questions, the feasibility of such studies has been demonstrated. The paleontological record of plant extinctions can be made useful to conservation biology.

3.2 CURRENT PLANT EXTINCTIONS: CHIAROSCURO IN SHADES OF GREEN

Bruce A. Stein and Warren L. Wagner

THE LAVENDER BLOSSOMS of *Calochortus indecorus* once brightened rocky serpentine outcrops along the slopes of southwestern Oregon's Sexton Mountain. Known only from its type locality, the existence of this distinctive species of mariposa lily was not brought to light until 1948. Yet less than 15 years later, the species had vanished, literally run over by construction of Interstate 5, a major north-south highway stretching from Canada to Mexico. Researchers from Southern Oregon University have scoured similar habitat in nearby areas searching for surviving populations of the diminutive wildflower, but with no luck. The Sexton Mountain mariposa lily has passed into darkness, existing now only as a herbarium specimen and on lists of extinct species.

Given the small range of *C. indecorus,* destruction of its only known habitat, and the intensity of searches carried out to locate remaining populations, it is possible with some confidence to declare the species extinct. Even in this relatively straightforward instance, however, making such a determination uses indirect evidence, relying on what was not found, rather than what was. The basic paradox of "documenting" extinctions is that absence of evidence is not necessarily evidence of absence.

The skilled use of light and dark to define volume and shape was a major contribution to western art of the Italian Renaissance painters. Artists such as Leonardo da Vinci and later Caravaggio were particularly adept at creating form through the contrast between light and shadow, a technique known as chiaroscuro. Much like these 15th- and 16th-century masters, modern botanists must use the uninhabited shadows in their efforts to discern whether a species has ceased to exist. One can think of the task of documenting plant extinctions as chiaroscuro in shades of green.

While the previous section focused on historical extinctions as indicated by the fossil record, this section reviews the current understanding of contemporary plant extinctions. What are the methods used to assess extinctions, and what does this say about the scale of contemporary extinctions? Botanical knowledge is very uneven across the globe, and so we focus on several regions for which relatively solid extinction data are now available, before moving on to consider the scale of the problem globally.

ASSESSING EXTINCTIONS

"Extinction is forever," proclaims the title of a seminal volume on the subject (Prance and Elias 1977). This declaration succinctly sums up why the problem of species extinctions is widely regarded in the scientific community as one of the most serious and irreversible environmental problems facing society. Yet despite the seeming finality of the concept, varying terminology can hinder efforts to gain a clear understanding of the problem. Quite simply, *extinct* means "no longer existing." In biological usage, however, *extinct* can refer to either no longer surviving anywhere on earth, or no longer existing in a given place. Because of the potential for confusion, we prefer to confine use of the term *extinct* to the global disappearance of a species, and apply the term *extirpated* to situations where a species has disappeared from a particular region or country, but still survives elsewhere.

A variety of related terms mark potential way stations on the road to extinction, including *endangered, threatened, imperiled,* and *vulnerable.* These all typically refer to an increase in risk of extinction, but some of these terms have both legal and vernacular definitions, and these definitions can vary considerably. For example, in the United States the term *endangered* has the legal meaning of "in danger of extinction within the foreseeable future throughout all or a significant portion of its range," while *threatened* refers to those species "likely to become endangered within the foreseeable future" (USFWS 1988). In contrast, the World Conservation Union's (IUCN) Species Survival Commission applies the term *threatened* generically to any species with a heightened risk of extinction, including its categories of critically endangered, endangered, and vulnerable (IUCN 2001).

Because of the paradox involved in documenting that which no longer exists, determining current levels of extinction is not an easy or straightforward exercise. Two major approaches have been employed in developing methods and criteria for assessing the extinction of particular species. The first focuses on a long absence in the observational record for a particular species. A 50-year absence was long regarded as a threshold for ascertaining extinction, and served as the basis for the pre-1994 IUCN Red List extinction category. The major problem with using a time-based criterion is that it does not take into account the highly variable level of inventory effort that is targeted toward different species and different places. A second problem is that it does not account for instances of undeniable extinctions that occur within such a time period.

The second major approach focuses on the absence of the species from suitable habitat, based on unsuccessful attempts to relocate the species. Issues with this approach include determining what constitutes suitable habitat, especially for very poorly known species, and what constitutes a sufficiently comprehensive search of this habitat from which to draw conclusions as to the likelihood of continued survival. The ability of seeds from some species to remain viable in the soil for long periods of time can further confound such efforts.

Three major systems for assessing extinction currently are in use: categories and criteria developed by NatureServe and its natural heritage member programs in the

Western Hemisphere (Stein et al. 2000); criteria developed by the IUCN Species Survival Commission for use globally in its Red List Programme (IUCN 2001); and criteria developed as part of an effort based at the American Museum of Natural History, known as CREO (Committee on Recently Extinct Organisms) (MacPhee and Flemming 1999; Harrison and Stiassny 1999), which has so far been applied just to mammals and fishes. These systems each tend to combine the two general approaches discussed above—time and search effort. However, there are some key distinctions.

The current IUCN Red List extinct category states that "[a] taxon is presumed Extinct when exhaustive surveys in known and/or expected habitat, at appropriate times (diurnal, seasonal, annual), throughout its historic range have failed to record an individual" (IUCN 2001, 14). Species that do not meet these strict criteria are placed in the critically endangered category, regardless of length of absence or other evidence that suggests a possible extinction. In contrast, NatureServe distinguishes between two categories: presumed extinct (GX for globally extinct) is applied to those species for which there is virtually no likelihood of survival; possibly extinct (GH for globally historical) applies to those species that are missing yet for which sufficient searches have not been carried out to definitively determine likelihood of continued existence. CREO takes an approach similar to NatureServe's, terming these two classes of extinction resolved and unresolved. In recognition of the utility of distinguishing between these two levels of extinction certainty, IUCN is currently considering adoption of a possibly extinct "flag" for appropriate species in their critically endangered category.

Why should it matter whether species are classified as missing, yet not definitely extinct? Although it may sound counterintuitive, once species are listed as missing or extinct, many field biologists take it as a personal challenge to attempt their rediscovery. Directing their attention toward those species where doubts linger, and where additional surveys are required, can often be very productive. Indeed, such enthusiastic responses from fieldworkers is a major reason that lists of supposedly extinct species can have a relatively high rate of change, such as noted by Burgman (2002).

Yet another complication in categorizing extinctions relates to those species that are extinct in the wild yet survive in cultivation or captivity. A well-known example is the beautiful shrub *Franklinia alatamaha* (Theaceae), discovered in Georgia by the early American botanists John and William Bartram. The shrub was last seen in the wild in 1803, having disappeared for unknown reasons. Fortunately, Bartram had brought the plant into cultivation, and it is now widely cultivated.

A more extreme example is the bird-pollinated tree *Kokia cookei* (plate 3.2), a Hawaiian relative of cotton found on the island of Molokai. When this species was discovered in the 1860s, only three individual trees were found. By the turn of the century free-ranging livestock had largely decimated the plant's dryland forest habitat, and the species survived as a single cultivated plant, which persisted for decades. In the 1970s Keith Woolliams of the Waimea Arboretum collected seeds from this last tree. Although he was able to grow a single tree, its seeds were not viable, forcing him to propagate the species by grafting cuttings onto a related *Kokia* species from the island of Hawaii. Consequently, all remaining *Kokia cookei* plants are clones with neither genetic variability nor rootstock. Currently there are only 15 individuals of this species in the world—all grafts.

EXTINCTION PORTFOLIOS: WHAT DO WE KNOW?

To assess current plant extinctions worldwide it is useful to start with an examination of some of the regions for which relatively comprehensive floristic inventories and conservation status assessments have been conducted. The United States is fortunate in having a long history of botanical exploration, resulting in a relatively well known flora. In addition, strong laws to protect endangered species have encouraged considerable work in documenting and assessing the status of wild plant species. This work has been carried out by a wide array of institutions in the governmental, academic, and nonprofit sectors. The first comprehensive assessment of the status of U.S. plants was carried out by the Smithsonian Institution as part of its landmark report to Congress on the threatened and endangered species of the United States (Ayensu and DeFilipps 1978). A great deal of additional inventory has been carried out since that time, and NatureServe and its state natural heritage member programs now maintain what generally is regarded as the most current assessment of U.S. plant conservation status.

Eleven presumed extinctions and another 130 possible extinctions have been documented from among the nearly 16,000 species of native U.S. vascular plants, for an overall extinction level of approximately 0.9% (Stein et al. 2000; see also NatureServe 2004). Interestingly, none of these are gymnosperms, and only four are ferns. Because of the ability of plants to persist in seed banks and in vegetative states, and to elude even determined field biologists, these figures reflect a conservative approach to assigning species to the different extinction categories (i.e., GX versus GH).

How does our understanding of plant extinctions compare with that of other groups of organisms? By calculating an extinction resolution index (ERI) we can gain insight into the relative depth of extinction knowledge across various groups (fig. 3.2). This index can be defined as the ratio of resolved extinctions to the total of all extinctions, presumed and possible (ERI = GX/[GX + GH]). Not surprisingly, mammals and birds, both well-known groups that have been the focus of intensive field inventories, have very high ratios reflecting few if any species for which questions about continued existence linger. Most well-studied invertebrate groups fall into a midrange, including freshwater mussels and snails, both species groups with quite high numbers of extinct species. U.S. flowering plants, however, have a very low extinction resolution index (0.08), demonstrating that even in one of the best-studied countries on earth, considerable uncertainty exists as to status of plant extinctions.

While plant extinctions have touched many parts of the United States, most are concentrated in the West, Southeast, and especially Hawaii. Indeed, Hawaii is com-

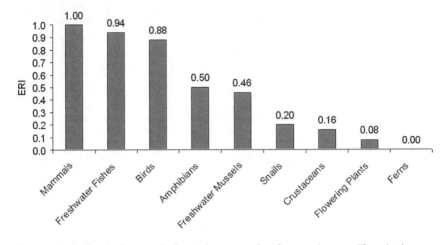

Figure 3.2 Ratio of resolved to unresolved extinctions across selected taxonomic groups. The extinction resolution index (ERI) indicates the proportion of resolved extinctions (GX) to the total number of presumed or possibly extinct species (GX + GH). Note the relatively low ERI for flowering plants and ferns relative to other groups of organisms.

monly referred to as the extinction capital of the nation, and with good reason. A volcanic archipelago separated from the nearest significant landmass by thousands of miles, Hawaii's plants evolved in almost complete isolation. The result is a flora that has one of the highest rates of endemism on earth. The Hawaiian flora also evolved in absence of humans and large herbivores with the result that many of its plants lack defensive adaptations, such as spines or thorns, commonly seen in continental relatives. Human colonization of the archipelago, first by Polynesians and later by Europeans, brought tremendous changes to the islands, sending many of its unique plant species into a downward spiral.

A recent assessment of endangerment and extinction in the Hawaiian flora regarded 82 species as presumed or possibly extinct, representing 8.0% of the native vascular flora (Wagner et al. 1999; see also Wagner et al. 2001–present). This represents nearly an order of magnitude increase over the plant extinction levels for the United States as a whole (0.9%), and a 27-fold increase over the extinction levels of the continental U.S. flora alone (0.3%). Work on the flora of Hawaii also illustrates the powerful influence that listing species as extinct has on field botanists. Since the publication of the *Manual of the Flowering Plants of Hawai'i* (Wagner et al. 1990), 35 species listed as extinct in that volume have been rediscovered, although most of these survive as just a few individuals. Conversely, fieldwork has failed to confirm the continued survival of another 22 species, which now have been included on the list of presumed or possibly extinct plants.

Several other relatively well studied countries provide additional context for extinction levels. Among New Zealand's approximately 2,400 species of native plants, 12 species currently are regarded as being extinct (D. Given, personal communication, 2003) for an extinction level of 0.5%. Similarly, in Australia 64 plant species are regarded as extinct out of a flora of about 15,970 (Burgman 2002) for an extinction level of 0.4%. Well-documented levels of extinctions are scarce for species-rich tropical countries. A recent analysis of endemic plants of Ecuador identifies 40 plant species as possibly extinct out of a total of 4,011 species assessed (Valencia et al. 2000), for an extinction level of 1.0%. Focusing just on Ecuador's Galápagos Islands—a well-studied but relatively species-poor archipelago—that same study noted 3 out of 560 species as having been lost, for an extinction level of 0.5%.

PLANT EXTINCTIONS WORLDWIDE

Data on plant extinctions across most of the globe are woefully inadequate. The most comprehensive figures currently available derive from the *1997 IUCN Red List of Threatened Plants* (Walter and Gillett 1998). Of the more than 30,000 species covered by this volume, only 380 were listed as extinct in the wild, and another 371 as extinct/endangered, a category defined as "taxa that are suspected of having recently become Extinct." As a proportion of a global flora of some 270,000 species (the estimate contained in that volume), the 1997 Red List documents a plant extinction level worldwide of about 0.3%, about the same level as for the continental United States. This first-ever global assessment of plant status was a monumental accomplishment, but there are several important limitations to its use. First, the assessments are based on pre-1994 IUCN Red List categories, and although reflecting the contributions of more than a dozen institutions and hundreds of scientists worldwide, documentation for individual submissions is often lacking. Perhaps more significant, however, the assessment clearly underrepresents many parts of the world where botanical knowledge is sparse, especially in the species-rich tropics.

More recent IUCN Red Lists represent a significant departure from the 1997 report by including only plant species that have been reassessed using the recently revised evaluation criteria and categories (IUCN 2001; see also chapter 11). The 2003 IUCN Red List contains just 81 plant species in the extinct category and lists another 25 as extinct in the wild (see IUCN 2003). Clearly, this does not reflect an actual reduction in the number of plant extinctions since 1997, but rather a change in the process by which species are assessed, documented, and listed.

As highlighted elsewhere in this volume, many parts of the world are very poorly known botanically, and this is especially true of many species-rich tropical regions. Unfortunately, development pressures and habitat conversion in many of these regions are far outstripping the rate of botanical exploration and documentation. For this reason, many projections of extinction risk worldwide rely on the empirical relationship between size of an area and the number of species it contains (MacArthur and Wilson 1967), and subsequent assumptions about the loss of species based on reduction in habitat area.

Centinela Ridge in western Ecuador serves as one of the most famous examples of the relationship between tropical habitat loss and plant extinctions. This relatively low ridge protrudes from the surrounding lowlands just before the start of the Andean foothills and was covered by a localized cloud forest. Alwyn Gentry and Calaway Dodson surveyed this site prior to its nearly complete deforestation and estimated that between 38 and 90 plants were endemic to the ridge (Dodson and Gentry 1991). Novel species in the family Gesneriaceae were especially well represented; the genus *Gasteranthus* alone had a reported four Centinela endemics. Clearing of this tiny ridge for agriculture, then, was thought to lead directly to the loss of up to 90 narrowly restricted endemics. Centinela, an unprepossessing hill in western Ecuador, was thereby transformed into a poster child for tropical plant extinctions (Wilson 1992).

How well has the Centinela story withstood the test of time? A recent review of the status of Gesneriaceae in western Ecuador (L. P. Kvist, L. E. Skog, J. L. Clark, and R. W. Dunn, unpublished data) provides an opportunity to review the original extinction predictions made by Dodson and Gentry. Kvist et al. confirmed that Centinela was indeed unique in its species diversity, having more endangered and endemic Gesneriaceae than any other correspondingly small forest in western Ecuador. Of six *Gasteranthus* species documented from the ridge, however, they found that only one is likely to have been entirely restricted to Centinela. Summarizing data for all 107 Gesneriaceae species found in western Ecuador, Kvist et al. consider seven species to be extinct (of which two survive in cultivation), yielding a regional extinction level of 6.5%.

The conclusion one can draw is that while Centinela Ridge was indeed an extraordinary site and its deforestation represents a considerable loss of biological diversity, a number of the seemingly site-specific plants have now turned up in surrounding areas. As a result, clearing of the ridge may not have caused quite as many extinctions as originally presumed. Nonetheless, assuming Gesneriaceae are somewhat representative of the western Ecuadorian flora as a whole, the region does exhibit a decidedly high level of extinctions.

FROM THE KNOWN TO THE UNKNOWN

From our review of documented extinctions in better-known regions, we can make some extrapolations of the potential range in species extinctions worldwide. One problem with making any such extrapolation is the widely divergent estimates of the total number of vascular plants on earth, from the widely used figure of 270,000, to an estimate of 310,000 by Prance et al. (2000), to independent estimates of 422,000 by Govaerts (2001) and Bramwell (2002). More recently, Scotland and Wortley (2003) have suggested that large-scale inclusion of synonyms has resulted in these higher estimates, and they offer the far more conservative figure of 223,000. Assuming a midpoint estimate of 310,000 species, and using a range in extinction levels from a low of 0.3% (e.g., continental United States, Galápagos Islands) to a high of 8.0% (Hawaii), yields a range in potential plant extinctions from about 1,000 to nearly 25,000 species. Clearly, Hawaii represents an unusual case, and it is hard to imagine that 25,000 species already have vanished, yet only 750 (Walter and Gillett 1998) have come to the attention of the botanical community worldwide. A conservative approach would be to assume current global extinction levels of between 0.5% and 1% totaling between 1,500 and 3,000 species. This represents between two and three times the number of currently recognized extinctions.

Because of the inherent difficulty in documenting extinctions, and the very poor level of knowledge in many parts of the world subject to the most rapid habitat losses, the relatively few recognized plant extinctions represent just a partial view of the situation. Of even greater concern is the way in which these known extinctions presage a far greater problem—species that are still extant, yet because of reduced population sizes and external pressures may already be committed to extinction. Threatened species are reviewed elsewhere in this volume (chapter 11), but the most conservative figure for threatened species levels worldwide is 12.5% (Walter and Gillett 1998), with more aggressive estimates ranging from 22 to 47% of the world's plant species (Pitman and Jørgensen 2002).

Of particular concern are those species referred to by Daniel Janzen (1986) as the "living dead." These are species whose future viability has already been compromised, either through a vast reduction in their populations, or the loss of an essential link in their reproduction, such as pollinators or seed dispersers. The island of Mauritius harbors one such example of the living dead, the palm *Hyophorbe amaricaulis*. This palm exists as just a single tree, living out its days in the confines of a botanical garden. Although the plant occasionally flowers and fruits, its seeds are not viable and there is no reproduction. When this tree eventually dies, the species will join Mauritius's most famous resident—the dodo—on the list of extinct species.

Hyophorbe amaricaulis offers a stark example of the bounds between life and death. For many of the plant species at the brink of extinction, the shading between life and death is far more nuanced and difficult to discern. Interpreting and documenting this chiaroscuro in green, however, will be essential if society hopes to reverse current trends in species declines and halt the loss through extinction of the earth's diversity of plant life.

LITERATURE CITED

Alvarez, L., Alvarez, W., Asaro, F., and Michel, H. 1980. Extraterrestrial cause for the Cretaceous-Tertiary extinctions. Science 208:1095–1108.

Ayensu, E. S., and DeFilipps, R. A. 1978. Endangered and Threatened Plants of the United States. Smithsonian Institution Press, Washington, DC.

Behrensmeyer, A. K., Hook, R. W., Badgley, C. E., Boy, J. A., Chapman, R. E., Dodson, P., Gastaldo, R. A., Graham, R. W., Martin, L. D., Olsen, P. E., Spicer, R. A., Taggart, R. E., and Wilson, W. V. H. 1992. Paleoenvironmental contexts and taphonomic modes. Pp. 15–138 in Behrensmeyer, A. K., Damuth, J. D., DiMichele, W. A., Potts, R. A., Sues, H.-D., and Wing, S. L., eds., Terrestrial Ecosystems through Time. University of Chicago Press, Chicago.

Behrensmeyer, A. K., Kidwell, S., and Gastaldo, R. A. 2000. Taphonomy and paleobiology. Pp. 103–147 in Erwin, D. H., and Wing, S. L., eds., Deep Time: Paleobiology's Perspective, vol. 25. Paleontological Society, Lawrence, Kansas.

Boulter, M. C., Spicer, R. A., and Thomas, B. A. 1988. Patterns of plant extinction from some paleobotanical evidence. Pp. 1–36 in Larwood, G. P., ed., Extinction and Survival in the Fossil Record, vol. 34. Clarendon Press, Oxford.

Bramwell, D. 2002. How many plant species are there? Plant Talk 28:32–34.

Burgman, M. A. 2002. Turner review no. 5: are listed threatened plant species actually at risk? Australian Journal of Botany 50:1–13.

Claeys, P., Kiessling, W., and Alvarez, W. 2002. Distribution of Chicxulub ejecta at the Cretaceous-Tertiary boundary. Pp. 55–68 in Koeberl, C., and MacLeod, K. G., eds., Catastrophic Events and Mass Extinctions: Impacts and Beyond. Geological Society of America Special Paper 356. Geological Society of America, Boulder, Colorado.

Collinson, M. E., Boulter, M. C., and Holmes, P. L. 1993. Magnoliophyta ("Angiospermae"). Pp. 809–841 in Benton, M. J., ed., The Fossil Record, vol. 2. Chapman and Hall, London.

DiMichele, W. A., and Phillips, T. L. 1996. Climate change, plant extinctions, and vegetational recovery during the Middle-Late Pennsylvanian transition: the case of tropical peat-forming environments in North America. Pp. 201–221 in Hart, M. L., ed., Biotic Recovery from Mass Extinctions. Geological Society of London, London.

Dodson, C. H., and Gentry, A. H. 1991. Biological extinction in western Ecuador. Annals of the Missouri Botanical Garden 78:273–295.

Erwin, D. H., Bowring, S. A., and Yugan, J. 2002. End-Permian mass extinctions: a review. Pp. 363–383 in Koeberl, C., and MacLeod, K. G., eds., Catastrophic Events and Mass Extinctions: Impacts and Beyond. Geological Society of America Special Paper 356. Geological Society of America, Boulder, Colorado.

Foote, M. 1994. Temporal variation in extinction risk and temporal scaling of extinction metrics. Paleobiology 20:424–444.

Gastaldo, R. 1988. Conspectus of phytotaphonomy. Pp. 14–28 in DiMichele, W. A., and Wing, S. L., eds., Methods and Applications of Plant Paleoecology. Paleontological Society Special Publication 3. Paleontological Society, Knoxville.

Govaerts, R. 2001. How many species of seed plants are there? Taxon 50:1085–1090.

Harrison, I. J., and Stiassny, M. L. J. 1999. The quiet crisis: a preliminary listing of the freshwater fishes of the world that are extinct or "missing in action." Pp. 271–332 in MacPhee, R. D. E., ed., Extinctions in Near Time: Causes, Contexts, and Consequences. Kluwer Academic and Plenum Publishers, New York.

Hotton, C. L. 2002. Palynology of the Cretaceous-Tertiary boundary in central Montana: evidence for extraterrestrial impact as a cause of the terminal Cretaceous extinctions. Pp. 473–502 in Hartman, J. H., Johnson, K. R., and Nichols, D. J., eds., The Hell Creek Formation and the Cretaceous-Tertiary Boundary in the Northern Great Plains: An Integrated Continental Record of the End of the Cretaceous. Geological Society of America Special Paper 361. Geological Society of America, Boulder, Colorado.

IUCN (World Conservation Union). 2001. IUCN Red List Categories and Criteria, Version 3.1. IUCN Species Survival Commission, Gland, Switzerland, and Cambridge.

———. 2003. 2003 IUCN Red List of Threatened Species. IUCN Species Survival Commission. http://www.redlist.org/.

Janzen, D. H. 1986. The future of tropical ecology. Annual Review of Ecology and Systematics 17:305–324.

Johnson, K. R. 1992. Leaf-fossil evidence for extensive floral extinction at the Cretaceous-Tertiary boundary, North Dakota, USA. Cretaceous Research 13:91–117.

———. 2002. Megaflora of the Hell Creek and lower Fort Union formations in the western Dakotas: vegetational response to climate change, the Cretaceous-Tertiary boundary event, and rapid marine transgression. Pp. 329–392 in Hartman, J. H., Johnson, K. R., and Nichols, D. J., eds., The Hell Creek Formation and the Cretaceous-Tertiary Boundary in the Northern Great Plains: An Integrated Continental Record of the End of the Cretaceous. Geological Society of America Special Paper 361. Geological Society of America, Boulder, Colorado.

Johnson, K. R., and Hickey, L. J. 1990. Megafloral change across the Cretaceous/Tertiary boundary in the northern Great Plains and Rocky Mountains, U.S.A. Pp. 433–444 in Sharpton, V. L., and Ward, P. D., eds., Global Catastrophes in Earth History: An Interdisciplinary Conference on Impacts, Volcanism, and Mass Mortality. Geological Society of America Special Paper 247. Geological Society of America, Boulder, Colorado.

Johnson, K. R., Nichols, D., Attrep, M. J., and Orth, C. 1989. High-resolution leaf-fossil record spanning the Cretaceous-Tertiary boundary. Nature 340:708–711.

Kerp, H. 2000. The modernization of landscapes during the late Paleozoic–early Mesozoic. Pp. 79–113 in Gastaldo, R. A., and DiMichele, W. A., eds., Terrestrial Ecosystems: A Short Course. Paleontological Society Papers 6. Paleontological Society, New Haven.

Knoll, A. H. 1984. Patterns of extinction in the fossil record of vascular plants. Pp. 21–68 in Nitecki, M. H., ed., Extinctions. University of Chicago Press, Chicago.

Lidgard, S., and Crane, P. 1990. Angiosperm diversification and Cretaceous floristic trends: a comparison of palynofloras and leaf macrofloras. Paleobiology 16:77–93.

Looy, C. V., Brugman, W. A., Dilcher, D. L., and Visscher, H. 1999. The delayed resurgence of equatorial forests after the Permian-Triassic ecologic crisis. Proceedings of the National Academy of Science USA 96:13857–13862.

Lupia, R., Lidgard, S., and Crane, P. R. 1999. Comparing palynological abundance and diversity: implications for biotic replacement during the Cretaceous angiosperm radiation. Paleobiology 25:305–340.

MacArthur, R. H., and Wilson, E. O. 1967. The Theory of Island Biogeography. Princeton: Princeton University Press, Princeton, New Jersey.

MacPhee, R. D. E., and Flemming, C. 1999. Requiem aeternam: the last five hundred years of mammalian species extinctions. Pp. 333–372 in MacPhee, R. D. E., ed., Extinctions in Near Time: Causes, Contexts, and Consequences. Kluwer Academic and Plenum Publishers, New York.

Magallon, S., Crane, P. R., and Herendeen, P. S. 1999. Phylogenetic pattern, diversity, and diversification of eudicots. Annals of the Missouri Botanical Garden 86:297–372.

Manchester, S. 1986. Vegetative and reproductive morphology of an extinct plane tree (Platanaceae) from the Eocene of western North America. Botanical Gazette 147:200–226.

NatureServe. 2004. NatureServe Explorer: An Online Encyclopedia of Life. NatureServe. http://www.natureserve.org/explorer/.

Nichols, D. J. 2002. Palynology and palynostratigraphy of the Hell Creek Formation in North Dakota: a microfossil record of plants at the end of Cretaceous time. Pp. 393–456 in

Hartman, J. H., Johnson, K. R., and Nichols, D. J., eds., The Hell Creek Formation and the Cretaceous-Tertiary Boundary in the Northern Great Plains: An Integrated Continental Record of the End of the Cretaceous. Geological Society of America Special Paper 361. Geological Society of America, Boulder, Colorado.

Nichols, D. J., and Johnson, K. R. 2002. Palynology and microstratigraphy of Cretaceous-Tertiary boundary sections in southwestern North Dakota. Pp. 95–144 in Hartman, J. H., Johnson, K. R., and Nichols, D. J., eds., The Hell Creek Formation and the Cretaceous-Tertiary Boundary in the Northern Great Plains: An Integrated Continental Record of the End of the Cretaceous. Geological Society of America Special Paper 361. Geological Society of America, Boulder, Colorado.

Niklas, K. J., Tiffney, B. H., and Knoll, A. H. 1980. Apparent changes in the diversity of fossil plants. Evolutionary Biology 12:1–89.

———. 1985. Patterns in vascular land plant diversification: an analysis at the species level. Pp. 97–128 in Valentine, J. W., ed., Phanerozoic Diversity Patterns: Profiles in Macroevolution. Princeton University Press, Princeton, New Jersey.

Norris, R. D., and Firth, J. V. 2002. Mass wasting of Atlantic continental margins following the Chicxulub impact event. Pp. 79–95 in Koeberl, C., and MacLeod, K. G., eds., Catastrophic Events and Mass Extinctions: Impacts and Beyond. Geological Society of America Special Paper 356. Geological Society of America, Boulder, Colorado.

Orth, C., Gilmore, J., Knight, J., Pillmore, C., Tschudy, R., and Fassett, J. 1981. An iridium abundance anomaly at the palynological Cretaceous-Tertiary boundary in northern New Mexico. Science 214:1341–1342.

Pitman, N. C. A., and Jørgensen, P. M. 2002. Estimating the size of the world's threatened flora. Science 298:989.

Pope, K. O. 2002. Impact dust not the cause of the Cretaceous-Tertiary mass extinction. Geology 30:99–102.

Prance, G. T., Beentje, H., Dransfield, J., and Johns, R. 2000. The tropical flora remains undercollected. Annals of the Missouri Botanical Garden 87:67–71.

Prance, G. T., and Elias, T. S. 1977. Extinction Is Forever. New York Botanical Garden, New York.

Rees, P. M. 2002. Land-plant diversity and the end-Permian mass extinction. Geology 30:827–830.

Regan, H. M., Lupia, R., Drinnan, A. N., and Burgman, M. A. 2001. The currency and tempo of extinction. American Naturalist 157:1–10.

Retallack, G. J. 1995. Permian-Triassic life crisis on land. Science 267:77–80.

Retallack, G. J., Veevers, J. J., and Morante, R. 1996. Global coal gap between Permian-Triassic extinction and Middle Triassic recovery of peat-forming plants. Geological Society of America Bulletin 108:195–207.

Scheihing, M. H. 1980. Reduction of wind velocity by the forest canopy and the rarity of non-arborescent plants in the Upper Carboniferous fossil record. Argumenta Palaeobotanica 6:133–138.

Schultz, P. H., and D'Hondt, S. 1996. Cretaceous-Tertiary (Chicxulub) impact angle and its consequences. Geology 24:963–967.

Scotland, R. W., and Wortley, A. H. 2003. How many species of seed plants are there? Taxon 52:101–104.

Spicer, R. 1981. The sorting and deposition of allochthonous plant material in a modern environment at Silwood Lake, Silwood Park, Berkshire, England. United States Geological Survey Professional Paper 1143. U.S. Geological Survey, Washington, DC.

Stein, B. A., Kutner, L. S., and Adams, J. S., eds. 2000. Precious Heritage: The Status of Biodiversity in the United States. Oxford University Press, New York.

Sweet, A. R. 2001. Plants: a yardstick for measuring the environmental consequences of the Cretaceous-Tertiary boundary event. Geoscience Canada 28:127–138.

Traverse, A. 1988. Plant evolution dances to a different beat: plant and animal evolutionary mechanisms compared. Historical Biology 1:277–301.

Tschudy, R., Pillmore, C., Orth, C., Gilmore, J., and Knight, J. 1984. Disruption of the terrestrial plant ecosystem at the Cretaceous-Tertiary boundary, western Interior. Science 225:1030–1032.

Twitchett, R. J., Looy, C. V., Morante, R., Visscher, H., and Wignall, P. B. 2001. Rapid and synchronous collapse of marine and terrestrial ecosystems during the end-Permian biotic crisis. Geology 29:351–354.

Upchurch, G. R. J. 1989. Dispersed angiosperm cuticles. Pp. 65–92 in Tiffney, B. H., ed., Phytodebris: Notes for a Workshop on the Study of Fragmentary Plant Remains. Paleobotanical Section of the Botanical Society of America, Toronto.

USFWS (U.S. Fish and Wildlife Service). 1988. Endangered Species Act of 1973, as Amended through the 100th Congress. U.S. Department of the Interior, U.S. Fish and Wildlife Service, Washington, DC.

Vajda, V., Raine, J. I., and Hollis, C. J. 2001. Indication of global deforestation at the Cretaceous-Tertiary boundary by New Zealand fern spike. Science 294:1700–1702.

Valencia, R., Pitman, N., León-Yánez, S., and Jørgensen, P. M., eds. 2000. Libro Rojo de las Plantas Endemicas del Ecuador. Pontifica Universidad Católica del Ecuador, Quito.

Valentine, J. W., Tiffney, B. H. and Sepkoski, J. J. Jr. 1991. Evolutionary dynamics of plants and animals: A comparative approach. Palaios 6:81–88.

Visscher, H., Brinkhuis, H., Dilcher, D. L., Elsik, W. C., Yoram, E., Looy, C. V., Rampino, M. R., and Traverse, A. 1996. The terminal Paleozoic fungal event: evidence of terrestrial ecosystem destabilization and collapse. Proceedings of the National Academy of Sciences USA 93:2155–2158.

Wagner, W. L., Bruegmann, M., Herbst, D. R., and Lau, J. Q. 1999. Hawaiian vascular plants at risk: 1999. Bishop Museum Occasional Papers 60:1–64.

Wagner, W. L., Herbst, D. R., and Palmer, D. D. 2001–present. Flora of the Hawaiian Islands. Smithsonian Institution. http://ravenel.si.edu/botany/pacificislandbiodiversity/hawaiianflora/index.htm. January 2004.

Wagner, W. L., Herbst, D. R., and Sohmer, S. H. 1990. Manual of the Flowering Plants of Hawai'i. Bishop Museum Special Publication 83. University of Hawaii Press and Bishop Museum Press, Honolulu.

Walter, K. S., and Gillett, H. J., eds. 1998. 1997 IUCN Red List of Threatened Plants. IUCN, Gland, Switzerland.

Wilson, E. O. 1992. The Diversity of Life. Harvard University Press, Cambridge.

Wing, S. L. 2000. Evolution and expansion of flowering plants. Pp. 209–231 in Gastaldo, R. A., and DiMichele, W. A., eds., Phanerozoic Terrestrial Ecosystems, vol. 6. Paleontological Society, New Haven.

66 | *Chapter 3*

Wolbach, W. S., Gilmour, I., Anders, E., Orth, C. J., and Brooks, R. R. 1988. Global fire at the Cretaceous-Tertiary boundary. Nature 334:665–669.

Wolfe, J. A. 1990. Palaeobotanical evidence for a marked temperature increase following the Cretaceous-Tertiary boundary. Nature 343:153–156.

Wolfe, J. A., and Upchurch, G. R. Jr. 1986. Vegetation, climatic, and floral changes at the Cretaceous-Tertiary boundary. Nature 324:148–152.

PART II

PLANT DIVERSITY

HABITATS AND TAXONOMIC GROUPS

CHAPTER 4

CASE STUDIES IN SELECT TROPICAL
AND SUBTROPICAL HABITATS

Most biologists and conservationists now agree that few undisturbed, pristine habitats remain on the earth. Perhaps no habitats have entirely escaped the hand of humans either past or present. However, scattered across the globe are areas of various sizes that are rich in plant diversity with high rates of endemism, are relatively undisturbed, but are constantly under threat of alteration. Most of these high-diversity habitats are found in the tropical and subtropical zones. Some are major wilderness areas that are still largely intact, others have been offered some protection by local or national governments, whereas other sites are faced with constant pressure of heavy exploitation and fragmentation as the result of logging, agriculture, and industrial exploration, for example oil, gas, gem, and mineral exploration and mining. This chapter highlights five case studies of regions of prime botanical importance; many other examples could have been selected. For each area, floristic diversity and botanical importance is described, current threats are outlined, and the consequences associated with the loss of their diversity are addressed.

4.1 THE ECUADORIAN ANDES

John L. Clark

ALTHOUGH MUCH ATTENTION has recently been focused on the conservation of Amazonian forests, the tropical portion of the Andes is rated as one of the most important regions on earth in terms of the number of plant and animal species and the high percentage of these species not found in other regions (Mittermeier et al. 1999; Myers et al. 2000). The Ecuadorian Andes, or the Sierra, corresponds to areas above 1,000 m and is flanked by the Oriente to the east and Costa to the west. There are 22 provinces in Ecuador, and those from north to south that primarily comprise Andean regions are Carchi, Imbabura, Pichincha, Cotopaxi, Tungurahua, Chimborazo, Bolívar, Cañar, Azuay, and Loja. Many of these provinces are not restricted to the Andes and contain significant area in the Costa, or western lowlands below 1,000 m.

The Andes Mountains run north to south and are the dominant topographic feature of Ecuador. In central and northern Ecuador, there are two parallel mountain chains connected by a series of transverse east-west ridges referred to as *nudos* (knots) that divide the intermountain plateau into roughly ten basins, or *hoyas*. The Nudo del Azuay is at 4,500 m and divides the country into northern and southern regions. The northern area is of modern volcanism and is topped by Quaternary volcanoes that reach 6,310 m (Volcan Chimborazo) and are capped by glaciers. These mountains are generally wetter and higher than the southern region, and their vegetation resembles that of Colombia. The region south of the Nudo del Azuay (Cañar, Azuay, and Loja provinces) barely reaches 4,000 m, is drier, and has vegetation that resembles that of northern Peru.

A third range east of the two main Andean ranges is tectonically related to the Cordillera Oriental of Colombia. This range is not continuous but is formed by a series of Cretaceous and Tertiary sediments including the cordilleras of Galeras, Cutucú, and Cóndor. The Reventador and Sumaco volcanoes are also part of this range (Sauer 1965).

Despite being one of the smaller countries in South America (with less area than Colorado), Ecuador harbors more than 16,000 plant species or roughly 8% of the plant diversity of the entire world (Jørgensen and León-Yánez 1999). The Andes harbor 9,865 species, or 64.4% of the total species found in Ecuador (Jørgensen and León-Yánez 1999). The number of endemic species in Ecuador is about 4,011 species, or roughly 26% of the native flora (Pitman et al. 2000), and these are concentrated in the Andes (2,965 endemics, or roughly 75% of the entire endemic flora).

Glacial erosion has had a major impact on the Andean landscape, producing wide valleys, ridges, and variable soil conditions that presumably have profound implications for current plant distributions. The variable soils, coupled by phenomena such as the Humboldt Current, are important factors that are responsible for the high degree of restricted local endemism in the Andes (Jørgensen et al. 1995). In addition to the Interandean Valley that separates the eastern and western ranges in central and northern Ecuador, the other significant barrier is the Girón-Paute valley that acts as a barrier between the northern and southern Andes (Jørgensen et al. 1995). The Girón-Paute valley has been a migration barrier both during the ice ages and at present, because of differences in geologic deposits and lower precipitation with a pronounced dry season in the southwestern part of the country. The Girón-Paute valley is an important factor in explaining why the vegetation of Loja is significantly different from that of the northern Andes, and why Parque Nacional Podocarpus in the south contains 211 endemic species, more than twice the number found in any other protected area in Ecuador.

Human influence on the Andean vegetation has been estimated to exist as far back as 10,000 years ago (Engel 1976). The name Andes is probably from the Spanish term *andén*, meaning "platform," because of the various methods used by the Inca Empire to build platforms or terraces for farming (Schjellerup 1992). After the Spanish conquest, a more agro-pastoral type of agriculture replaced terrace agriculture, intensifying soil erosion. This switch in agricultural practices, coupled with cattle production for beef and milk and deforestation primarily for coal production and cooking, has eliminated most forest. Threatening the integrity of the remaining forests is the significant increase in human population; between 1957 and 2004, the population of Ecuador increased from less than 4 million to 13.4 million (Population Reference Bureau 2003). Remnants of forest can still be found on the eastern and western slopes of the Andes, but virtually no forest exists in the most densely populated sections of the Interandean regions, except in steep ravines, or *quebradas*. Based on satellite images of remnant forests from 1996 and estimates of original forest cover (Sierra 1999), Andean forests have decreased by 43%. This estimate includes a large range of vegetation types; the Interandean forests and western slope forests are the most critically endangered because they are the least protected, most populated, and most deforested. For example, Interandean forests located between 2,000 and 3,000 m north of the Girón-Paute valley are over 75% deforested (Sierra 1999) and have mostly been replaced by cultivated groves of *Eucalyptus globulus* from Australia.

The highest concentration of protected forests in the Andes is on the eastern slopes, many of which are included in the Ecuadorian government's system of protected areas, Sistemas Nacional de Areas Protegidas del Ecuador (SNAP). Some of the larger of these protected areas, from north to south, are Reserva Ecológica Cayambe-Coca (403,000 ha), Reserva Ecológica Antisana (120,000 ha), Parque Nacional Llanganates (219,707 ha), Parque Nacional Sangay (517,725 ha), and Parque Nacional Podocarpus (146,280 ha). Along the western slopes some of the larger SNAP reserves are Reserva Ecológica Cotacachi-Cayapas (204,000 ha), Reserva Geobotánica Pululahua (3,383 ha), Parque Nacional Cotopaxi (33,393 ha), Reserva Ecológica Los Ilinizas (149,900 ha), Reserva de Producción de Fauna Chimborazo (58,560 ha), and Parque Nacional El Cajas (28,808 ha). The only Interandean forest within Ecuador's SNAP system is in Refugio de Vida Silvestre Pasochoa (500 ha).

Ecuador's Andean forests are recognized internationally and nationally as important resources that warrant protection. This is especially evident in the recent creation of new parks and protected areas by the Ecuadorian government. Despite the attention these forests receive, the reality is that many of these protected areas exist only on paper. Local populations still depend on cutting down and living in what remains of these forests. The future integrity of the Andes will depend on balancing local resources in a manner that will help sustain local communities while protecting what little remains of these forests.

4.2 THE RAMAL DE GUARAMACAL
IN THE VENEZUELAN ANDES

Laurence J. Dorr, Basil Stergios, and S. Miguel Niño

THE RAMAL DE GUARAMACAL is a northeastern outlier of the Sierra Nevada de Mérida, one of the two main mountain chains composing the Venezuelan Andes. The other chain is the Sierra del Norte, or Sierra de La Culata, lying to the north of the Sierra Nevada de Mérida. Combined, these mountain chains cover more than 32,000 km² and run 425 km from the Táchira depression near the Colombian border to the Barquisimeto-Carora depression in Lara State. They rise from foothills at about 250 m bordering the llanos (plains) in the east and Lake Maracaibo in the west to a permanently snow-capped Pico Bolívar (5,007 m) in Mérida State. Guaramacal National Park, which protects the rugged ridge of the same name, covers a relatively insignificant part of the Venezuelan Andes—a mere 215 km² of undisturbed lower montane forest, montane (cloud) forest, elfin forest, and *páramo* (tropical alpine vegetation) situated between 1,200, 1,600, or 1,800 m (depending upon the park boundary) and 3,150 m. The Guaramacal ridge, almost entirely in Trujillo and partly in Portuguesa State, runs no more than 40 km in a northeast-trending direction. It is separated from the higher Ramal de Calderas to the southwest by the deep valley of the Río Boconó and from the higher páramos of Dinira National Park to the northwest by a relatively heavily settled and cultivated valley that includes the small city of Boconó. The steep, south-facing flank of Guaramacal is almost fully exposed to the prevailing winds coming across the llanos.

While it is estimated that the Venezuelan Andes contain between 4,500 and 5,000 species of vascular plants with as many as 10% of them endemic (Huber et al. 1998), precise figures are lacking. For a region that has been populated since pre-Columbian times and that was colonized by the Spanish in the late 16th century (the largest city, Mérida, was settled in 1558, and Boconó in 1592), this is somewhat surprising. Plant collectors have concentrated their efforts on the higher elevations surrounding Mérida and along the very few paved roads that traverse the mountains. The extensive, forested llanos-facing slopes that are found in Táchira, Mérida, Barinas, and Trujillo states are, with one notable exception, virtually unknown botanically.

The exception is Guaramacal National Park, and the relatively small ridge that it protects consequently becomes significant out of proportion to the area that it covers, because it is the only park in the Venezuelan Andes that has a well-documented flora. A combination of factors, some fortuitous and others planned, have made it well-known botanically. In the early 1960s, several orchid collectors were attracted to a road being cut to facilitate the construction of television towers on the summit of the Páramo de Guaramacal (3,150 m). In 1988, when the entire ridge became a national park, casual plant collecting was replaced by organized efforts to understand the flora (Ortega et al. 1987; Rivero and Ortega 1989; Dorr et al. 2000) and vegetation (Cuello 1999, 2002). Currently, 158 families, 588 genera, and 1,396 species of vascular plants are known to occur within the boundaries of Guaramacal National Park (Dorr et al. 2000, unpublished data). This is a relatively rich flora, containing almost one-third of the species estimated to occur in the Venezuelan Andes. By way of comparison, the native component of the vascular flora of the British Isles, an area of 248,000 km², is only about 1,400 species (Williamson 2002).

On the north-facing slope of Guaramacal, the most conspicuous element of the lower montane forest (1,800–2,600 m) is the 20-m-tall *yagrumo blanco, Cecropia telenitida* (Urticaceae). Its white leaves delimit a band at 1,800–2,000 m that is visible from a distance of several kilometers. Other emergent trees also reach 20 m or more in height. Tree ferns are common in the understory. As one moves up in elevation, there is more diversity in tree species, and in places one encounters dense colonies of the palm known locally as *mapora* (*Wettinia praemorsa*) and extensive patches of *Croizatia brevipetiolata* (Phyllanthaceae). Interestingly, in upper montane forest (2,350–2,600 m) there is a notable decrease in the size of the leaves found on trees. Examples of this phenomenon include *Weinmannia balbisiana* (Cunoniaceae) and *Podocarpus oleifolius* var. *macrostachyus* (Podocarpaceae). Here the canopy becomes lower (7–12 m tall), but emergents still reach 15–20 m in height. Epiphytes are abundant. Above 2,600 m, upper montane forest is replaced by elfin forest, which has a 5–8-m-tall canopy. *Ruilopezia paltonioides* (Asteraceae), representing one of several genera of *frailejón* (Asteraceae, Espeletiinae) endemic to the Venezuelan Andes, is the most conspicuous arborescent species in this forest (plate 4.1). Its silhouette has been informally used as an emblem of the park.

The south-facing slope of Guaramacal is steeper than that of the north, and receives significantly more precipitation. The lower slopes (1,400–1,800 m), like their north-facing counterparts, have 25-m-tall canopy emergents such as *Chrysophyllum* and *Pouteria* (both Sapotaceae). The forest found from 1,800 to 2,100 m closely resembles in species composition that of the north-facing slope at similar elevations. At higher elevations (2,100–2,600 m), species of *Clusia* (Clusiaceae) and *Miconia* (Melastomataceae) begin to dominate, and the canopy reaches only 8–15 m in height.

The least-extensive vegetation type in the park is páramo. It is dominated by herbaceous plants, grasses, and isolated islands of shrubs. Conspicuous species in the Páramo de Guaramacal include *Blechnum colombianum* (Blechnaceae), *Puya aristeguietae* (Bromeliaceae), *Chusquea angustifolia* (Poaceae), and *Ruilopezia viridis* (Asteraceae). Two species of *Elaphoglossum* (Dryopteridaceae), *E. appressum* and *E. delicatulum,* appear to be endemic to the páramos of the park. The former also is found near the summit of the Fila de Agua Fría (ca. 2,800 m), an isolated peak that is very similar in species composition to the Páramo de Guaramacal. Although still poorly collected, the Páramo El Pumar (ca. 2,600 m) has genera not found in the other páramos of the park. These include *Isoetes* (Isoetaceae), *Eccremis* (Hemerocallidaceae), *Luzula* (Juncaceae), and various grasses such as *Aulonemia, Cinna,* and *Festuca.*

In the southwestern section of the park above the Río Amarillo there are extensive cave formations and karst topography. Karst, which is uncommon in the Andes, often harbors endemic plants. Preliminary botanical exploration of this area suggests that Guaramacal will not be an exception. A number of plant genera such as *Lastreopsis, Tectaria* (both Dryopteridaceae), *Dicranoglossum* (Polypodiaceae), *Monstera* (Araceae), *Peltastes* (Apocynaceae), *Discocactus* (Cactaceae), and *Clavija* (Theophrastaceae) are recorded for the park only from these calcareous slopes.

More than 30 species of vascular plant have been described from Guaramacal (Dorr et al. 2000, appendix I). Although most of them are known from nowhere else, our knowledge of the Andean slopes facing the llanos is still so poor that it is difficult to tell whether these truly represent endemic species or simply collecting bias. At least one remarkable disjunction has been discovered and deserves notice. *Koanophyllon isillumense* (Asteraceae), once thought to be endemic to Peru, has been collected in the park.

The faunal diversity of Guaramacal is unevenly known. Inventories of mammals (37 species, including 16 bat and 8 rodent species), birds (149 species), amphibians (15 species), and reptiles (11 species) have been made (Soriano et al. 1990; García-Pérez 1999; Ríos U. 1999; Utrera 1999), but information on fish and invertebrates is essentially nonexistent. Two conspicuous but rarely seen residents of Guaramacal are the spectacled bear (*Tremarctos ornatus*) and puma (*Felis concolor*) (Goldstein 1993, 1999; Utrera 1999). Both are endangered throughout the Andes, and preserving suitable habitat for them is essential for their continued survival in South America.

Guaramacal is one of the best-maintained national parks in Venezuela, with a superintendent and staff sympathetic to conservation and scientific research. Despite their vigilance, threats to the integrity of the park and its biota still can be identified. The most significant come from oil and gas exploration. Natural gas deposits have been found within 50 km of Guaramacal, and one proposal for exploiting them involves the construction of a highway from Barinas to Boconó through the valley of the Río Boconó. If built, this road would destroy the unique karst habitat in what is now one of the least accessible sectors of the park.

Agricultural activity, especially coffee production, is causing the southeastern buffer of the park to be deforested and will contribute to the isolation of the remaining undisturbed forest. Coffee, an important cash crop since colonial times, is intensively cultivated in the broad valley between the park and the low-lying Cerro Negro (1,200 m), which isolates the village of Guaramacal and the park from the llanos. Coffee is typically planted between 800 and 1,200 m, but there is a temptation to encroach on the boundaries of the park up to 1,600 m. Cattle grazing, common at lower elevations on the northwestern slopes of Guaramacal, has led to the establishment of numerous *potreros* (pastures), some of which extend up to the boundary of the park at 1,800 m. Cattle undoubtedly are to blame for the introduction of most of the alien grass species found in the park.

Hunting of small game, albeit illegal, is common in Guaramacal. Poaching of the larger mammals has not been documented, although it occurs in other Venezuelan parks where puma have killed cattle. The role of different animal species in dispersing seed and fruit is not well understood, but there is evidence (L. J. Dorr and S. M. Niño, unpublished data) that small rodents and frugivorous birds are important and their populations need to be maintained to keep the forests of the park healthy. Hunters also have a direct impact on the vegetation when they set fires to flush game. Burning can induce páramo plants to occur at lower elevations than expected. One example of this phenomenon can be seen at 2,400 m on the north-facing slopes below Páramo El Pumar.

Organized timber extraction ceased when the park was created. Until then, large *peineto*, or Spanish cedar trees (*Cedrela odorata*), were systematically removed. Limited extraction of *mapora*, and presumably other species, continues for local house construction. Generally, there is an appreciation that the park's forest cover needs to be maintained, if not for preserving biodiversity, at least for maintaining water resources. Most of the park drains into the Río Boconó that feeds into the Tucupido Reservoir, which in turn is the source of drinking water for Guanare, the capital of Portuguesa State, and through the reservoir's dam and turbines a supplier of hydroelectric power.

The television (and telecommunication) towers near the summit of the Páramo de Guaramacal, which mostly antedate the creation of the park, have had an impact on the vegetation, most notably through the building of the all-weather road that now bisects the park and a separate set of power lines to provide electricity. The road construction disturbed forest along the north-facing Quebrada Segovia and south-facing Quebrada Jirajara. Alien taxa have come in along this road, some, such as *Phytolacca rugosa* (Phytolaccaceae) and *Rumex acetosella* (Polygonaceae), arriving within the past few years. More undoubtedly will come, since the road is currently being paved. As yet, evidence of alien species invading undisturbed vegetation has not been seen.

When the clouds lift and provide the visitor with vistas of extensive forested slopes, it is tempting to think of Guaramacal as pristine, but on closer examination there is ample evidence that this mountain ridge was more intensively utilized by humans in the past. It is crisscrossed with mule trails that connect the llanos with the cities of Boconó and Trujillo. Now that most coffee planters rely on four-wheel-drive vehicles for transport, these trails have been abandoned, but their ruts are easily traced. Similarly, during a period of political instability in the 1960s, the Venezuelan military placed camps in several more or less level areas within what is now the park. The disturbance at these sites is still visible 40 years later. One of these *campamentos*, which created a permanently wet area in Quebrada Jirajara, is the only locality in the park for the widespread *Utricularia alpina* (Lentibulariaceae).

The majority of the park's visitors (18,000 in 2000) do little more than picnic near two artificial lakes, the Laguna Negra and the Laguna de Los Cedros. Both are located on the northwestern edge of the park, and both have large numbers of introduced and adventive plant species. As yet, few foreign tourists visit Guaramacal National Park. Those who do, however, seem most interested in birds. Ecotourism could probably be expanded with negligible impact on the park's biodiversity.

It is likely, given the enormous human pressure being exerted on all tropical forests, that relatively well studied natural areas such as Guaramacal National Park will become increasingly important as isolated exemplars of entire biomes. Our obligation is therefore not only to protect them, but also to understand them.

4.3 THE GUIANA SHIELD

Vicki A. Funk and Paul E. Berry

THE GUIANA SHIELD underlies the northeastern corner of South America and includes much of the area east and south of the Río Orinoco and east and north of the Rio Negro and Rio Amazonas. The area includes Bolívar, Amazonas, and Delta Amacuro states in Venezuela; most of Guyana, Surinam, and French Guiana; parts of northern Brazil; and parts of southeastern Colombia. The area, which includes the famous tepuis of Sir Arthur Conan Doyle's *Lost World,* is known to be rich in species diversity; for many groups, including some plant families, it serves as a center of species diversity. The Venezuelan portion of the Shield is the subject of a detailed flora (families A–M have been published), with the first volume dedicated to extensive background information on the Guiana Shield and its flora (Steyermark et al. 1995–present). The *Flora of the Guianas* (Guyana, Surinam, and French Guiana) is in progress, and a number of volumes have been published (Görts–van Rijn 1989–present).

The Guiana Shield refers to an ancient craton that was formed well before the breakup of the supercontinent of Gondwana. The Shield's igneous-metamorphic basement was laid down in several events from 0.8 to 3.6 billion years ago (Mendoza 1977; Schubert and Huber 1990). This granitic basement is easily observed in the many black "hills" of granite that dot the landscape across the Guiana Shield and is exposed on some of the mountains and massifs. Between 1 and 1.6 billion years ago, sedimentary covers of sand were successively laid down and cemented during thermal events (Huber 1995). The resulting quartzite and sandstone rocks are known today as the Roraima Formation. Some recent work has suggested that the eastern rocks are the oldest (Huber 1995), so this would make the Pakaraima Mountains and the eastern parts of Venezuela older than the rest of the Roraima Formation.

Over the last 200 to 600 million years, intrusions of diabases and granite have penetrated both the granitic basement and the quartzite sedimentary rocks.

The most distinctive feature of the Guiana Shield is the tepuis, the steep-walled table mountains prominently featured in photos and films. For many groups of organisms, they support a unique flora and fauna. Tepui elements begin to appear in the biota at around 300–1,000 m elevation, but predominate above 1,500–1,800 m. The easternmost peaks that reach heights of 2,000 m include Mount Ayanganna and Mount Wokomong in Guyana. The highest tepui, Sierra de la Neblina, just exceeds 3,000 m elevation and occurs in the western part of the Shield on the border of Venezuela and Brazil. Many of the remaining Venezuelan tepuis have summits between 2,000 and 2,400 m. There is a large, midelevation, sandy plateau between 400 and 1,500 m called La Gran Sabana that occupies southwestern Venezuela and adjacent parts of Guyana, and there are variously sized areas of lowland white-sand savannas scattered in different parts of the Shield.

The Guiana Shield is located slightly north of the equator and is affected by trade winds blowing off the Atlantic Ocean. The trade winds bring moisture and collide with other masses of air in what is known as the intertropical convergence zone (ITCZ). Over the course of the year, the ITCZ migrates north and south, leading to distinct wet and dry seasons in different parts of the equatorial zone of South America. The rainy part of the year in the Venezuelan part of the Guiana Shield is from May to August, and the driest part is from January to March (earlier to the south, later to the north). As one moves to the east, a second rainy season develops from December to January. This second rainy season is shorter, is less intense, and does not penetrate as far inland from the Atlantic Ocean. As a whole, the Guiana Shield has a tropical climate characterized by a high mean annual temperature exceeding 25°C at sea level (but decreasing with elevation); a diurnal range of temperature exceeding the annual range; and an annual temperature range of less than 5°C. Most of the Shield area is covered by evergreen forest, but scrub or savanna predominates in some lowland areas, particularly when soil conditions are extremely poor or clay hardpans prevent penetration by tree roots and lead to flooding during the rainy season. Slopes of mountains that face the incoming trade winds usually have higher precipitation than the leeward sides (Clarke et al. 2001).

Only a few parts of the Guiana Shield have well-known floras, such as Mount Roraima (Venezuela), Iwokrama–Mabura Hill and Kaieteur (Guyana), Tafelberg (Surinam), and Saül (French Guiana). Most areas, such as the lowland forests and tepui slopes in Venezuela, the Pakaraima Mountains and New River Triangle in Guyana, and southern Surinam and French Guiana, are very poorly known. Hollowell et al. (2001) recently published a checklist for about half the flowering plants of the Guiana Shield (families from A to L). Using this publication and subsequent online updates (see Biological Diversity of the Guiana Shield 2004) as a guide, we estimate there are 12,500 species of flowering plants currently known for the Guiana Shield (excluding Brazil and Colombia). The ten largest groups are the Fabaceae (sensu lato), ferns or pteridophytes (all families), Orchidaceae, Rubiaceae, Poaceae,

Cyperaceae, Melastomataceae, Euphorbiaceae, Myrtaceae, and Asteraceae. The Venezuelan portion of the Shield includes about 9,000 species, Guyana 6,200, Surinam 4,500, and French Guiana 5,000 (these species partially overlap between countries). If the southern part of Surinam were better explored, the number of species for that country would surely increase. The Venezuelan Guayana includes 672 species of ferns and fern allies (Berry et al. 1995), while there are 638 species in the three Guianas. The estimated overlap of these two areas is 74%, so that the total number of ferns is predicted to be close to 1,000 species (not including the Brazilian and Colombian parts of the Shield). Thus, a conservative estimate of the total number of vascular-plant species from the Guiana Shield is 13,500 species, perhaps closer to 15,000 with the Colombian and Brazilian parts of the Shield included. Berry et al. (1995) calculated that 40% of the plant species occurring in the Guiana Shield do not occur outside this area. A closer analysis of the flora of Kaieteur Falls, Guyana (Kelloff and Funk 2004; plate 4.2), shows that 42% of the plant species are endemic to the Guiana Shield. Consequently, we can conclude that about 6,000 species of vascular plants are restricted to the Guiana Shield. Considering just the plants of the Pantepui area (over 1,500 m elevation), Berry et al. (1995) found that the percentage of endemics increases to 65%. Notable families with high levels of endemism include Asteraceae, Bonnetiaceae, Bromeliaceae, Eriocaulaceae, Podostemaceae, Rapateaceae, Rubiaceae, Tepuianthaceae, and Xyridaceae. This level of endemism must be one of the highest for noninsular floras, and documents the existence of a "Guiana Shield flora" that is separate from other floras such as the Andean, Amazonian, and Brazilian Shield floras.

With a few exceptions, such as cities along the Río Orinoco, the Rupununi savanna (Guyana), and the coastal areas of the Guianas, the Guiana Shield has benefited from its isolation and low population density, and much of the vegetation is still relatively undisturbed by human activities. This has led to its designation as a "tropical wilderness" (Mittermeier et al. 1998). Unfortunately, the pace of disturbance has accelerated greatly in recent times because of logging by Asian and local companies, gold and diamond mining, oil drilling, bauxite mining, dams for hydroelectric power, wildlife trade, burning, grazing, and agriculture. If this pace of activity continues, the Guiana Shield will lose its place as part of one of the three remaining "tropical wilderness" areas in the world.

Efforts to conserve this interesting and unique region vary by country. Since 1962, Venezuela has set up 7 national parks, 29 natural monuments, and 2 biosphere reserves covering 142,280 km², or almost 31% of the Shield that lies in the country, and about 15% of the country. In Guyana the totals are much less, with only one major national park, the expanded Kaieteur National Park (627 km² or about 3% of the country). Several other areas have been proposed, but the boundaries are ambiguous and no legislation has been passed. Surinam has 18 areas of nature or forest reserves or national parks that total 7,290 km² (1,310 km² of which is proposed; Lindeman and Mori 1989) and that make up 4.4% of the country. French Guiana has no designated protected areas, but there are 18 proposed sites that total 6,710 km² and make up 7.5% of the country (Lindeman and Mori 1989). However, just because an area is marked on a map as a park or reserve does not mean that the area is actually protected. As with many countries in the tropics, areas that are designated as parks are often only "paper" parks because they lack the infrastructure and financial backing to effectively protect the areas. As a result parks often host gold miners, hunting, wildlife trade, and other disruptive activities. Currently, Venezuela and Guyana have the most pressure being put on their respective biodiversities, while French Guiana is probably the least threatened of the countries and therefore has the best chance to protect its environment.

The Guiana Shield encompasses parts of five countries, each with a different administrative structure and official language; there are a number of border disputes, and the borders are porous to drug, gold, and wildlife trafficking; and there are serious issues concerning native peoples. All of these issues will have to be overcome before a viable reserve system for the Guiana Shield can be designed and maintained.

ACKNOWLEDGMENTS

This is number 86 in the Smithsonian's Biological Diversity of the Guiana Shield Program publication series.

4.4 PACIFIC OCEANIC ISLANDS

Warren L. Wagner, Denise Mix, and Jonathan Price

THE PACIFIC OCEAN is the earth's largest feature, covering about one-third of the surface in an expanse of 155,000,000 km². It is interspersed with thousands of islands totaling roughly 1,250,000 km² of land, which is equal to the size of Peru and about one-half the size of Greenland. Many of the islands are situated on the Pacific plate, where the land arises from volcanic activity and has never been connected to a continental landmass. These islands are considered to be truly oceanic. The remainder of the Pacific Islands, which make up a greater proportion of the land area, are on the Indo-Australian plate. These islands generally have a continental origin with some volcanic composition, and will not be addressed here. Fiji is included, although it formerly was considered to have originated on the northeastern edge of the Indo-Australian plate. More recent syntheses of data suggest that Fiji's origin was on the Pacific plate and that it is an oceanic island that made contact with a fragment of Gondwana, now part of Tonga.

Two types of islands can be distinguished throughout the Pacific oceanic islands:

high volcanic islands and low coral islands. High volcanic islands have two forms and a wide variation in age (table 4.1). Steep-coned, explosive volcanoes form at the boundaries between plates and are laid out in sweeping arcs, as seen in the younger northern Mariana Islands (Mueller-Dombois and Fosberg 1998). Less-explosive shield volcanoes with gentler slopes (plate 4.3) arise from stationary hot spots at intraplate sites and are arranged in linear or cluster formations, as seen in the Hawaiian, Society, and Marquesas islands. Low coral islands develop around a volcanic base as the volcanic material erodes and the volcano eventually submerges back into the water, as seen in the older southern Mariana Islands. When the coral dies, a limestone island remains, as seen in the Cook Islands. Some of the oldest islands were formed 60–85 million years ago (Davis et al. 1995; Clague 1996), with the oldest submerged former islands of the Hawaiian-Emperor volcanic chain dated at 85 million years ago. New islands are continually being formed, and many have not yet broken the water surface, such as Lohihi off the coast of the island of Hawaii.

Island environments range from tropical rain forest to sand deserts. The most common vegetation zones within the Pacific include lowland, montane and limestone forests, scrublands and grasslands, and littoral zones. Sapotaceae, Rutaceae, and Rubiaceae are among the more widespread families found in lowland forests, which are found at the lower elevations of high volcanic islands. Most lowland forests have disappeared due to logging and agriculture, although Fiji and Samoa still have some lowland forest areas. *Metrosideros* is the dominant genus of the higher-elevation montane forests, as seen in the Hawaiian Islands. Wet montane forests of the eastern islands are dominated by tree ferns. Dry montane forests have *Pandanus* in addition to tree ferns, and the larger islands also include *Acacia* and *Santalum*. Human activities are generally deterred because of the steep slopes, but other threats exist, such as grazing feral animals and invasive plant species. Limestone forests, like those in Fiji, consist of specialized vegetation growing from coral-reef beds. Many limestone forests have been cleared. Scrub and grasslands are most typically seen as secondary growth as a result of the clearing of forests. Littoral zones are frequently cut down for coconut cultivation and tourist developments. Littoral zones serve as bird-nesting areas and prevent coastal erosion.

Geographically, the Pacific oceanic islands are divided into three regions: Melanesia to the southwest (Fiji group); Micronesia to the northwest (with the Mariana, Caroline, Marshall, and western Kiribati groups); and Polynesia to the east (from north to southwest it includes the Hawaiian, eastern Kiribati, Samoa, and Tonga groups; southwest to southeast it includes the Niue, Cook, French Polynesia, Pitcairn, and Easter groups). Floristically, these regions are divided slightly differently (Takhtajan 1986). The Fijian region includes the island groups of Fiji, Rotuma, Samoa, Tonga, and Niue. The Polynesian region is broken into the Micronesian Province, which includes the Caroline, Mariana, Marshall, Gilbert, and Phoenix island groups, and the Polynesian Province, which includes the Line, Society, Cook, Austral, Marquesas, Pitcairn, and Easter groups. The Hawaiian Islands are distinct enough to be their own region, which also includes Johnston Atoll.

Many factors associated with the distinctive evolutionary patterns in the native

Table 4.1 Attributes of Pacific oceanic islands

Archipelago	Area of archipelago (km²)	Number of high islands (>300 m)	Nearest high islands	Distance to nearest high islands (km)	Ages of present high islands (Ma)	Ages of eroded former high islands (Ma)
Hawaii	16,641	8	Marquesas	3,500	forming–5.1[a]	30[a]
Marquesas	1,275	10	Society	1,400	1.3–6[b]	—
Society	1,598	7	Austral	500	forming–3.8[c]	>4.5[c]
Rapa	40	1	Austral	600	5[d]	—
Austral	135	3	Rapa	600	5.5–12[e]	—
Caroline (FSM)	1,170	5	Bismarck	1,000	forming–continental[f]	—
Samoa	447	6	Tonga	900	forming–1.5[g]	13.5[g]
Tonga	699	2	Fiji	700	forming–continental[f]	—
Cook	230	1	Austral	900	1.1–2.1[d]	18.0–19.3[d]
Fiji	18,270	15	Tonga	700	40–50[f]	—

[a]Clague 1996.
[b]Duncan and MacDougall 1974.
[c]Duncan and MacDougall 1976.
[d]Krummenacher and Noetztlin 1966.
[e]Turner and Jarrard 1982.
[f]Mueller-Dombois and Fosberg 1998.
[g]Nunn 1994.

Hawaiian biota may also be related to the vulnerability of the flora. The islands have a high diversity of habitats over a small geographic area, with isolation limiting colonization and leading to a high proportion of endemism (table 4.2). Factors affecting successful colonization of oceanic islands are the geologic age of the island and the distance from the source area (table 4.1). Remote oceanic islands tend to have less variety of species per island when compared to islands near continental landmasses; however, the number of endemic species per island is much greater. Endemic families are rare among oceanic islands, because generally the islands are not old enough for endemic families to have evolved (Clague 1996; Wagner 1991). Because successful colonists of remote insular environments populate a decreasingly open set of communities, unusual morphology often accompanies rapid speciation. These atypical morphologies have led to the recognition of many endemic genera. With the advent of molecular and morphological phylogenetic studies during the past two decades, many of these divergent lineages are found to be divergent species groups that arose within continental genera (e.g., *Hesperomannia, Haplostachys, Phyllostegia,* and *Stenogyne* from the Hawaiian Islands and *Plakothira* from the Marquesas Islands), resulting in the need for revised classification of either the endemic or the splitting of the continental genera. Thus, endemism in oceanic island plants is primarily at the species level.

One exception is Fiji, which boasts a single endemic family, the Degeneriaceae. This small family with two species is one of the basal flowering plant families in the Magnoliales. At 40 million years, the flora of the Fijian region is considerably older and richer than that of many of the other islands (Smith 1979–1996; Mueller-Dombois and Fosberg 1998; Takhtajan 1986) and is much closer geographically to source areas. Species endemism in Fiji approaches 50% (Davis et al. 1995). Its flora is most closely affiliated with the Malesian flora (loosely equivalent to the flora of Indonesia, Malaysia, and the Philippines), particularly from New Guinea and the Solomon Islands (Smith 1979–1996; Takhtajan 1986), but also contains a small percentage from the Australian area (Mueller-Dombois and Fosberg 1998).

The Polynesian region (not including Hawaiian Islands) has about 11 endemic genera (Takhtajan 1986). The Micronesian province contains a large number of endemic species, the majority of which are from the Caroline Islands, followed by the Marianas. The flora most resembles that of East Asia in the northern Polynesian region, while the southern part of the region has associations with the Philippines, Southeast Asia, and Indonesia (Mueller-Dombois and Fosberg 1998). Its western taxa are related to those of the Papuan–Indo-Malaysian and Australian areas (e.g., *Coprosma, Meryta,* and *Scaevola*). Most of the eastern taxa are from Australia, but some are from American (e.g., *Cordia, Heliotropium,* and *Plakothira*) and Hawaiian (e.g., *Cheirodendron*) sources.

The Hawaiian region, the most isolated and one of the smallest floristic regions in the world (Takhtajan 1986), has the highest level of generic and species endemism—about 33 endemic vascular-plant genera and 1,183 endemic vascular-plant taxa (Wagner et al. 1999; Wagner et al. 2001; Wagner 1991). Species endemism is exceptionally high, approaching 90% for the vascular plants. The Marquesas

Table 4.2 Statistics on the vascular-plant taxa of Pacific oceanic islands

Archipelago	Native angiosperms	Native pteridophytes	Native gymnosperms	Endemic vascular plants	Total native vascular plants	Naturalized vascular plants
Hawaii	1,159[a]	181[b]	0	1,183[b]	1,340[a]	1,159[a]
Marquesas	236[a]	105[a]	0	156[a]	341[a]	243[a]
Society	416[c]	207[c]	0	273[c]	623[c]	~400
Rapa	115[c]	73[c]	0	71[c]	188[c]	?
Austral	121[c]	81[c]	0	87[c]	202[c]	?
Caroline (FSM)	992[d]	201[d]	1[d]	293[d]	1,194[d]	380[d]
Samoa	536[d]	228[d]	0	214[d]	764[d]	>157[d]
Tonga	360[d]	102[d]	1[d]	25[d]	463[d]	374[d]
Cook	184[d]	100[d]	0	33[d]	284[d]	273[d]
Fiji	1,307[d]	310[d]	11[d]	812[d]	1,628[d]	1,000[d]

[a]Wagner et al. 2002.
[b]Wagner et al. 2001.
[c]Florence 1987.
[d]Davis et al. 1995.

archipelago, which is positioned closer to other land areas compared to the Hawaiian Islands, has a considerably lower species endemism rate of 45% (Wagner and Lorence 2002; Wagner 1991). Most of the Hawaiian flora affiliations are with the Indo-Malesian area, but there are also affiliations with the Australian and American areas (Fosberg 1948a, 1948b; Takhtajan 1986; Wagner et al. 1990, 1999). Recent molecular studies are increasing the proportion of colonists from the Americas (e.g., *Schiedea,* Nepokroeff et al., in press; *Sanicula,* Vargas et al. 1998; *Viola,* Ballard and Sytsma 2000; *Rubus,* Alice and Campbell 1999).

Attempting to determine how colonization of oceanic islands took place has been problematic, especially when considering islands that are more remote. Because oceanic islands have never been contiguous with continental land and may be distant from one another, colonizers have to find ways to travel to these islands. Not only must a colonizing species arrive in good shape, but it must also land in an area suitable for growth. Studies have suggested that the most likely dispersal method for remote islands is on the feathers or feet of birds (Carlquist 1974, 1996), with water and air dispersal somewhat less likely. A significant and underestimated (but difficult to quantify) dispersal type is large storm systems, such as hurricanes (cyclones).

Island ecosystems, including the Hawaiian Islands, are well known for the insights they have provided about speciation and endemism, and more recently, about extinction (e.g., Carlquist 1974; Wagner and Funk 1995; Craddock 2000; Myers et al. 2000). Over half of the 25 global biodiversity hot spots are islands or island groups (e.g., New Caledonia, New Zealand, Madagascar, the Philippines) or ecological islands (e.g., Cape Floristic Province, southwestern Australia, Caucasus) (Myers et al. 2000). The Polynesia/Micronesia hot spot includes the Hawaiian Islands and ranks fourth in the world for the number of endemic plant species per area (after only the Eastern Arc and coastal forests of Tanzania/Kenya, the Philippines, and New Caledonia). Less than a quarter of the original area of primary vegetation of Polynesia/Micronesia remains (Myers et al. 2000). Within the United States, the State of Hawaii has the highest number of extinct plant and animal species (Stein et al. 2000), and all four Hawaiian counties rank in the top five counties nationally for numbers of federally listed plant and animal species (Rutledge et al. 2001).

Endemic Pacific island plants are usually adapted to very specialized insular habitats, and typically have very restricted distributions, often limited to only a portion of a single island. They are especially vulnerable to extinction from habitat destruction (e.g., fire, agriculture, development, and deforestation) and competition from introduced organisms.

Few plants and animals have been able to successfully cross the Pacific Ocean, colonize, and establish reproducing populations on the islands. Because native plant species have evolved in the absence of large grazing or browsing mammals as well as many plant competitors, they have been particularly susceptible to extinction, especially with the introduction of alien species by the Polynesians as they colonized this vast region and by Europeans several hundred years ago (Wagner et al.

1985). Unique adaptations such as secondary woodiness (Carlquist 1974) can evolve in island habitats, but these adaptations may also result in species that cannot compete well in highly perturbed systems, especially with introduced large mammals.

Native plants of the Pacific Islands are also at great risk from encroachment into their habitat. On Fiji, human activities including logging, mining, slash-and-burn agriculture, expanding population, tourism, and trade in endangered species are the biggest threats to the native plant populations. On the Hawaiian Islands and several of the Society and Micronesian islands, the greatest threats include clearing of the land for agriculture, introduced plants and animals, logging, fire, tourism, and urban development in coastal and lowland areas.

In general, preservation programs throughout the Pacific are severely inadequate, with poor policies and few designated protected areas. Some of the best-developed programs are in the Hawaiian Islands, where 12% of the land is designated as protected areas (Davis et al. 1995). Many islands have less than 1% of their total land area designated as protected. In many cases where land has been set aside as a natural preserve or national park, the plants most at risk are located outside the protected area. The Phoenix, Line, and Cook islands are among the few in the central Pacific that have been found to have a fair percentage of their flora protected within a park or preserve. In French Polynesia, where a high level of endemism occurs, there are few parks or preserves. Ecological and botanical research is urgently needed to provide a solid foundation for preservation and management programs. There are many organizations and research programs promoting environmental conservation, but few of the recommendations are put into practice. Acting on recommendations is not so simple, since the local people, not the government, own the land on many islands. This is problematic, because the local people need to develop the land for agriculture and to accommodate expanding population. The success of preservation programs then rests on the local people to work in conjunction with development needs and the need to manage land use. With rapidly decreasing native plant populations, if protected areas are not designated soon and management programs established, the habitats that hold the most vulnerable plants could be lost forever. Steps must be taken quickly to enact and enforce conservation programs in order to preserve the rich biodiversity unique to this part of the world.

4.5 THE GAOLIGONG MOUNTAINS OF SOUTHWEST CHINA AND NORTHEAST MYANMAR

Ai-Zhong Liu and W. John Kress

THE GAOLIGONG MOUNTAINS and adjacent regions in northeastern Myanmar (Burma), together an extension of the east Himalayas, have recently drawn exceptional attention from biologists and conservationists because of their rich biodiversity. The Gaoligong Mountains, rising between the great Salween (Nujiang) and Irrawaddy (Dulongjiang) rivers, lie between latitude 24°40′ and 28°30′ N in the border area between southwestern China and northeastern Myanmar (Li et al. 2000; plate 4.4). These mountains cover a total area of 111,000 km², half of which lies in northwestern Yunnan Province of China and the other half in eastern Kachin State of Myanmar.

The Gaoligong Mountains extend from north to south, with the highest peak at 4,640 m in the north and the lowest elevation at 930 m in the south. The diverse mountain climates and microenvironments found in this region are due in part to these extremes in altitude over a very short distance. Geologically, these mountains lie along the border of the Gondwanan and Laurasian supercontinents, with uplifting due to the collision between the Indian and Yangtze landmasses since the Cretaceous (Li 1994; Li et al. 1999). This region is part of the Burma-Malaya geological zone that extends to the south (Bande and Prakash 1986; Li 1994; Li et al. 1999; Li et al. 2000). Because of its geologic history as well as its ecological and climatic diversity, the Gaoligong Mountains have been a refuge for ancient floristic elements as well as recent evolutionary diversification. The tropical and temperate floristic elements fuse in this area, with the tropical species distributed in the southern part at low elevations and the temperate species in the northern part and the higher elevation zones (Li et al. 2000).

Plant species density in the Gaoligong Mountains is among the highest in China. Based on a preliminary survey in this part of southwestern China, 4,303 seed-plant species have been recorded. More specifically, 10 of the 16 families of seed plants endemic to East Asia are found in this region (Li et al. 2000), and 434 species are endemic. Of the 36 species of bamboo recorded, 11 species are endemic to the Gaoligong Mountains and another 11 species are endemic to Yunnan. The genus *Gaoligongshania*, the only epiphytic bamboo in the world, is found only in these mountains. The number of endemic taxa will be increased without question when similar surveys are carried out in the Gaoligong region of northeastern Myanmar, which has been little explored by biologists.

The Gaoligong Mountains are considered to be one of the cradles of the East

Asian flora and recognized as a key area for understanding the origin and diversification of the plants of the eastern Himalayas (Li et al. 1999; Li et al. 2000). They are extremely rich in biological resources, such as timber, wildlife, and edible, medicinal, and ornamental plants. For example, *Taxus yunnanensis*, a source of the anticancer drug taxol, has its geographic center in the Gaoligong Mountains. Many new herbaceous taxa of value in traditional Chinese medicine, such as *Gentiana, Impatiens, Coptis,* and especially *Fritillaria,* have been discovered in this region. In addition, many wild species of orchids and rhododendrons, renowned for their horticultural importance, have been described from these forests.

The Gaoligong Mountains not only are special as a natural environment, but also display rich indigenous human cultures. Fifteen different religious groups reside in the China section, as well as various unique cultures. Before 400 BC, the Gaoligong Mountains were a key part of the old Silk Road that formed a commercial and cultural link between China and South Asia. The nearly inaccessible river valleys of the eastern slopes of the Gaoligong Mountains in Myanmar are also the home of various Kachin tribal people.

Based on a preliminary investigation, 34 species of rare or threatened plants, which have been regarded at the national level as key protected plants in China, occur in the Gaoligong area. In particular, species such as *Alsophila spinulosa, Taiwania flousiana, Euptelea pleiosperma, Tetracentron sinense, Circaeaster agrestis, Sinopodophyllum emodi, Acrocarpus fraxinifolius, Dipentodon sinicus, Eurycorymbus cavaleriei, Emmenopterys henryi, Gochnatia decora,* and *Nouelia insignis* are endangered in China. Many of these taxa continue to maintain healthy populations in the Gaoligong Mountains. Other rare species, however, such as *Panax zingiberensis* and *Tacca chantrieri,* which can grow only in low-elevation areas, are threatened in the wild. Herbarium specimens indicate that they were present in the Gaoligong Mountains several decades ago, but currently they have become very difficult to find because of habitat degradation.

Like other diversity centers, the loss of biodiversity in this area is increasing. Two main factors are involved. The first factor is that this region has low agricultural productivity, with most of the local people pursuing subsistence agriculture. With increases in population, many natural habitats are being quickly fragmented by agricultural land use, in particular in the low-elevation area. The second factor is that important individual taxa (e.g., *Codonopsis convolvulacea* var. *forrestosa* Ballard, *Maianthemum oleraceum,* and various orchids) are being overexploited for the medicinal and horticultural markets.

Fortunately, unlike other montane ecosystems in southwestern China, forests in the Gaoligong Mountains have been protected by virtue of their remote location. Currently, approximately half of the area remains undisturbed. It is more difficult to determine the extent of primary forest remaining on the eastern slopes of the Gaoligong Mountains in Myanmar. However, because of its inaccessibility and low human population density, much of the forest probably remains intact and is threatened only by locally based hunting and subsistence agriculture.

Because of the unique biological features of the Gaoligong Mountains, the region has drawn the interest of many botanists since the 1860s. On the China side, collection activity remained limited until the 1980s because of poor transportation and accessibility. During the last two decades, Chinese botanists in collaboration with botanists from the United Kingdom, the United States, and Australia have made comprehensive collections of plant taxa in this region. An inventory of the plants collected during this time has resulted in the recent publication of the *Flora of Gaoligong Mountains,* containing 4,303 species of seed plants (Li et al. 2000). This volume provides a sound basis for initiating biodiversity research and conservation in this region.

In contrast, neither a modern flora of Myanmar nor the area encompassing the Gaoligong Mountains exists. An outdated checklist of the plants of Myanmar (Hundley 1987) was recently revised (Kress et al. 2003), increasing the number of species recorded from the country by 67%, from 7,000 to nearly 12,000. Concurrently, new investigations of plant diversity in the Gaoligong region of the country have also begun. Previously, the few specimens available for study were collected by the early British explorers of the region in the late 1800s and early 1900s. The great plant explorer Frank Kingdon Ward visited the eastern edges of the Gaoligong Mountains, and much of our knowledge of the flora of these mountains is recorded in his writings (Kingdon Ward 1941, 1944–1945) and the botanical collection of George Forrest in the herbarium at the Royal Botanic Gardens in Edinburgh, Scotland. With renewed botanical efforts by staff of the Myanmar Forest Department and the botany department of Yangon University, as well as curators from the National Museum of Natural History at the Smithsonian Institution, the Kunming Institute of Botany in Yunnan, the California Academy of Sciences, Harvard University, and the Royal Botanic Gardens at Edinburgh, a better understanding of the plant diversity will be forthcoming.

In 1995, the Gaoligong Mountains were identified by the Global Environment Facility (GEF) and the United Nations Environment Programme (UNEP) as a global biodiversity "hot spot." Partly because of this international recognition, the Gaoligong National Nature Reserve was established in China in 1983. It includes a total area of 430,000 ha (Xue 1995; Xu 1998). Since the establishment of the Gaoligong National Nature Reserve in China, conservation regulations and boundaries have been established, and specific management plans have been organized. However, these activities must be strengthened at the grassroots level if they are to be successful. The biodiversity conservation ethic of the local people in this region is still weak, and the primary conflict, as in other parts of the developing world, is between local use by subsistence farmers and protection by regional and national governments. Economic development in the area remains poor, and related educational activities to improve agricultural practices are necessary to heighten the conservation awareness of local residents. In Myanmar, none of the Gaoligong region has been designated as a wildlife sanctuary or protected area. However, plans for a transborder international nature reserve are under consideration.

LITERATURE CITED

Alice, L. A., and Campbell, C. S. 1999. Phylogeny of *Rubus* (Rosaceae) based on nuclear ribosomal DNA internal transcribed spacer region sequences. American Journal of Botany 86:81–97.

Ballard, H. E., and Sytsma, K. J. 2000. Evolution and biogeography of the woody Hawaiian violets (*Viola*, Violaceae): arctic origins, herbaceous ancestry, and bird dispersal. Evolution 54:1521–1532.

Bande, M. B., and Prakash, U. 1986. The tertiary flora of Southeast Asia with remarks on its palaeoenvironment and phytogeography of the Indo-Malayan region. Review of Palaeobotany and Palynology 49:203–233.

Berry, P. E., Huber, O., and Holst, B. K. 1995. Floristic analysis and phytogeography. Pp. 161–191 in Berry, P. E., Holst, B. K., and Yatskievych, K., eds., Flora of the Venezuelan Guayana, vol. 1, Introduction. Missouri Botanical Garden, St. Louis.

Biological Diversity of the Guiana Shield. 2004. Smithsonian National Museum of Natural History. http://www.mnh.si.edu/biodiversity/bdg/.

Carlquist, S. 1974. Island Biology. Columbia University Press, New York.

———. 1996. Plant dispersal and the origin of Pacific island floras. Pp. 153–164 in Keast, A., and Miller, S. E., eds., The Origin and Evolution of Pacific Island Biotas, New Guinea to Eastern Polynesia: Patterns and Processes. SPB Academic Publishing, Amsterdam.

Clague, D. A. 1996. The growth and subsistence of the Hawaiian-Emperor volcanic chain. Pp. 35–50 in Keast, A., and Miller, S. E., eds., The Origin and Evolution of Pacific Island Biotas, New Guinea to Eastern Polynesia: Patterns and Processes. SPB Academic Publishing, Amsterdam.

Clarke, H. D., Funk, V. A., and Hollowell, T. 2001. Using Checklists and Collections Data to Investigate Plant Diversity. Part 1, A Comparative Checklist of the Plant Diversity of the Iwokrama Forest, Guyana. Sida, Botanical Miscellany 21. Botanical Research Institute of Texas, Fort Worth.

Craddock, E. M. 2000. Speciation processes in the adaptive radiation of Hawaiian plants and animals. Evolutionary Biology 31:1–53.

Cuello A., N. L., ed. 1999. Parque Nacional Guaramacal. UNELLEZ-Fundación Polar, Caracas.

———. 2002. Altitudinal changes of forest diversity and composition in the Ramal de Guaramacal in the Venezuelan Andes. Ecotropicos 15:160–176.

Davis, S. D., Heywood, V. H., and Hamilton, A. C., eds. 1995. Centres of Plant Diversity: A Guide and Strategy for Their Conservation. Vol. 2, Pacific Ocean Islands. Information Press, Oxford.

Dorr, L. J., Stergios, B., Smith, A. R., and Cuello A., N. L. 2000. Catalogue of the vascular plants of Guaramacal National Park, Portuguesa and Trujillo states, Venezuela. Contributions from the U.S. National Herbarium 40:1–155.

Duncan, R. A., and MacDougall, I. 1974. Migration of volcanism with time in the Marquesas Islands, French Polynesia. Earth Planetary Science Papers 21:414–420.

———. 1976. Linear volcanism in French Polynesia. Journal of Volcanology and Geothermal Research 1:197–227.

Engel, F. A. 1976. An Ancient World Preserved: Relicts and Record of Prehistory in the Andes. Crown Publishers, New York.

Florence, J. 1987. Endémisme et évolution de la flore de la Polynesie Française. Bulletin de la Société Zoologique de France 112:369–380.

Fosberg, F. R. 1948a. Derivation of the flora of the Hawaiian Islands. Pp. 107–119 in Zimmerman, E. C., ed., Insects of Hawaii, vol. 1. University of Hawaii Press, Honolulu.

———. 1948b. Immigrant plants in the Hawaiian Islands. Part 2. Occasional Papers University of Hawaii 46:1–17.

García-Pérez, J. E. 1999. La herpetofauna del Parque Nacional Guaramacal. Pp. 127–137 in Cuello A., N. L., ed., Parque Nacional Guaramacal. UNELLEZ-Fundación Polar, Caracas.

Goldstein, I. 1993. Distribución, presencia, y conservación del oso frontino en Venezuela. BioLlania 9:171–181.

———. 1999. El oso frontino en el Parque Nacional Guaramacal. Pp. 161–164 in Cuello A., N. L., ed., Parque Nacional Guaramacal. UNELLEZ-Fundación Polar, Caracas.

Görts–van Rijn, A. R. A., ed. 1989–present. Flora of the Guianas. Koeltz Scientific Books, Utrecht, Netherlands; and Royal Botanic Gardens, Kew, London.

Hollowell, T., Berry, P. E., Funk, V. A., and Kelloff, C. 2001. Preliminary Checklist of the Plants of the Guiana Shield. Vol. 1. Smithsonian Institution Press, Washington, DC.

Huber, O. 1995. Geographical and physical features. Pp. 1–61 in Steyermark, J. A., Berry, P. E., and Holst, B. K., eds., Flora of the Venezuelan Guayana, vol. 1, Introduction. Missouri Botanical Garden, St. Louis.

Huber, O., Duno, R., Riina, R., Stauffer, F., Pappaterra, L., Jiménez, A., Llamozas, S., and Orsini, G. 1998. Estado Actual del Conocimiento de la Flora en Venezuela. Ministerio del Ambiente y de los Recursos Naturales Renovables, Caracas.

Hundley, H. G. 1987. List of Trees, Shrubs, Herbs, and Principal Climbers, etc., Recorded from Burma with Vernacular Names. 4th revised edition. Government Printing Press, Rangoon, Burma.

Jørgensen, P. M., and León-Yánez, S. 1999. Catalogue of the Vascular Plants of Ecuador. Monographs in Systematic Botany from the Missouri Botanical Garden, vol. 75. Missouri Botanical Garden Press, St. Louis.

Jørgensen, P. M., Ulloa Ulloa, C., Madsen, J. E., and Valencia, R. 1995. A floristic analysis of the high Andes of Ecuador. Pp. 221–249 in Churchill, S. P., Balslev, H., Forero, E., and Luteyn, J. L., eds., Biodiversity and Conservation of Neotropical Montane Forests. New York Botanical Garden, New York.

Kelloff, C., and Funk, V. A. 2004. Phytogeography of Kaieteur Falls, Potaro Plateau, Guyana: flora distributions and affinities. Journal of Biogeography 31:501–513.

Kingdon Ward, F. 1941. The Vernay-Cutting Expedition, November, 1938, to April, 1939: Report on the vegetation and flora of the Hpimaw and Htawgaw Hills, northern Burma. Brittonia 4:1–19.

———. 1944–1945. A sketch of the botany and geography of North Burma. Journal of the Bombay Natural History Society 44:550–574, 45:16–30, 45:133–148.

Kress, W. J., DeFilipps, R. A., Farr, E., and Daw Yin Yin Kyi. 2003. A checklist of the trees, shrubs, herbs, and climbers of Myanmar. Contributions from the U.S. National Herbarium 45:1–590.

Krummenacher, D., and Noetztlin, J. 1966. Ages isotopiques K-Ar de roches prélevées dans les possessions françaises du Pacific. Bulletin du Société Géologique de France 8:173–175.

Li H. 1994. The biological effect to the flora of Dulongjiang caused by the movement of Burman-Malaya Geoblock. Acta Botanica Yunnanica supplement 4:113–120.

Li H., Guo H.-J., and Dao Zh.-L. 2000. Flora of Gaoligong Mountains. Science Press, Beijing.

Li H., He D.-M., Bartholomew, B., and Long C.-L. 1999. Re-examination of the biological effect of plate movement: impact of Shan-Malay plate displacement (the movement of Burma-Malaya Geoblock) on the biota of the Gaoligong Mountains. Acta Botanica Yunnanica 21:407–425.

Lindeman, J. C., and Mori, S. A. 1989. Floristic inventory of tropical countries. Pp. 376–390 in Campbell, D. G., and Hammond, H. D., The Guianas. New York Botanical Garden, Bronx.

Mendoza, V. 1977. Evolución tectónica del Escudo de Guayana. Boletín de Geología, publicación especial 7:2237–2270.

Mittermeier, R. A., Myers, N., Robles Gil, P., and Mittermeier, C. G. 1999. Hotspots: Earth's Biologically Richest and Most Endangered Terrestrial Ecoregions. Cemex, Mexico City.

Mittermeier, R. A., Myers, N., Thomsen, J. B., da Fonseca, G. A., and Olivieri, S. 1998. Biodiversity hotspots and major tropical wilderness areas: approaches to setting conservation priorities. Conservation Biology 12:516–520.

Mueller-Dombois, D., and Fosberg, F. R. 1998. Vegetation of the Tropical Pacific Islands. Springer-Verlag, New York.

Myers, N., Mittermeier, R. A., Mittermeier, C. G., da Fonseca, G. A. B., and Kent, J. 2000. Biodiversity hotspots for conservation priorities. Nature 403:853–858.

Nepokroeff, M., Wagner, W. L., Soltis, P. S., Weller, S. G., Sakai, A. K., and Zimmer, E. A. In press. Monograph of *Schiedea* (Caryophyllaceae subfam. Alsinoideae). Systematic Botany Monographs.

Nunn, P. D. 1994. Oceanic Islands. Blackwell Publishers, Oxford.

Ortega, F., Aymard, G., and Stergios, B. 1987. Aproximación al conocimiento de la flora de las montañas de Guaramacal, Estado Trujillo, Venezuela. BioLlania 5:1–60.

Pitman, N., Valencia, R., and León-Yánez, S. 2000. Resultados. Pp. 15–23 in Valencia, R., Pitman, N., León-Yánez, S., and Jørgensen, P. M., eds., Libro Rojo de las Plantas Endémicas del Ecuador 2000. Publicaciones del Herbario QCA, Pontificia Universidad Católica del Ecuador, Quito.

Population Reference Bureau. 2003. 2004 World Population Data Sheet of the Population Reference Bureau. Washington, DC. http://www.prb.org.

Ríos U., G. A. 1999. Avifauna del Parque Nacional Guaramacal. Pp. 139–151 in Cuello A., N. L., ed., Parque Nacional Guaramacal. UNELLEZ-Fundación Polar, Caracas.

Rivero, R., and Ortega, F. 1989. Notas fitogeográficas y adiciones a la pteridoflora de las montañas y páramo de Guaramacal, Estado Trujillo, Venezuela. BioLlania 6:133–142.

Rutledge, D. T., Lepczyk, C. A., Jialong, X., and Liu, J. 2001. Spatiotemporal dynamics of endangered species hotspots in the United States. Conservation Biology 15:475–487.

Sauer, W. 1965. Geología del Ecuador. Ministerio de Educación, Quito.

Schjellerup, I. 1992. Pre-Columbian field systems and vegetation in the jalca of northeastern Peru. Pp. 137–150 in Balslev, H., and Luteyn, J. L., eds., Páramo: An Andean Ecosystem under Human Influence. Academic Press, London.

Schubert, C., and Huber, O. 1990. The Gran Sabana: Panorama of a Region. Lagoven Booklets, Caracas.

Sierra, R., ed. 1999. Propuesta Preliminar de un Sistema de Clasificación de Vegetación para el Ecuador Continental. Proyecto INEFAN/GEF-BIRF y EcoCiencia, Quito.

Smith, A. C. 1979–1996. Flora Vitiensis Nova: A New Flora of Fiji (Spermatophytes Only). Vols. 1–5 and comprehensive indexes. Pacific Tropical Botanical Garden, Lawai, Hawaii.

Soriano, P. J., Utrera, A., and Sosa, M. 1990. Inventario preliminar de los mamíferos del Parque Nacional General Cruz Carillo (Guaramacal), Estado Trujillo, Venezuela. BioLlania 7:85–99.

Stein, B. A., Kutner, L. S., and Adams, J. S. 2000. Precious Heritage: The Status of Biodiversity in the United States. Oxford University Press, Oxford.

Steyermark, J. A., Berry, P. E., and Holst, B. K., eds. 1995–present. Flora of the Venezuelan Guayana. Missouri Botanical Garden, St. Louis.

Takhtajan, A. 1986. Floristic Regions of the World. University of California Press, Berkeley.

Turner, D. L., and Jarrard, R. D. 1982. K-Ar dating of the Cook-Austral island chain: a test of the hot-spot hypothesis. Journal of Volcanology and Geothermal Research 12:187–220.

Utrera, A. 1999. Mastofauna del Parque Nacional Guaramacal. Pp. 153–159 in Cuello A., N. L., ed., Parque Nacional Guaramacal. UNELLEZ-Fundación Polar, Caracas.

Vargas, P., Baldwin, B. G., and Constance, L. 1998. Nuclear ribosomal DNA evidence for a western North American origin of Hawaiian and South American species of *Sanicula* (Apiaceae). Proceedings of the National Academy of Science USA 95:235–240.

Wagner, W. L. 1991. Evolution of waif floras: a comparison of the Hawaiian and Marquesan archipelagos. Pp. 267–284 in Dudley, E. C., ed., The Unity of Evolutionary Biology: The Proceedings of the Fourth International Congress of Sytematics and Evolutionary Biology. Dioscorides Press, Portland, Oregon.

Wagner, W. L., and Funk, V. A., eds. 1995. Hawaiian Biogeography: Evolution on a Hot Spot Archipelago. Smithsonian Institution Press, Washington, DC.

Wagner, W. L., Herbst, D. R., and Palmer, D. D. 2001–present. Flora of the Hawaiian Islands. Smithsonian Institution. http://ravenel.si.edu/botany/pacificislandbiodiversity/hawaiianflora/index.htm.

Wagner, W. L., Herbst, D. R., and Sohmer, S. H. 1990. Manual of the Flowering Plants of Hawai'i. University of Hawaii Press and Bishop Museum Press, Honolulu.

———. 1999. Manual of the Flowering Plants of Hawai'i. Revised edition. University of Hawaii Press and Bishop Museum Press, Honolulu.

Wagner, W. L., Herbst, D. R., and Yee, R. S. 1985. Status of the native flowering plants of the Hawaiian Islands. Pp. 23–74 in Stone, C. P., and Scott, J. M., eds., Hawai'i's Terrestrial Ecosystems: Protection and Management. Cooperative National Park Resources Studies Unit, University of Hawaii, Honolulu.

Wagner, W. L., and Lorence, D. H. 2002. Flora of the Marquesas Islands. Smithsonian Institution. http://ravenel.si.edu/botany/pacificislandbiodiversity/marquesasflora/index.htm.

Williamson, M. 2002. Costs and consequences of non-indigenous plants in the British Isles. Botanical Electronic News (BEN), no. 281. http://www.ou.edu/cas/botany-micro/ben/ben281.html.

Xu Zh.-H. 1998. Nujiang Nature Reserve. Yunnan Art Publishing House, Kunming.

Xue J.-R. 1995. The Gaoligong National Nature Reserve. Chinese Forestry Press, Beijing.

CHAPTER 5

CASE STUDIES AMONG SELECT
TAXONOMIC GROUPS

Any effective strategy for conserving biological diversity requires a good estimate of how many species exist and where they are found. Yet the number of species of seed plants present in the world today is still unknown; estimates range from about 223,000 (Scotland and Wortley 2003) to over 420,000 species (Govaerts 2001; Bramwell 2002). With such a great diversity of plants, both in marine and terrestrial environments, different taxonomic groups with different life-history strategies will react to habitat disturbance in very different ways.

There are many different definitions of what constitutes the plant kingdom. Here we take a broad perspective to encompass most organisms that photosynthesize in some fashion, including algae and dinoflagellates, lichens, mosses, and hepatics, as well as ferns, gymnosperms, and flowering plants. Each group has unique features and characteristics, and each may require individual strategies for conservation assessment, management, and protection. The nine examples in this chapter document the diversity, characteristics, distribution, ecology, and conservation status for a spectrum of plant types. The details on taxonomy and classification offer an insider's view to the complexity of our understanding of these plants and provide a broad taxonomic perspective of the consequences of habitat alteration and degradation.

5.1 DINOFLAGELLATES:
PHYLUM DINOFLAGELLATA

Maria A. Faust

DINOFLAGELLATES ARE microscopic unicellular microorganisms in aquatic environments. They are an ancient and successful group; the first evidence in the fossil record dates back to the Silurian age (Fensome et al. 1996). They are primarily single-celled, free-swimming, cosmopolitan, eukaryotic organisms distributed worldwide in fresh and marine, temperate and tropical waters. Cells have features common to both plants and animals. Photosynthetic and nonphotosynthetic species are known in this diverse and unusual polyphyletic group. The same morphospecies occur around the world within broad latitudinal limits (Steidinger and

Tangen 1996). Cells exhibit a wide divergence in shape from round to discoid to heart-shaped, and range in size from 5 μm to 1 mm long. Some cells are armored with plates, while others are enclosed in an organic membranous sack. Cell surface can be smooth or ornate, laced with spines, pores, or grooves. Cell nutrition varies from autotrophic (via photosynthesis) to heterotrophic (via consumption of small microscopic organisms) (Steidinger and Tangen 1996).

Some dinoflagellate species produce resting cysts that can survive in sediments for an extended period of time and germinate to initiate blooms, some of which can form red tides (Anderson et al. 1996). When harmful, these blooms affect the surrounding sea life and their consumers, and cause mass mortalities of fish, shellfish, birds, and mammals. Tainted seafood may cause human illness (Smayda and Shimizu 1993). For example, two toxin-producing species have been identified from Belize; *Prorocentrum hoffmannianum* (plate 5.1, panel 1; Faust 1990) and *P. belizeanum* (plate 5.1, panel 2; Faust 1993) are known to cause ciguatera fish poisoning in humans (Morton and Bomber 1994; Morton et al. 1998). Socioeconomic stresses can also result from closing commercial fisheries and aquaculture until the harmful bloom dissipates.

There are 130 dinoflagellate genera recognized today, with about 4,500 species. As many as 11,000 species are estimated to exist (Sournia et al. 1991). Our knowledge of species number is imprecise because inconspicuous species have not received their proper share of taxonomic attention. Cells are microscopic and difficult to collect, which has particularly hampered our knowledge of the number of species found in aquatic habitats. Coastal marine environments are still an enormous frontier for biological diversity studies (Primack 2000). It is a little-known fact that only a small percentage of fauna and flora of the species-rich Caribbean coral-reef mangrove ecosystems has been described (Ruetzler and Feller 1996).

DINOFLAGELLATES IN A BELIZEAN MANGROVE SWAMP

Mangrove swamps exhibit high productivity of microscopic organisms, which are a rich food source for fish and shellfish, and are known as the nursery grounds of marine fish. This unique ecosystem, however, is increasingly threatened by human activity (e.g., garbage dumping) and natural events (e.g., hurricanes). The factors responsible for regulating species biodiversity are not well understood for unicellular organisms (Hillebrand and Sommer 2000). Coastal aquatic ecosystems can be subjected to nutrient enrichment and subsequent loss of biodiversity (Cederwall and Elmer 1990). One example of a threatened mangrove swamp is the Lair Channel at Twin Cays, Belize, in which disposal of household waste caused the disappearance of dinoflagellates and other microscopic organisms for several years. In 1996, Twin Cays became part of a protected region, the South Water Marine Reserve. This area is a World Heritage Site under the UNESCO World Heritage Convention, a significant and protected natural habitat of outstanding global importance.

The topography of Twin Cays reflects several thousand years of natural history, and provides a testament to the power of the countless storms and hurricanes that have buffeted this island. The island's periphery and channel banks are bordered by tall red mangroves, *Rhizophora mangle,* that extend stiltlike roots into 1–3 m of deep water, beyond the peat bank that supports the trees (Ruetzler and Macintyre 1982). Successful colonizers of the red mangrove stilt roots include various species of macroalgae, sponges, and anemones, which offer food and refuge to a variety of fauna, such as fish, oysters, and crabs. Here, dinoflagellates are free-floating or attached to macroalgal surfaces, sediments on coral rubble or mud, or sandy substrates mixed with mangrove detritus that is rich in organic matter. Dinoflagellates are the primary food source for zooplankton, the primary consumers, including filter feeders and juvenile fish (Fenchel 1969). Dinoflagellates and zooplankton proliferate in response to their unique physical, chemical, and biological needs (Faust and Gulledge 1995).

VULNERABILITY OF DINOFLAGELLATES

Investigations conducted over the last 20 years recognized the forces that shape and maintain the biology of mangrove environments and control the nutrient processes and the participation of microorganisms in marine food webs (Alongi 1998). Because mangroves, like coral reefs and other tropical shallow-water communities, exist near the limit of ecological tolerance of their inhabitants, they are significant indicators of environmental quality. Species association is another important indicator of certain stability in mangrove communities that are constantly threatened. Studies targeting processes in the Caribbean have examined benthic and epiphytic dinoflagellates from shore to forereef in the Virgin Islands (Tindall and Morton 1998). Similar individual species were found in a wide range of specific habitats, even within a small reef or reef flat of mangroves. Most species tend to show preference for one habitat: on sessile macroalgae, associated with sand, or free-floating in the water column.

The unique features of Twin Cays habitats are conducive to rich and diverse dinoflagellate populations. With slow flushing rates and high organic nutrient levels from red mangrove leaf litter, the detrital mat formation is extensive, resulting in a diverse benthos-associated dinoflagellate community (Faust 1996). Dinoflagellate blooms are not uncommon: *Prorocentrum elegans* (80,230 cells/liter; plate 5.1, panel 3) often forms red tide (Faust 1993). In these waters dinoflagellates are the most abundant (50–90%) microorganisms, followed by diatoms (5–15%), and cyanobacteria (3–25%). Here 22 dinoflagellate taxa were identified, representing 30 harmful and nonharmful species. Heterotrophic organisms identified included ciliates, nematodes, copepods, and invertebrates (Faust and Gulledge 1995). Long-term biodiversity and ecological studies of phytoplankton, aquatic invertebrates, and insects on Twin Cays have revealed an astonishing number of new species. Ten percent of the species have proved to be novel. But perhaps as many as 20–30% of the phytoplankton, algae, sponges, and worms living in Twin Cays may be from undescribed species (Ruetzler and Feller 1996).

Microbial communities can be damaged and species driven to extinction by external factors that do not change the structure of higher-plant communities, so that the damage is not immediately apparent to the human eye. One of the most subtle and universal forms of environmental degradation involves released liquid waste. The effects of pollution on water quality are a cause of great concern, not only as threats to the biological diversity of organisms, but also to aquatic productivity. This kind of water pollution has occurred via waste disposal in the waters at the South Water Marine Reserve, Belize.

Illegal dumping of domestic waste in black plastic bags in the shallow waters of the Lair Channel was discovered during field studies. In 1995 waste disposal started, and it continued for three years. By 2000 pollutants released during the decomposition process resulted in the dramatic loss of phytoplankton and zooplankton diversity in these waters. The first sign of population recovery occurred in May 2001; benthic dinoflagellates were observed attached to floating detritus. The composition of dinoflagellate species, however, was significantly altered. Some significant observations were that photosynthetic species were the major component in the population and heterotrophic species were the minor component. The most abundant photosynthetic species were less than 20 μm in size, for example, *Plagodinium belizeanum* (plate 5.1, panel 4) and a new *Prorocentrum* species (plate 5.1, panel 5). Only 5 dinoflagellate genera and 18 species were present, numbers greatly reduced from 1995 (Faust 1996). Diatoms, cyanobacteria, and zooplankton were minor components in the collected samples.

CONSERVATION AT THE POPULATION AND SPECIES LEVELS

While dinoflagellates have little economic value, indirectly they may signal change in the health of an undisturbed aquatic ecosystem as illustrated in Twin Cays. Mangrove swamps indeed constitute an extremely delicate natural system. Three years of waste disposal in Twin Cays significantly altered dinoflagellate assemblages and greatly changed the microscopic food web of this habitat. Continuation of waste disposal would probably have caused irreversible harm. The effect of household waste ultimately could have resulted in a loss of habitat for organisms in microscopic food webs, and in turn, the loss of a food source for marine fish and shellfish.

SUMMARY

In Twin Cays, microscopic organisms are the primary producers and primary consumers within the system. Our knowledge is limited in understanding the effects of household waste on the microscopic biodiversity of shallow slow-flowing coastal water environments. Long-term studies of species of these ecosystems are necessary to distinguish normal year-to-year fluctuations from long-term trends. The fact that adverse environmental effects may lag for many years behind their initial causes creates a challenge in understanding change in ecosystems. The South Water Marine Reserve with its rich dinoflagellate population is an ideal research site that can provide an early warning system for disruption or decline of ecosystem functions.

ACKNOWLEDGMENTS

The author thanks Dr. Klaus Ruetzler at the National Museum of Natural History, Smith-sonian Institution, for his encouragement and use of the field station's facilities at Carrie Bow Cay, Belize. This investigation was supported by grants from the Caribbean Coral Reef Eco-system Program (CCRE), Smithsonian Institution. This paper is contribution no. 651 from the CCRE Program.

5.2 LICHENS: PHYLUM ASCOMYCOTA

Rebecca Yahr and Paula T. DePriest

LICHENS ARE obligate, intimate symbioses between lichen-forming fungi and their photosynthetic partners, algae or cyanobacteria. Their conservation poses some unique challenges although overall it may be similar to and fruitfully borrow from that of vascular plants. Like plants, lichens are sessile, photosynthetic, and closely tied to a particular ecological context, such as a plant community or successional stage. Two species of lichen-forming fungi are listed as endangered in the United States, *Cladonia perforata* Evans and *Cetradonia linearis* [Evans] J. C. Wei & Ahti (≡ *Gymnoderma lineare* [Evans] Yoshimura & Sharp). *Cladonia perforata* was listed in 1993 (CFR 58 FR 25746 25755), and a recovery plan was published in 1999. In the species recovery plan, the U.S. Fish and Wildlife Service (USFWS) targeted five general conservation actions: (1) to determine the current distribu-tion, (2) to protect, manage, and restore known populations, (3) to conduct research on life history and demography, including responses to management and disturbance, (4) to monitor populations, and (5) to provide public information (USFWS 1999). This article follows the USFWS framework to discuss important themes in lichen conservation, using the case of the relatively well studied lichen *Cladonia perforata*.

CURRENT DISTRIBUTION

Many lichen species are hard to identify, and museum collections often give an in-complete picture of range or abundance. In contrast, *Cladonia perforata* is easily recognized in the field (plate 5.2), and its distribution is well documented (fig. 5.1; USFWS 1999). It grows directly on bare sand and with a suite of other *Cladonia* spe-cies sometimes dominates open areas of xeric sandy soils in Florida rosemary scrub. Although such microhabitats can be found on knolls in open scrub sites through-out Florida and adjacent states, only 30 such scrub sites in Florida are currently in-habited by *C. perforata* (USFWS 1999).

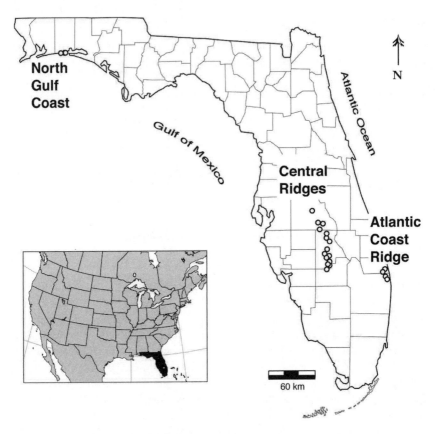

Figure 5.1 Distribution of *Cladonia perforata*. Most populations occur on central Florida's Lake Wales Ridge, and only a few on coastal dunes along both Florida's Gulf and Atlantic coasts. Populations have a patchy distribution, occupy a small area, and often occur as metapopulations within a matrix of occupied and unoccupied sites.

PROTECTING, RESTORING, AND MANAGING KNOWN POPULATIONS

Habitat loss is the most common threat to biodiversity in the United States and abroad. In Florida scrub, habitat loss due to suburban development and citriculture (citrus plantations) was the driving force in the listing of *C. perforata* and several other scrub endemics (USFWS 1999). Less than 15% of the original extent of Florida scrub remains (Peroni and Abrahamson 1985), and habitat loss continues to accelerate. Because of this ever increasing habitat loss, a system of reserves on Florida's biodiverse and imperiled Lake Wales Ridge (e.g., Dobson et al. 1997) targets protection of xeric scrub in peninsular Florida. This system of reserves boasts the only national wildlife refuge set aside primarily for the protection of plant—including lichen—resources (USFWS 1999). Close to two-thirds of *C. perforata* populations are located within reserve boundaries or on federal lands. Rare and endangered lichens should continue to be considered in such multispecies conservation efforts.

Conservation of species is more than just acquisition of habitat and protection from development. The direct results of habitat destruction, reduced population sizes or reduced distributions, can mean that natural threats or periodic demographic fluctuations may result in local or global extinction of species. For instance, a small or lone population is more likely to be extirpated by a local disturbance, such as a fire, that might have impacted only a portion of a larger population or metapopulation prior to habitat loss. Long-term survival in disturbance-prone habitats may depend on the potential to move between patches as successional stages change. Managing for a mosaic of successional stages is important for *C. perforata* (Yahr 2000) as well as many Lake Wales Ridge plants that depend on temporally transient bare-sand habitats (Hawkes and Menges 1996; Menges and Hawkes 1998). Similar successional requirements are known for other lichen species (e.g., Hedenas and Ericson 2000).

The impacts of other stochastic disturbances such as hurricanes can be more difficult to manage, due to the potential for severe and extremely far-ranging effects. For example, as a result of Hurricane Opal in 1995 all three northern Florida populations of *C. perforata,* which occur at most 16 km apart on a single barrier island, were severely damaged. The two small populations were extirpated, and the largest population was reduced by over 70% (Yahr 1997). To restore populations after this storm, a series of small-scale experiments were undertaken to evaluate sites and to refine protocols for successful reintroduction (Yahr 1997). Based on these experiments, a large-scale population reintroduction was conducted in cooperation with USFWS and the landowner Eglin Air Force Base. Two new and apparently stable populations of *C. perforata* were established in its pre-Opal range (Yahr 2001).

Still, all three *C. perforata* populations in north Florida could be extirpated by the direct hit of another large storm with higher winds and storm surge. Mitigation of past, and even future, large-scale stochastic events may be best practiced at a regional level, by establishing multiple populations in widely disjunct locales. In north Florida, for instance, several barrier island reserves with suitable habitats for *C. perforata* already exist that could be used for future introductions. Regional planning should be a part of any conservation scheme, especially for species with poor dispersal or reduced ranges.

LIFE HISTORY, DEMOGRAPHY, AND GENETICS

Even where lichens can be identified, ranges outlined, and habitats managed and protected, their conservation may be hampered by incomplete knowledge of their life-history characteristics. Life-history studies, including basic status surveys and demography, can be challenging because of lichens' slow observable responses to environmental changes and lack of determinate life stages. In addition, numeric counts are impossible for some lichens and often uninformative, since "individuals" are frequently vegetatively derived fragments, and these may be too small and too numerous to count (Will-Wolf et al. 2003). Finally, the vagrant habit of some lichens, including *C. perforata*, means that the fragments are free to drift, unattached to any substrate and carried by wind, water, or animals. Demographic studies of vagrant species require more difficult and error-prone methods such as mark/recapture. In the cases where standard demographic methods are problematic, changes in abundance or cover can supply useful measures for characterizing populations and monitoring conservation efforts (e.g., Rosso et al. 2000).

An important part of lichen population demography is estimating dispersal of various propagules including spores, vegetative fragments, or specialized structures (e.g., soredia, isidia). Colonization of disjunct habitat patches may be via spores or specialized long-distance dispersal units in some lichens, but *C. perforata* has only large, bulky vegetative fragments that are poor dispersers. Limits to dispersal may be the most important demographic feature of this species, since survival of *C. perforata* transplants in unoccupied or recently burned sites is nearly 100% (Yahr 2000). Similarly, in experiments with several other lichens, transplants using soredia or other small thallus fragments were successful in terms of survival, attachment, or establishment (Heinken 1999; Sillett et al. 2000; Zoller et al. 2000). Such transplant studies show that recruitment is not limited by microclimatic factors, but rather by poor dispersal (see also Gauslaa 1997; Dettki et al. 2000; Sillett and Goslin 1999). Furthermore, these studies demonstrate that transplants can be a viable strategy for managers interested in conservation through population augmentation or relocation.

Identification of transplant source material, selection of target populations for conservation action, and delimitation of reserve boundaries can be based on many disparate types of criteria; preserving maximal genetic variability is one of the most common. Genetic variation is viewed as the source of a species' adaptive potential (Ellstrand and Elam 1993), and should be part of an integrated conservation program that targets the ecological and evolutionary processes that maintain genetic diversity (Bowen 1999). In the endangered lichen *Lobaria pulmonaria* (Zoller et al. 1999), genetic variability within Swiss populations (as a result of sexuality) was used to prioritize sites for preservation. Conservation priorities for the apparently asexual *C. perforata* can also be compared in this way, although each population is probably predominantly clonal. In clonal populations, variability can only be protected by protecting multiple, genetically different, populations. Nonetheless, for *C. perforata*, despite the low number of genotypes and strong spatial structure (Yahr, unpublished data), populations are likely to be stable under natural disturbance regimes. Overall, risks from demographic factors seem low in comparison to those challenges posed by loss of habitat and improper management.

MONITORING

By monitoring changes in species abundance, community composition, or environmental conditions, and by comparing current and desired conditions, managers can determine appropriate actions to attain conservation goals. Small populations that are stable in terms of abundance require less immediate attention than large populations that are declining. However, detecting population changes in lichens can be technically difficult, because growth of lichens is slow and difficult to estimate (e.g., simultaneous growth and death in different parts of a lichen, non-modular growth, etc.). This slow growth can result in abundance estimates with higher error terms than real population changes even over a period of several years. For example, after three years of postfire monitoring of *C. perforata,* no significant increases in abundance were detected although the population had clearly expanded outside the sampling area (Yahr 2000). Such informal monitoring can be useful when expected impacts (e.g., trampling effects) can be efficiently detected by casual qualitative assessments. Such an evaluation led to the recent construction of a fence to limit trampling of *C. perforata* at Eglin Air Force Base (D. Teague, personal communication) in an area with many unauthorized trails.

Sometimes when it is difficult to detect changes in abundance in a target species, these changes can be tracked indirectly by monitoring surrogate species or habitat characteristics. For example, in long-unburned Florida scrub, two reindeer lichens (*C. subtenuis* and *C. evansii*) tend to replace *C. perforata* as openings in the understory close and become more shaded. Therefore, measuring increases in abundance of *C. subtenuis* and *C. evansii* (or closing of the understories) may predict the decline of *C. perforata.* Similarly, habitat changes such as increased air pollution (e.g., Nimis and Purvis 2002) or loss of preferred microhabitats may predict the decline of target species.

PROVIDING PUBLIC INFORMATION

One of the most important foundations of conservation is public education. Without basic knowledge of local or regional ecosystems, public sentiment and therefore political support may resist appropriate management techniques. For example, prescribed fire or mechanical treatments are often used to open canopies or reduce fuel loads, but these may be misconstrued as solely destructive activities rather than critical habitat maintenance tools. On the other hand, providing information may carry its own risks of intentional destruction or overcollecting. In the case of *Cetradonia linearis,* localities have not been publicly disclosed, with the intention of protecting the lichen.

OTHER CONSIDERATIONS

Although lessons from plant conservation can be applied to lichens, their unique physiological features require additional conservation considerations. First, many lichens are sensitive to air pollution, especially nitrogen oxides, sulfur oxides, and ozone, and they accumulate heavy metals and radionuclides (James 1973). This sensitivity to such pollutants can be both a conservation challenge and a potential boon. On the one hand, pollution sensitivity creates a particularly difficult management issue since air- and windborne pollutants can cross management or jurisdictional boundaries. On the other hand, this same sensitivity led to the development of a monitoring system using lichen species with differing pollution tolerance as indicators of ecosystem health (Geiser and Reynolds 2002; Nimis and Purvis 2002). Additionally, some lichens may be sensitive to other chemical impacts, such as saltwater storm surge from hurricanes in the case of *C. perforata*.

Second, unlike many vascular plants, lichens cannot be propagated effectively in their natural (symbiotic) state, and they lack long-lived propagules (seeds) and belowground stems or storage organs. Lichen fungi and algae can be successfully cultured axenically and in symbiotic associations ex situ, but they cannot be routinely reintroduced into the wild after they have been brought into cultivation. Some wild-harvested propagules and fragments have been used successfully for transplants, reintroductions, and augmentations, but ex situ propagation via culturing is not a practical conservation strategy. Therefore, resources would be better spent on alternative conservation measures such as purchasing and protecting refugia or preserving maximal genetic diversity across the range of a species.

CONCLUSIONS

To conserve lichens, both (or all) symbiotic partners must be conserved in situ. For *Cladonia perforata* and its photosynthetic partners, the most important threats are ongoing habitat loss and periodic natural disturbances such as fires and hurricanes. Recovery from stochastic fires and hurricanes is hampered by the combined effects of low population sizes, low dispersal, and limited metapopulation structure (few available sites, fewer still occupied). Historical population bottlenecks and resulting low genetic diversity are a concern in efforts to conserve populations of *C. perforata*. Despite this suite of challenges, however, land managers have implemented an integrated and successful management regime, which suggests that there are many effective tools for conserving lichens and other symbiotic associations.

5.3 MOSSES: PHYLUM BRYOPHYTA

Harold E. Robinson

THE BRYOPHYTA (mosses and hepatics) form one of the two major groups of "land plants" or "Embryophyta." The group is smaller in number of species and mostly smaller in size than the vascular plants (plants with xylem and phloem), but they are coequal in age. Both groups date back to the Devonian age. The Bryophyta are most easily distinguished by the prominence of the gametophyte generation in their life cycle, the generation that occurs between meiosis and recombination. The sporophyte generation, the postsexual generation that dominates in other land plants, is comparatively small, with limited unbranched growth, reproducing only by spores. The group consists of between 20,000 and 25,000 species.

Bryophyta are comparatively small in stature, and survive readily in microhabitats. They also seem to evolve comparatively slowly, and a few modern species seem to differ little from entities fossilized in the Late Devonian. Nevertheless, many Bryophyta seem to have originated in the last 60 million years. At present, Bryophyta occur on all the continents, with over 60 species in the Antarctica, where only one or two vascular plants occur. They also inhabit all kinds of habitats from tundra, temperate forests, and tropical jungle to deserts. They can grow on bare granite rock or have their whole life cycle on the surfaces of leaves of tropical trees. They form extensive turfs in tundra and in the high Andean páramos.

Bryophyta are autotrophs, and thus contribute to the production of biomass and breathable air. Few forms are as useful as *Sphagnum* (peat moss), which has remarkable absorptive and antiseptic properties. Still, mechanical difficulties have prevented any serious attempt to use *Sphagnum* in surgery since World War I. Bryophyte gametophytes seem mostly immune to insect damage, but this seems mostly because of poor nutritional value. Anobiid beetle larvae have been seen in moss specimens where they have starved to death at the end of a trail of consumed material. Some hepatics have adverse effects due to presence of complex sesquiterpene lactones that induce skin reactions and photosensitivity. Bryophytes are important indicator plants, both for ecological and phytogeographic studies. Bryophyte species often show distributions similar to genera of vascular plants. Bryophyte distribution is often most important as an indicator of the quality of a habitat.

Bryophyta are easily preserved as dead specimens, and have lent themselves to many floristic studies. This is especially true of the mosses, with floras prepared for many temperate areas and some tropical ones such as Hawaii (plate 5.3), Guatemala, the West Indies, Colombia, Surinam, the Philippines, and parts of India and Indonesia. All needed studies are not complete. Even major groups have gone unrecognized until the last 50 years. The order Takakiales was not recognized until the 1950s. Even more recent is the Ambuchananiales of Tasmania, described in the 1990s. The full distribution of some supposedly rare bryophytes might still be questioned. Some seemingly rare ephemeral mosses are actually much undercollected, including the spring mosses *Pleuridium* Rabh., *Astomum* Hampe, and *Acaulon* Müll. Hal. and the mid- or late-summer *Ephemerum* Hampe and *Micromitrium* Aust. Additions to the known distributions of some species are exceptional. The hepatic *Gymnomitrium laceratum* (Steph.) Horik., once known only from alpine habitats in Japan, is now known from similar habitats in central and south Africa, the Great Smoky Mountains of Tennessee, Mexico, Peru, North Borneo, and the Himalayas (Arnell 1962; Ando 1972). *Didymodon michiganensis* (Steere) Saito, once believed endemic to the south shore of Lake Superior, has since been found in Assam, India, and Okinawa. *Forsstroemia ohioensis* (Sull.) Lindb., once considered endemic to southeastern Ohio, has since been found in Virginia, southern Missouri, and central Mexico, and seems conspecific with *F. producta* (Hornsch.) Par. of South America, Africa, Australia, and Asia (Robinson 1964). A genus once thought restricted to the Andes, *Tristichium* Müll. Hal., is now known from Mexico (Sharp et al. 1994). The entity now known as *Neosharpiella* H. Rob. & Delgad. was once thought restricted to the Andes, but is now known from Mexico and South Africa (Robinson et al. 2001). The necessary proper study of Bryophyta on a worldwide basis remains very incomplete.

Survival of some types of bryophytes is not in danger. *Bryum argenteum* Hedw. and *Ceratodon purpureus* (Hedw.) Brid. are species found on every continent. Some other species such as *Tortula pagorum* (Milde) DeNot. may have increased their distribution in smoggy areas where ultraviolet light exposure is reduced. In the infamous Polk County, Tennessee, where copper smelting has killed trees and herbs down to the bare red earth, *Scopelophila ligulata* (Spruce) Spruce, the "copper moss," has become more common.

The most vulnerable bryophytes are those that grow as epiphytes. *Forsstroemia nitida* Lindb. is known only from two specimens from southern Ontario collected over 130 years ago. The species is apparently extinct. The species was recorded from an unexceptional locality. Other mosses seem much more vulnerable, such as *Philophyllum* Müll. Hal., known only from the water tanks of epiphytic bromeliads. Even less specialized epiphytic bryophytes are endangered by the sweeping destruction of tropical forests. Deforestation is also accompanied by climatic changes that could destroy hundreds of species. Areas of southern Ecuador and northern Peru were once forested, with regular afternoon or evening showers, but since deforestation and the loss of tree transpiration, they have become deserts with cacti.

5.4 GRASSES: FAMILY POACEAE

Paul M. Peterson

THE GRASS FAMILY (Poaceae or Gramineae) is the fourth-largest flowering-plant family and contains about 11,000 species in 800 genera worldwide. Twenty-three genera contain 100 or more species or about half of all grass species, and almost half of the 800 genera are monotypic or diatypic, that is, with only one or two species (Watson and Dallwitz 1992).

The most important feature of this family is a one-seeded indehiscent fruit (the seed coat is fused with the ovary wall), known as a caryopsis, or grain (fig. 5.2). The grain is rich in endosperm starch and can contain protein and small quantities of fat. The embryo is located on the basal portion of the caryopsis and contains high levels of protein, fats, and vitamins. The stems are referred to as culms and the roots are fibrous, principally adventitious or arising from lower portions of the culms. Many perennial grasses have rhizomes (underground stems) or stolons (horizontal

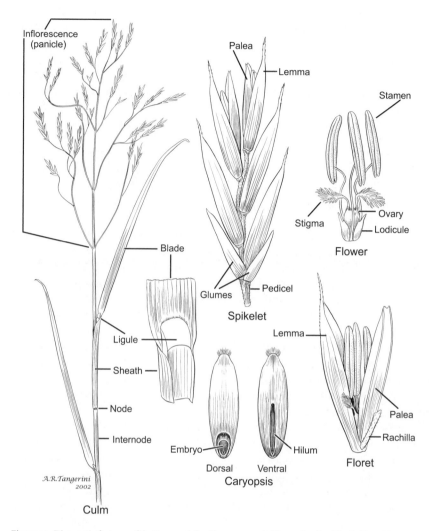

Inflorescence
(panicle)

Blade

Ligule

Sheath

Node

Internode

A.R.Tangerini
2002

Culm

Palea

Lemma

Glumes — Pedicel

Spikelet

Stamen

Stigma

Ovary

Lodicule

Flower

Lemma

Palea

Rachilla

Floret

Embryo

Hilum

Dorsal   Ventral

Caryopsis

Figure 5.2 Diagnostic features of the Poaceae (after Peterson 2003). Illustration by Alice R. Tangerini.

aboveground branches) that allow for vegetative reproduction. Silica is a conspicuous component of the epidermis and is stored in silica cells. Another important
feature of grasses is intercalary meristems, which allow growth well below the apex,
typically near the base of the plant. The leaves are parallel-veined and two-ranked
with the basal portion forming cylindrical sheaths and the upper portions referred
to as a blade. A ligule, located on the upper surface at the junction of the blade and
sheath, commonly consists of flaps of tissue or hairs. The primary inflorescence
is referred to as a spikelet with one to many two-ranked bracts inserted along the
floral axis or rachilla. The lowest two bracts of each spikelet, inserted opposite each

other, are glumes, above which along the rachilla are borne pairs of bracts termed florets. Each floret consists of a lemma (lower bract) and palea (upper bract). Within each palea the highly reduced flowers can be found. Each grass flower usually consists of two or three small scales at the base called lodicules, an ovary with a style and two plumose stigmas, and one to six but most commonly three stamens with basifixed anthers that contain single-pored, wind-dispersed pollen grains. Lodicules open the florets during flowering and possibly represent reduced perianth (sepals and petals) segments. Since the morphological features are often cryptic or lacking, identification to species is often very difficult and requires a trained specialist.

The grass family has probably been characterized as a distinct entity in most cultures. Three hundred years before the Christian era, Theophrastus, a Greek scholar, recognized the grass family and began to teach his students the concepts of plant morphology. The first scientific subdivision of the family was made by Robert Brown (1814), who recognized two different spikelet types that differentiate subfamilies Panicoideae and Pooideae (Festucoideae). Bentham (1881) recognized 13 tribes grouped in the 2 major subfamilies. Hitchcock (1935) and Hitchcock and Chase (1951) in their treatments of the grasses of the United States recognized 14 tribes in these 2 major subfamilies. The two subfamily classification was used by most agrostologists for almost 150 years until more modern syntheses. With the infusion of molecular data our present concept and classification of the grasses is changing at a rapid rate. We currently recognize 12 subfamilies: Bambusoideae, Anomochlooideae, Pharoideae, Puelioideae, Ehrhartoideae, Centothecoideae, Pooideae, Arundinoideae, Danthonioideae, Aristidoideae, Chloridoideae (plate 5.4), and Panicoideae (GPWG 2001), and in these subfamilies we recognize 40 tribes and 76 subtribes (Soreng et al. 2000).

The highly reduced floral structure and wind pollination in the grasses has enabled the family to be extremely successful at planetwide radiation and colonization. Grasses are well adapted to open, marginal, and frequently disturbed habitats, and can be found on every continent, including Antarctica. Two major photosynthetic or carbon dioxide (CO_2) assimilation pathways can be found in the grasses (C_3-fixing of CO_2 by ribulose 1,5-diphosphate [the Calvin-Benson cycle, found in all vascular plants] and C_4-fixing of CO_2 by an additional enzyme [phosphoenolpyruvate] to form four carbon molecules [oxaloacetate or malate]) and there are anatomical, physiological, phytogeographic, and ecological differences between these two types. The C_3 grasses are well adapted to temperate climates with winter precipitation whereas C_4 grasses are well suited to tropical environments with summer/fall precipitation. The addition of C_4 photosynthesis has allowed the grasses to outcompete other plants in warm, tropical environments by lowering the oxidation levels (photorespiration) of photosynthetic products. All of these features have led to the family's ability to occupy nearly one-quarter of the earth's surface in various climatic environments as the dominant component, the grasslands. Historically, the grassland biome has been maintained by a myriad of biotic, climatic, and edaphic effects. First, there must be a dry season in which grasses and adjacent forest border dry out and become inflammable (Axelrod 1985). Repeated fires favor grasses over most tree and shrub species since they very easily resprout from the base. Second, large herbivorous mammals (e.g., bison, antelope, yaks, guanacos, vicuñas, and llamas) are instrumental at maintaining and further opening up grassland communities (Axelrod 1985). An often overlooked consequence of grazing animals is their effect on soil compaction, which again favors sod-forming grasses over trees and shrubs.

Economically, the grasses are the most important plant family for food production. Rice (*Oryza sativa* L.), wheat (*Triticum aestivum* L.), corn (*Zea mays* L.), barley (*Hordeum vulgare* L.), rye (*Secale cereale* L.), oats (*Avena sativa* L.), sorghum (*Sorghum bicolor* [L.] Moench), pearl millet (*Pennisetum glaucum* [L.] R. Br.), finger millet (*Eleusine coracana* [L.] Gaertn.), and tef (*Eragrostis tef* [Zucc.] Trotter) are widely cultivated grains. Grains were one of the first plants domesticated by humans and are the basis for all early civilizations. Corn, or maize, originated in Mexico and formed the basis for the Aztec, Inca, and Mayan civilizations. Wheat and barley originated in the Middle East or Fertile Crescent and were used to make breads, pastas, and beer. Rice originated in southeastern Asia and can be called the world's most important crop because it is estimated that over 200 million tons of rice are consumed each year by 1.6 billion people (Simpson and Conner-Ogorzaly 1986). In wheat two proteins, gliadin and glutein, combine with starch to form gluten. When these proteins in bread flour are mixed with water and kneaded, the result is an elastic dough perfectly suited for baking, with the addition of yeast. The domestication of modern wheat, a hexaploid ($6n = 42$), is an interesting story in the coevolution of humans and food, since early domesticates of wheat, the einkorns (*Triticum monococcum* L., $2n = 14$) and emmers (*Triticum turgidum* L., $4n = 28$), were harder to harvest, that is, to separate the grains from florets by threshing, and when baked produced inferior breads. Other notable economic uses of grasses include landscaping, construction (primarily bamboos), and sugarcane (*Saccharum officinarum* L.) production.

Although the grasses seem well suited to ecosystem margins and moderately disturbed sites, the major threat to extant species is loss of habitat. Grasses have been very successful in an evolutionary sense and in their ability to adapt to human needs as major food sources for all complex civilizations. It is important for us to maintain and manage these critical resources for future generations. One way to accomplish this goal is to preserve as many of the wild relatives as possible, because they may carry the genetic code for improved production or pest resistance. Once these genes are introduced into a crop species, the hybrid vigor of a crop can be dramatically improved, increasing pest resistance and productivity.

Worldwide, Walter and Gillett (1997) list 776 grass species as threatened, which includes possibly extinct, endangered, rare, vulnerable, and indeterminate categories. In North America 215 species are listed as threatened, with 40 (19%) of these occurring in California alone. In comparison, only 80 species are listed as threatened in South America. South America is only three-quarters the size of North America, but North America has about twice as many threatened grass species per square mile as South America does. Obviously, a major factor in determining the status of these threatened grass species is the extent of botanical knowledge scientists have gathered on these plants. We still need to learn more about the grasses in South America, especially in the tropics.

5.5 DAY FLOWERS:
FAMILY COMMELINACEAE

Robert B. Faden

THE COMMELINACEAE is a family of small to medium-sized herbaceous plants. The family comprises about 650 species in 41 genera (Faden 1998). Commelinaceae are mainly tropical, but species occur nearly worldwide. The three major centers of diversity are tropical Africa, Mexico to northern Central America, and peninsular India to southwestern China. The largest genera are *Commelina, Tradescantia, Aneilema, Murdannia,* and *Cyanotis.*

Commelinaceae are of limited importance economically. In the tropics they are used locally as medicines, in rituals, or as fodder for domestic animals. Ecologically, they are important components of the understory in lowland African rain forests. In temperate regions species are used mainly as garden and greenhouse ornamentals. Some Commelinaceae are weeds in crops and waste places.

The conservation status of species of Commelinaceae is generally difficult to ascertain. This is due in large measure to taxonomic uncertainties and to our very incomplete knowledge of the distributions of most species. The family is undercollected, and herbarium specimens are particularly difficult to work with.

Commelinaceae species are often overlooked by collectors. Similar congeneric species may occur at the same site, and the diversity goes unrecognized. For example, up to five small, orange-flowered, annual species of *Commelina* were found on roadside banks in southern Tanzania (R. B. Faden, unpublished data), but such local species richness had not been reported previously or documented through collections. Five species of *Aneilema* have been observed growing together in bushland in Kenya (Faden 1983), five of *Murdannia* in old rice paddies in northern Sri Lanka, and six of *Palisota* in the forests of Cameroon (R. B. Faden, unpublished data). Naturally, a specialist tends to find more species of his group than the general collector does, but collections of the latter provide most of the available specimens.

Dried specimens of Commelinaceae are difficult to study and identify because of the general absence of flowers and other diagnostic characters. Flowers of all species are at anthesis (the period during which a flower is fully open and functional) for only a few hours, after which they collapse and liquefy. Thus collections have to be made at the right time of day in order to encounter flowers. Then the flowers have to be pressed immediately in the field or preserved in liquid. Even when pressed promptly, the flowers often do not preserve well, either shriveling, sticking to the newsprint, or both.

The overwhelming majority of herbarium specimens are incomplete. Almost invariably they lack flowers or have just fragments thereof. Depending upon the species—small annuals are exceptions—most specimens also lack mature fruits and seeds. In species for which the subterranean organs can be diagnostic, for example, many *Cyanotis* and some *Commelina,* these parts are commonly left in the ground by collectors.

The final difficulty in working with herbarium specimens is in the usual inadequacy of collectors' notes. The only useful detail that is commonly included is the flower color, and even that can be misleading. Very few collectors add further data about the flower, and almost no one mentions the hours of flowering, perhaps because its significance as a species-specific character is not known.

Murdannia clarkeana Brenan and *Commelina nairobiensis* Faden are two species of Commelinaceae that occur in seasonally wet soils mainly in and near Nairobi, Kenya (fig. 5.3). *Murdannia clarkeana* was described 50 years ago (Brenan 1952), but *C. nairobiensis* was not formally named until recently (Faden 1994b). Both species remained unnamed for a considerable period after they were first collected, both have some taxonomic loose ends, and both are threatened by habitat destruction.

Murdannia clarkeana was first collected in 1902. According to Brenan (1952), C. B. Clarke annotated a collection as "Aneilema delicatulum" but never published this name. The name was picked up, however, and used without a formal description by Jex-Blake (1948). It was described as *M. clarkeana* by Brenan (1952) in a revision of the genus *Murdannia* in Africa.

Murdannia clarkeana is a slender, rhizomatous perennial usually under 30 cm tall, with linear, subterete, onionlike leaves. The small, blue to lavender flowers open about 11 a.m. and fade about 4 p.m. A study of leaf anatomy showed that all the stomata have four subsidiary cells, instead of the normal six for the genus (Faden and Inman 1996). A chromosome number of $2n = 24$ was obtained by Faden and Suda (1980), showing that the species is tetraploid. Plants grow in rock pools, in seasonally waterlogged shallow soil on granitic outcrops, and occasionally near streams.

Murdannia clarkeana ranges in central Kenya from the Nairobi area north to the southern edge of Samburu District, southeast of Maralal, a distance of about 270 km (fig. 5.3B). All but 7 of the 22 collections seen have come from the Nairobi area. The plant appears to have been extirpated from the locality in Thika, ca. 40 km to the northeast (R. B. Faden, personal observation, 1996). None of the other localities outside of Nairobi has been checked in recent years.

This species has also been recorded from Chad and the Central African Republic by Lebrun et al. (1972). Those specimens, which are more than 2,000 km disjunct from the Kenyan populations, have been examined and can be separated from the Kenyan plants only by their annual habit. Perhaps additional characters might be

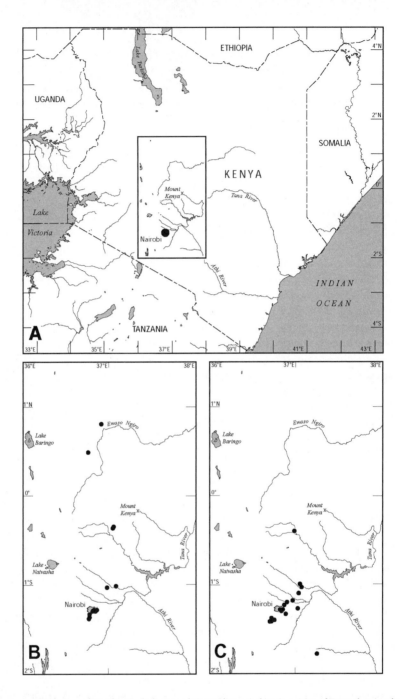

Figure 5.3 Distribution of *Murdannia clarkeana* and *Commelina nairobiensis*. A, Map of Kenya showing the overall distributions of *M. clarkeana* and *C. nairobiensis* (box). The solid circle represents the position of Nairobi, the capital and largest city in Kenya. B, Inset from A, showing distribution of *M. clarkeana*. C, Inset from A, showing distribution of *C. nairobiensis*.

found in the flowers, but no flowers are preserved. However, the difference in habit would be sufficient to treat the Central African populations at least as a distinct subspecies.

Commelina nairobiensis apparently was first collected in 1964, which seems remarkably late, considering its frequency in Kenya's most populous city and along the main road to Thika (fig. 5.3C). Possibly there were older collections in an ill-fated 1976 loan of *Commelina* from the East African Herbarium that was lost, but, if so, there is no record of them. *Commelina nairobiensis* was recognized as a distinct species by 1970 and designated as *Commelina* sp. "A" in Faden 1974. It was formally described in Faden (1994b), using an epithet that had been in common use for more than two decades.

Commelina nairobiensis is endemic to central Kenya with a range of about 170 km in diameter (fig. 5.3C). It is a perennial with tufted, erect or ascending stems up to 60 cm tall. This dark blue–flowered species is characterized by swollen shoot bases that somewhat resemble orchid pseudobulbs and by a squarish capsule that contains four spherical, warty seeds (plate 5.5). A chromosome count of $2n = 24$, obtained from two collections, is a highly unusual number in *Commelina*. Plants grow in seasonally wet grassland, especially in black-cotton soil (Faden 1994b).

The uncertainties about this species are twofold. One concerns the relationship between its morphology and ecology. The other has to do with the similarity of its capsules and seeds to those of two other *Commelina* taxa in Kenya. Swollen shoot bases, which characterize *C. nairobiensis,* have also been observed in other *Commelina* species, notably *C. benghalensis* L. and *C. latifolia* A. Rich., but only when these species grow in black-cotton soil. Could *C. nairobiensis* just be a black-cotton soil ecotype of another species? The obvious species to consider are the two other Kenyan taxa, *C. eckloniana* Kunth and *C. echinosperma* K. Schum., with similar fruits and seeds (Faden 1994a). These taxa are sufficiently distinct morphologically, so *C. nairobiensis* could not just be an ecotype of either one, and therefore it has to be a distinct species.

Both *M. clarkeana* and *C. nairobiensis* must be considered threatened because of their local distributions in a densely populated area with a rapidly expanding human population. The rock formations that hold the pools in which *M. clarkeana* grows make excellent building sites. The black-cotton soil in which *C. nairobiensis* grows, once marginal for farming, has increasingly been utilized for crops and grazing. *Murdannia clarkeana* appears more vulnerable to habitat destruction than *C. nairobiensis* because its habitat is more specialized and much rarer. Both species are common on a rocky hillside adjacent to the Carnivore Restaurant in Nairobi, just outside Nairobi National Park. This may be the largest remaining population of *M. clarkeana.* This area has been of great interest to conservationists, but it has been repeatedly threatened by development. These two species provide additional reasons why the area should be preserved.

5.6 ACANTHUS: FAMILY ACANTHACEAE

Dieter C. Wasshausen

THE ACANTHACEAE is a pantropical family of approximately 240 genera and 3,240 species. In the New World, there are approximately 85 genera and 2,000 known species, with 75 genera and 1,780 species in tropical America. About 8 genera, including the two largest, *Justicia* (450–600 species) and *Ruellia* (ca. 300 species), are pantropical. Of the remaining 77 genera, 33 are monotypic and 14 others possess only two or three species each. Genera of Acanthaceae are often distinguished on relatively minor characters such as anther configuration and pollen sculpture, and therefore there are numerous taxonomic problems at the generic level (Cronquist 1981). Until these problems are resolved, it is difficult to infer phylogenetic relationships and geographic patterns in the family.

Species of Acanthaceae occur almost exclusively in the tropics and subtropics, with five major centers of diversity (Lindau 1895) and richness. The Indo-Malesian region is distinguished mainly by abundant taxa in the genus *Strobilanthes* and the tribe Andographideae (Bremekamp 1981). From this center radiate regions that possess few endemics, so that Japan and China have mostly widely distributed species. In Afghanistan and still farther removed in the Mediterranean region are found only xerophytic types (e.g., *Acanthus* with 50 species). Finally, Australia has only 57 native and naturalized species (Barker 1986). A second center of diversity, which is at least as rich in taxa as Asia, is tropical Africa and Madagascar. This region possesses a large number of endemic genera.

The remaining three centers of diversity occur in the New World. The Andes of South America are especially rich in Acanthaceae. Examples of genera with extensive species diversity in this region are *Aphelandra* (180 species), *Justicia* (450–600 species), *Sanchezia* (50 species), and *Pachystachys* (13 species) as well as a number of monotypic genera. The second South American center of diversity is Brazil. Its vast and diverse flora (38 genera and 490 species of Acanthaceae) is especially rich in species of *Mendoncia* (24 species) and *Ruellia* sections *Dipteracanthus* and *Physiruellia* (100 species). Genera and species from this center of diversity radiate to the north into the Guianas and to the south into northern Argentina. The third center of diversification in the New World is tropical Mexico and Central America. More than 350 species in 40 genera occur in Mexico alone (Daniel 1993). Representatives of the tribes Aphelandreae and Odontonemeae reach their highest level of development in this region.

Species of Acanthaceae occur in almost every type of habitat—in forests, swamps, and meadows. Some species are found on rocky outcrops and others in deserts. At the higher elevations (i.e., above 3,000 m) the family is found only in habitats that are protected from the cold and have sufficient rainfall, such as the south slope of the Himalayas or the eastern slope of the Cordillera in Bolivia.

A great array of diverse floral forms is evident in Acanthaceae. Characteristically, the flowers are proterogynous (the pistil matures before the stamens), nectar-yielding, brightly colored and often arranged in large and showy inflorescences. At anthesis the receptive stigma lies directly above the already dehiscent anther. Usually, insects, hummingbirds, and bats strike the downward directed tips of the pollen sacs and are dusted with pollen. The pollinator then transfers the pollen to the projecting stigmas of other flowers. As only one or two flowers are open per inflorescence daily, many pollinator-mediated pollen transfers must result in cross-fertilization. Toward early afternoon, the stigmas of all flowers, whether pollinated or not, characteristically arch downward, thus making direct contact with the inwardly dehiscing anthers and ensuring self-pollination in the absence of previous cross-pollination. In essence, in Acanthaceae speciation appears to be pollinator-driven. Floral visitors observed and reported include carpenter bees (on *Thunbergia grandiflora*), euglossine bees (on *Justicia stipitata*), bats (on *Ruellia exostema* and *R. verbasciformis*), butterflies (on *Tetramerium* and *Pseuderanthemum* [80 species]) and, commonly in the long tubular red flowers, hummingbirds (on *Pachystachys spicata* [plate 5.6] and *Justicia secunda*). Furthermore, the spicate inflorescence of numerous species (*Ruellia proxima*, *Stenostephanus longistaminus*, and *Pachystachys ossolaea*) is extensively covered with scale insects and ants. Floral morphology appears to preclude myrmecophily as a factor in the pollination system, however (Long 1970); it is suspected that an exudation of plant juices immediately surrounding the scale insects attracts the ants. The ants feed on this exudate and, in return, protect the plants from herbivorous predators.

A developing capsule is normally evident a few days after pollination and fertilization. In most genera the seeds are dispersed for up to 5 m from the parent by the explosive dehiscence of the ripe capsule. This explosive mechanism, unique to the Acanthaceae, has been attributed to the progressive loss of water from the maturing fruit. The seeds in numerous species are covered with minute hairs that, when moistened, become erect and mucilaginous; possibly this is an adaptation that helps to disperse the seeds and to anchor them to a suitable substrate. It is suspected that the seeds of some widespread disjunct species such as *Dicliptera sexangularis* stick to the webbed feet of waterfowl and other birds and thus have been transported over long distances.

The most common use of the Acanthaceae is in cultivation as ornamentals, especially through modern technology, which has enabled horticulture to adopt many of these tropical species. Currently species of no less than 36 genera are ornamental plants in tropical and subtropical regions of the New World, many of which persist or escape. The large decorative leaves of the genus *Acanthus* have held an immortal position in both the arts and ancient horticulture. The "acanthus" ornamental leaf pattern of *A. spinosus* was a favorite decoration in classical sculpture that was first exhibited on Corinthian columns. According to legend, the sculptural use of the leaf originated about 430 BC when a Greek sculptor adopted it as a distinct element in the decoration of temple columns. Virgil described an acanthus design embroidered on the robe of Helen of Troy. Species of *Acanthus* have also been used medicinally. An extract of the boiled leaves of *Acanthus ebracteatus*, sea holly, is used as a cough medicine in parts of Malay, while the roots of *A. mollis,* bear's breech, are used to treat diarrhea in some parts of Europe.

Justicia pectoralis is reported to have numerous medicinal, hallucinogenic, and economic uses throughout its wide range. The sweetly fragrant dried leaves are pulverized by some South American Indians and are added to their *Virola*-based hallucinogenic snuff. It is suspected that *J. pectoralis* is employed as a flavoring rather than as a hallucinogenic agent. Recently species of *Ruellia* and other Acanthaceae species have been reported in pharmaceutical journals as possessing anti-infective phytomedicines and antimicrobial activities. These findings are proving that higher plants used as anti-infective phytomedicines may serve as promising new sources of antibiotics.

Human activities in the regions where Acanthaceae are commonly found are often extractive or destructive; a few have left lasting gains. At the lower elevations, between 600 and 1,800 m, there are great demands for agricultural land both illegal (extensive coca plantations) and legal (coffee, fruits, maize, rice). It is presently estimated that anywhere between 10 and 15% of the known species of Acanthaceae are either rare or presently in danger of becoming extinct. Two examples of this type of habitat destruction come to mind. In 1975, a new species of *Aphelandra, A. pepeparodii,* was collected in Peru's department of Ayacucho, La Mar Province, in a lowland rain forest along the Río Catute. Two years later, this locality was revisited because of its high degree of Acanthaceae endemism (37 species in a 30-km radius). The area was totally cleared for coffee growth. What prompted the clearing was a rapid rise in the price of coffee. To date, this species is known only from the type locality. Similarly, in the summer of 1987, a number of endemic genera of Acanthaceae (*Achyrocalyx, Pseudoruellia,* and *Pseudodicliptera*) were collected in the dry forest of Sakaraha, near Ilhosy, Madagascar. This region along the edge of Madagascar's Central Plateau was also a classical locality for a number of succulents such as *Euphorbia enterophora* ssp. *enterophora.* Two years after the visit a photograph was received with the inscription, "This was the forest of Sakaraha." The photograph depicted a smoldering and denuded scene of what once was a large area of natural vegetation. Farmers of this region practice shifting agriculture, known locally as *tavy* cultivation. They had cut down the forest and cleared the land by burning the dried wood. The soil is enriched with nutrients through this process, and the farmers can cultivate maize, dry rice, manioc (cassava), or batatas (sweet potatoes). The land usually can produce only a single or, perhaps, a second crop, then a new *tavy* must be established in a different locality.

The *1997 IUCN Red List of Threatened Plants* (Walter and Gillett 1997) lists 429 threatened species of Acanthaceae, 17% of the recorded species. Four species are listed as extinct, 63 as endangered, 197 as rare, 108 as intermediate, and 68 as vulnerable. The overall percentage more or less accurately reflects the status of endangered Acanthaceae in the New World, but the breakdown of the individual categories probably differs.

5.7 DAISIES AND SUNFLOWERS: FAMILY ASTERACEAE

Vicki A. Funk and Harold E. Robinson

THE COMPOSITAE (Asteraceae) family contains the largest number of described species of any plant family, approximately 25,000 distributed in over 2,200 genera, and they occur on all continents except Antarctica. Estimates vary, but assuming that there are 200,000–300,000 species of flowering plants, then one out of every 8–12 species of flowering plants is in the Compositae (about 10%). The family is monophyletic, characterized by florets arranged on a receptacle in heads (plate 5.7) and by anthers fused in a ring with the pollen pushed out by a style that acts as a plunger (fig. 5.4). Although the family is well-defined, there is a great deal of variation among the members. The habit varies from annual and perennial herbs to shrubs, vines, or trees, although a few are true epiphytes. The heads can have 1 to more than 1,000 florets. Chromosome numbers range from $2n = 4$ to high-level polyploidy with $2n = 228$. Species grow in just about every type of habitat from forests to páramo and puna; however, they are less common in tropical wet forests and more common in open areas. They can be showy or obscure, fragrant or foul, breathtakingly beautiful or nondescript. However, the general perception of this family as "weedy" is not correct. Certainly some members benefit from disturbance, such as a few species of dandelions, goldenrods, and thistles, but most species have a restricted distribution, and just about every "at risk" habitat in the world contains members of this family that are an important part of the flora. In fact, the *1997 IUCN Red List of Threatened Plants* (Walter and Gillett 1997) lists 2,553 species of Compositae. Particularly vulnerable are previously unappreciated epiphytic members of the family in tropical forests, island floras, and the páramo and puna floras of the Andes.

That the family is monophyletic has never been in question. Every early worker in plant classification recognized the Compositae as a group at some level. The compound inflorescence with florets, often found in the striking ray/disk arrangement (plate 5.7), enabled everyone to accurately delimit this family (e.g., Bentham in Bentham and Hooker 1873). There are several general references that treat the family (Bremer 1994; Heywood et al. 1977; Hind 1996); the most up-to-date is the Kubitzki volume (in press) that provides keys to and descriptions for all the genera of every tribe. In every molecular analysis that contains more than one member of this family, the results show that the family is monophyletic (i.e., Jansen and Palmer 1987; Bremer et al. 1992; Hansen 1992). Within the family, however, things are not as clear-cut. The family is traditionally divided into three subfamilies: Barnadesioideae, Cichorioideae, and Asterioideae (fig. 5.5). The Cichorioideae has been

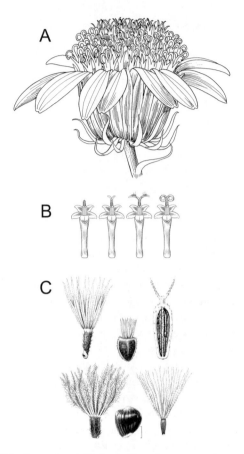

Figure 5.4 Characters of the Compositae. A, The head with ray florets arranged around the perimeter, disk florets in the center, and involucral bracts surrounding the outer florets. B, The pollen is released when the style pushes out through the anthers, which are fused at the margins. C, A few of the achene (cypsela) and pappus types found in the Compositae. The scale bar in C is 1 mm long.

divided into additional subfamilies (Panero and Funk 2002); for this treatment three of these are used. The Asterioideae is subdivided into two supertribes; each of these is divided into tribes (Baldwin et al. 2002). The tribes are the subunits of the family used in most treatments.

Most tribes contain some useful and noxious species as well as some that are common and some that are rare. No simplistic answers are available for management or protection of these types of plants in the various tribes other than the preservation of ample representations of the richest overall habitats.

The subfamily Barnadesioideae lacks the complex chemistry and twin hairs that characterize most of the rest of the family: their terpenoids are mostly diterpenes. Most members have paired axillary spines. The pollen lacks the characteristic form

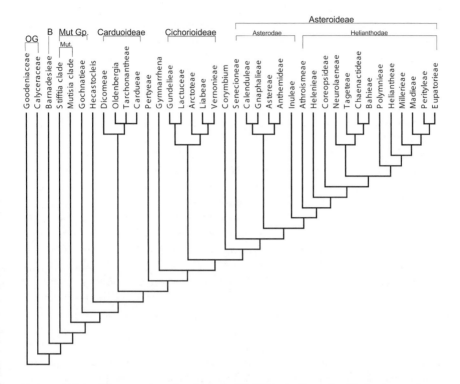

Figure 5.5 Relationships among the major tribes and subfamilies in the Compositae. The basal group, the monophyletic subfamily Barnadesioideae, contains less than 1% of the species in the family. The monophyletic terminal subfamily, the Asterioideae, contains ca. 65% of the species in the family. This large and diverse subfamily contains many of the tribes that are widespread and easily recognized, as well as several smaller ones. The third subfamily, the Cichorioideae (ca. 35% of the species in the family), vary in their morphological and molecular characters. Current thinking on the number of tribes ranges from 17 to 22; here 19 tribes are recognized (Funk et al., in press). B, subfamily Barnadesioideae; Mut., tribe Mutisieae; Mut. Gp., mutisioid group.

of higher Asteraceae. There is only one tribe, Barnadesieae, consisting of 9 genera and 90–95 species. Members are restricted to the Andes except for *Dasyphyllum* Kunth, which extends into southern Brazil, and the monotypic *Schlechtendahlia* Less. in lowland southern South America. Their habit ranges from small herbs to trees. Although many of the species are rare, most of them grow in habitats that are relatively inhospitable, and only a few of these habitats are threatened.

The former paraphyletic subfamily Cichorioideae can be divided into three groups: the complex and variable Mutisieae, called the mutisioid group, and two monophyletic subfamilies, the Carduoideae and Cichorioideae. In addition there are several genera that do not fit in any of these groups, and they have been placed in subfamilies of their own (Panero and Funk 2002).

The paraphyletic mutisioid group is notable for its variable floret structure, generally showy involucral bracts, absence of much of the specialized chemistry of higher Compositae, and a nonspecialized pollen type. Members of the former large tribe Mutisieae are now found at the base of the Carduoideae and between the Mutisieae and Carduoideae and the Carduoideae and Cichorioideae.

The tribe Mutisieae sensu strictu consists of ca. 70 genera and nearly 1,000 species. The members occur mainly in Central America and South America. Most are Andean, on the Brazilian Shield, or on the Guiana Shield; however, there are also genera in Africa and Madagascar, Asia, and one monotypic genus in Australia, *Amblysperma* Benth. Numerous members of the tribe are of limited distribution and at risk of extinction because of habitat destruction, such as several species of the rare small tree *Wunderlichia* Riedel ex Benth. from the Brazilian *planalto*. This tribe includes some of the most interesting plants found in the unusual tepui flora found on the Guiana Shield in northeastern South America.

Two unusual dioecious genera that used to be in the Mutisieae from southern and tropical Africa and Madagascar (*Tarchonanthus* L. and *Barchylaena* R. Br.) have been placed in their own tribe, Tarchonantheae (Keeley and Jansen 1991). They are now in the Carduoideae.

The tribe Cynareae (Cardueae) has 74 genera and ca. 2,500 species. Centers of distribution are primarily Eurasia and North Africa; some are successful weeds. Notable members are the thistles (*Cirsium* L.), the artichoke (*Cynara* L.), and various *Centaurea* L. (e.g., bachelor's button). These often spiny-leaved plants are sometimes appreciated horticulturally, but can be unwelcome invaders outside of their normal range, as in western North America, the pampas of Argentina, and the fallow fields of Australia. Some taxa are restricted in their distribution, and most of these are endangered by habitat destruction and invasive species. For instance, *Centaurodendron* Johow and *Yunquea* Skottsb. are endemic to the Juan Fernandez islands, with the latter known only from a few plants from one summit. There are many small endemic genera in the Middle East (*Centaurothamnus* [Forssk.] Wagenitz & Dittrich), Central Asia (*Schmalhausenia* C. Winkl.), and Iran (*Aegopordon* Cass., *Karvandarina* Rech. f., *Myopordon* Cass.), all of which are difficult to find. Many endemic species of various genera, such as *Stemacantha australis* (Gaud.) M. Dittrich in Australia, *Cirsium* in the western United States, and *Centaurea* in the Mediterranean region, are rare and often endangered.

The subfamily Cichorioideae contains tribes with specialized asteraceous pollen, echinate or often lophate, and style branches with undivided areas of stigmatic papillae on their inside surfaces. The tribe Arctoteae is a small group of 15 genera almost totally restricted to southern Africa; one outlying genus, *Cymbonotus* Cass., is endemic to Australia, where it is rare. A number of the species are used in horticulture, especially members of *Gazania* Gaertn. and *Arctotis* L. Many of the species found in the ephemeral flora of Namaqualand (a recognized conservation hot spot) are in this tribe.

Vernonieae have nearly 120 genera with ca. 1,400 species. It is widespread in North America and South America, Africa south of the Sahara, and areas around the Indian Ocean. A few species are invasive weeds such as *Cyanthillium cinereum*

(L.) H. Rob. of Africa. The widespread *Struchium sparganophorum* (L.) Kuntze of possible American origin invades agricultural fields in Africa and India. *Baccharoides anthelmintica* (L.) Moench of Asia and Africa, and *Gymnanthemum amygdalinum* (Del.) Sch. Bip. ex Walp. of Africa have some medicinal value and have been introduced in a limited way into America. *Lychnophora* (Candelabra) is restricted to rocky outcrops in the Brazilian Shield and is endangered, even within the protected areas, by fire and grazing. Many species from Africa and Asia are threatened, such as the species of *Distephanus* in Madagascar and Reunion Islands. In addition, the Tanzania genus *Hystrichophora* Mattf. is still known from only one collection that was destroyed in Berlin; a fragment exists at the British Museum. Located at the base of the Vernonieae is the small tribe Moquineae (Robinson 1994).

The tribe Liabeae contains 15 genera and ca. 180 species. Many genera are like the Cichorieae in having milky sap. Members of the Liabeae are found in Mexico, Central America, the Greater Antilles, and the Andes; however, the greatest diversity is in Ecuador and northern Peru. A few taxa are common, but most genera, such as *Pseudonoseris* H. Rob. & Brettell and *Chionopappus* Benth., are characterized by species that are difficult to find. One rare Peruvian genus, *Bishopanthus* H. Rob., is still known from only one collection, and the pygmy *Microliabum humile* (Cabrera) Cabrera of the Andes of northern Argentina has been collected only twice.

The tribe Cichorieae (Lactuceae) includes nearly 100 genera and several thousand species. Members of the tribe have milky sap and unusual ligulate florets. Distribution is nearly worldwide, but is concentrated in Eurasia and western North America. Notable members are lettuce (*Lactuca* L.) and *Cichorium* L., used as food and flavoring, and the dandelion, *Taraxacum* Weber, which is usually considered a weed, can be used as a salad green and for making wine. More localized and endangered elements are found in restricted habitats around the world such as California and Eurasia, and some are island endemics, such as *Dendroseris* D. Don of Juan Fernandez with 11 species, 2 of which are extinct and another which is known from only three specimens.

The tribe Gundelieae contains a single genus with one to two species distributed from Turkey to the Middle East and Iran. *Gundelia tournefortii* L. is believed to have supplied the pollen and the plant images found on the shroud of Turin, and it has been suggested that it was the "crown of thorns." In spite of its distribution, the species may be threatened because its leaves, stems, roots, and undeveloped flower buds are edible and it is harvested from the wild before it sets seed.

The subfamily Asteroideae is notable for the familiar echinate, tricolporate, asteraceous pollen and the stigmatic papillae of the style branches often separated into two separate bands. When rays are present, they are "true" rays, and the disk florets often have short lobes. It is divided into two supertribes, one paraphyletic, Asterodae, and one monophyletic, Helianthodae. The supertribe Asterodae has tribes without phytomelanin in the walls of their achenes. Six of the seven tribes in the supertribe Asterodae are discussed below (Panero and Funk 2002; Funk et al., in press).

The tribe Astereae has ca. 190 genera and nearly 3,000 species; it is nearly worldwide in distribution. Included are the common asters (e.g., *Aster* L.) and goldenrod (*Solidago* L.), the widespread *Erigeron* L., the weedy *Conyza* L., *Amellus* L. of South Africa, and the Bellidinae (the true daisies) of Europe. Some genera are very local and endangered, such as the tree genera *Commidendron* DC. and *Melanodendron* DC. of St. Helena, *Remya* Hildebr. ex Benth. of Hawaii, *Pleurophyllum* Hook. f. of the Subantarctic islands near New Zealand, *Pacifigeron* Nesom of Rapa Island, and the small bromeliad-like *Novenia* Freire of Peru and Bolivia. There are also a fairly high number of substrate endemics in the Andes and the western United States, particularly in *Townsendia* Hook. and *Erigeron* L., a number of which are threatened by mining and grazing.

The tribe Gnaphalieae (187 genera, 1,250 species) has a nearly cosmopolitan distribution, and has an especially rich diversity in Australia and the southern Africa–Madagascar region. It is notable for *Antennaria* Gaertn. (pussytoes), *Anaphalis* DC. (pearly-everlasting), and *Leontopodium* R. Br. ex Cass. (edelweiss). Most species have a restricted distribution; however, a few taxa such as *Gamochaeta* Wedd. turn up in sidewalks around the world. *Calocephalus* in Australia is restricted to one vulnerable area in Canberra. Two more recently described small-statured Andean genera, *Cuatrecasasiella* H. Rob. and *Jalcophila* Dillon & Sagást., may or may not be endangered.

Inuleae (including Plucheeae) seem to be a monophyletic group with 66 genera and over 700 species with a nearly cosmopolitan distribution but concentrated in Eurasia and the Neotropics. Cultivated members include *Inula helenium* L. (elecampane) and *Buphthalmum salidifolium* L. (yellow ox-eye). There are several weedy taxa, but most genera have only a few species nearly all of which have restricted distributions. Some are endemic to islands (Canary Islands: *Allagopappus* Cass.; Mauritius: *Monarrhenus* Cass.; Madagascar: *Neojeffreya* Cabrera), or Namibia (*Antiphiona* Merxm.), or Australia (*Allopterigeron* Dunlop). *Cylindrocline* Cass. has two very rare species on Mauritius, one of which is known from ten specimens at a single locality.

The tribe Senecioneae contains 138 genera and over 3,000 species in both the Eastern and Western Hemispheres (Nordenstam, personal communication). It is on record as the largest tribe, with *Senecio* L. alone containing over 1,000 species, particularly in Eurasia and America. The horticultural cineraria (actually *Pericallis* D. Don in Sweet), native to the Canary Islands, is now widely introduced; the South African true *Cineraria* L. is distinguished in the tribe by its obcompressed achenes. One narrowly endemic genus is *Dendrosenecio* (Hauman ex Hedberg) B. Nord., whose members are trees in the mountains of tropical east Africa. Other groups of particular interest are the Jamaican genus *Odontocline* B. Nord., which is still known mostly from 18th-century collections of Swartz, and *Hoehnephytum* Cabrera, a rare genus from Brazil. There are two remarkable trees on St. Helena, both very important for conservation efforts, *Lachanodes arborea* (Roxb.) B. Nord., which is critically endangered, twice thought to be extinct, and *Pladaroxylon leucadendron* (G. Forst.) Hook. f., also critically endangered with fewer than 50 plants (Cronk 2000). *Robinsonia* DC. has seven species endemic to the Juan Fernandez islands; two are extremely rare or possibly extinct, and the monotypic *Lordhowea* B. Nord. from Lord Howe Island is endemic and known only from a small population. The Mascarene Islands have two small genera as well, *Faujasiopsis* C. Jeffrey and *Parafaujasia* C. Jeffrey.

The tribe Anthemideae has over 100 genera and ca. 1,750 species that are found mostly in the Northern Hemisphere, especially in the Mediterranean region and Central Asia, with an additional center of diversity in South Africa. The chrysanthemums and daisies (*Chrysanthemum* L. with a conserved type and *Leucanthemum* Miller, respectively) are widely cultivated, as is *Tanacetum* L., which includes the tansies and pyrethrum. A few taxa, such as *Matricaria matricarioides* Porter ex Britton, *M. recutita* (L.) Rauschert, and the yarrow, *Achillea* L., are weeds. Members of the tribe are rich chemically, and are the source of the important insecticides named pyrethrins. *Artemisia* (sensu lato) is one of a handful of anemophilous genera in Asteraceae (Faegri and van der Pijl 1979), and its species are a major cause of allergies in humans (Lewis et al. 1983). All *Artemisia* (sensu lato) species produce aromatic oils, and several are culinary herbs or used as flavorings, hallucinogens, vermifuges, and pharmaceuticals (Lee and Geissman 1970; Heinrich et al. 1998; Burrows and Tyrl 2001). The habitats of the Mediterranean and South African Anthemideae are severely threatened.

The tribe Calenduleae is credited with 8 genera and ca. 110 species. Its native range is almost completely restricted to Africa. Today, genera such as *Calendula* L. and *Dimorphotheca* Moench are often planted in gardens or on maintained roadsides. The tribe is most notable for extreme differences in the form of the ray and disk achenes. A few of the species are local and potentially endangered.

The supertribe Helianthodae contains tribes that usually have phytomelanin in the walls of their achenes. There are 12 tribes in this supertribe; here 5 are discussed. The tribe Tageteae is an American group of ca. 16 genera and 250 species. *Pectis* L. is widespread, and one species of *Porophyllum* is weedy, but most species of this genus have restricted distributions and some are narrowly endemic. Some species of *Tagetes* L. are rare and local, and the Peruvian type species of *Schizotrichia* Benth. has been very rarely collected. Most notable in horticulture is the marigold (*Tagetes*), which is often planted among other garden plants that would benefit from its insect-repelling monoterpene chemistry.

The tribe Helenieae includes 900 species in 130 genera. It occurs prominently in western North America with some interesting disjunctions of some of the species in Chile and Argentina. The tribe contains the horticultural *Gaillardia* Foug., the weedy *Schkuhria* Roth, and *Helenium* L. (sneezeweed). Many of the North American species are rich in terpenoids, and they are problems in range management. The tarweeds (e.g., *Madia* Molina) of California and their relatives, including the 19 species of the Hawaiian silversword alliance (*Argyroxiphium* DC., *Wilkesia* A. Gray, and *Dubautia* Gaudich.) are especially important not only because of their distinctive natural habitats but because they are threatened and endangered. *Lasthenia* Cass. and *Perityle* Benth. contain many isolated species in California or Mexico that are subject to extinction.

The tribe Coreopsideae contains ca. 18 genera and over 300 species. It is basically American; some species of *Bidens* L. have become pantropical with the help of the retrorse barbs on the pappus awns. Horticultural genera include *Coreopsis* L., *Cosmos* Cav., and *Dahlia* Cav. One particularly interesting Pacific endemic is the tree-like *Fitchia* Hook. f. with liguliform corollas, and a potentially endangered monotypic genus is the St. Helena endemic *Petrobium* R. Br. known only from ca. 250 plants.

The tribe Heliantheae has nearly 150 genera and 2,400 species. The tribe rather characteristically has the combination of persistent pales on the receptacles of the heads and slowly maturing achenes (Robinson 1981). It is almost wholly American, with some pantropical elements such as *Melanthera* L. and *Wedelia* Jacq. Crop plants include the sunflower (*Helianthus* L.) from North America and *Guizotia* Cass. (niger seed) of Africa and India. Horticultural plants include *Tithonia* Desf. (Mexican sunflower), *Rudbeckia* L. (black-eyed Susan), and *Echinacea* Moench (coneflower). Among the noxious weeds are *Xanthium* L., the cocklebur, and *Acanthospermum* Kunth, both of which have fruits with hooked or straight spines, and ragweed (*Ambrosia* L.), the source of allergenic pollen. Some genera are rare endemics such as *Faxonia* Brandeg., collected once in southernmost Baja California, *Espeletia* Mutis ex Humb. & Bonpl., providing a special forested aspect to the páramo habitats of the northern Andes, and *Exomiocarpon* Lawalrée, endemic in Madagascar. In spite of the appearance of great numbers in some species, the habitat is at risk, and many more-narrowly ranging species could easily become extinct. Some genera and species of restricted distribution, such as *Scalesia* Arn. and *Trigonopterum laricifolium* (Hook. f.) W. L. Wagner & H. Rob. of the Galápagos, may be more vulnerable than others.

The tribe Eupatorieae consists of ca. 173 genera and 2,400 species. It is almost exclusively American, with main centers of diversity in Mexico, the Andes, and Brazil. Members include the well-known *Ageratum* L., *Liatris* Gaertn. ex Schreb. (blazing star), and *Eupatorium* L. (bonesets and joe-pye weeds). Their chemistry prevents them from being used as food, and a number of species (e.g., *Chromolaena odorata* [L.] R. M. King & H. Rob.) have become weeds when introduced to Hawaii, Asia, Africa, or Australia. Many of the more than 2,000 species have narrow distributions, and one monotypic forest genus from Costa Rica, *Standleyanthus* R. M. King & H. Rob., known from one stem, may already be extinct. There are a number of narrow endemics from eastern Brazil, such as *Agrianthus* Mart. ex DC. The tribe has many genera of pygmy plants known mostly from only one or two collections such as *Iltisia* S. F. Blake from Costa Rica and Panama, *Monogerion* G. M. Barroso & R. M. King and *Cavalcantia* R. M. King & H. Rob. from the south edge of the Amazon basin, *Piqueriopsis* R. M. King from southwest Mexico, *Ferreyrella* S. F. Blake from northern Peru, *Piqueriella* R. M. King & H. Rob. from northeast Brazil, *Siapaea* Pruski from the Guayana Highland area of southern Venezuela, *Ciceronia* Urban and *Antillia* R. M. King & H. Rob. from Cuba, *Parapiqueria* R. M. King & H. Rob. from Pará in Brazil, and *Teixeiranthus* R. M. King & H. Rob. from the São Francisco area of eastern Brazil. Some of these have not been re-collected for more

than a century, and they may be extinct. Others like them may remain to be discovered or become extinct without ever having been known to science.

OVERVIEW

Two characteristics of the family bear examining in the context of conservation:
chemistry and pollination. The Compositae is notable for having a diverse secondary metabolite chemistry based especially on the presence of unique structural
types, sesquiterpene lactones, and a variety of types of polyenes (= polyacetylenes)
(Mabry and Bohlmann 1977; Hegnauer 1977). This chemistry is considered a major
factor in the success of the family. The basal element in the family, the Barnadesieae,
lacks complex polyenes and sesquiterpene lactones, but has secondary metabolite
chemistry based mostly on triterpenes. Most of the widespread weeds are not in
these lower branches, so the chemistry may play an important role in the success of
the more common taxa. Concerning pollination, insect and wind are the most
common vectors for the Compositae. Bird pollination also occurs, however, being
most common in the Barnadesioideae and Mutisieae and sporadic in tribes such as
Cardueae, although there is bird pollination scattered in other tribes such as the
Heliantheae and Senecioneae (Leppik 1977). The Barnadesioideae, Mutisieae, and
Cardueae are the basal branches of the Compositae phylogeny (fig. 5.5), and one
can surmise that the ancestor of the extant Compositae was bird pollinated. In addition, one could say that most of the species in these basal groups are rare and at-
risk taxa, and so one might surmise that it was the switch from bird pollination to
insect pollination, along with the chemistry, that led to the diversification and success found in the higher branches of the phylogeny.

Endemicity in the family can also be examined from a habitat standpoint. For
instance, within the area of rich diversity mentioned in the tribe Liabeae is a smaller
and even more critical area of endemism. This recently noticed area is in the eastern foothills of the Andes in Morona-Santiago and Zamora-Chinchipe Ecuador
and in closely adjacent northern Amazonas Peru. The area remains moist and
mostly forested to this day, although areas not far to the west and south are now deforested and desertified. Rarities in the area include *Stenopadus andicola* Pruski
(Mutisieae), the only member of that genus outside of the Guiana Shield;
Munnozia luyensis H. Rob. (Liabeae); three recently described species of *Pipto-
carpha* (Vernonieae), including one that has very unusual heads; *Pentacalia tilletii*
H. Rob. & Cuatrec. (Senecioneae); and some rarely recollected older species such as
Ayapanopsis mathewsii (B. L. Rob.) R. M. King & H. Rob. (Eupatorieae). The most
inexplicable endemic is the recently described new genus *Holoschkuhria* H. Rob.,
whose closest relatives are all from drier areas of North America and Mexico or
from southern South America.

Around the globe various habitats contain rare and at-risk members of the
Compositae. From the tree senecios of east Africa (*Dendrosenecio* [Hauman ex
Hedberg] B. Nord., Senecioneae), to the *Espeletia* Mutis ex Humb. & Bonpl. of the
northern Andes (Heliantheae), the hummingbird-pollinated trees of the tepuis of

the Guiana Shield (e.g., *Stenopadus* S. F. Blake, Mutisieae) and the *planalto* of Brazil (*Wunderlichia* Riedel ex Benth., Mutisieae), the daisy trees of Mexico to northern South America (*Montanoa* Cerv., Heliantheae), and the silverswords of Hawaii (*Argyroxiphium* DC. and *Wilkesia* A. Gray, Helenieae), members of the Compositae are threatened, endangered, or extinct. Numerous habitats that are dominated by the Compositae are also under threat, such as the fields of spring flowers in Namaqualand (South Africa), Western Australia, California, and the Mediterranean. While it is true that some members of the family are well-known weeds, the majority of species in the family are restricted in their distributions, and many are headed toward extinction. Clearly, this family does not deserve its reputation as "weedy," and indeed it might be used as an indicator of importance for many areas of the world that are not heavily forested.

5.8 AFRICAN VIOLETS:
FAMILY GESNERIACEAE

Laurence E. Skog

THE GESNERIACEAE, commonly known as the African violet family, is a moderately sized family of herbs, subshrubs, shrubs, lianas, and epiphytes, and mainly tropical. The family has over 2,500 species in about 135 genera usually grouped by seedling morphology and geography into two subfamilies of nearly equal size, the Gesnerioideae from the Neotropics and the Cyrtandroideae from the Paleotropics, and a small subfamily of few species in southern South America and the southwestern Pacific, the Coronantheroideae. Closely related to the Scrophulariaceae and distinguished from that family by placentation characters, the Gesneriaceae are commonly found in humid habitats, often in wet or cloud forests at upper elevations, and are relatively uncommon in low-elevation rain forests and drier habitats. The plants, particularly those in the Neotropics, sometimes have modified storage stems, such as tubers (e.g., *Sinningia* spp.), or aerial and subterranean rhizomes (e.g., *Gloxinia* spp.), or stolons (e.g., *Episcia* spp.), all of which allow them to be propagated vegetatively, as well as from leaves and other plant parts. This ease of propagation, in addition to the attractiveness of the flowers, has made the plant family especially popular with amateur and commercial plant growers (e.g., the florists' gloxinia [*Sinningia speciosa* cultivars] and the African violet [*Saintpaulia* spp. and cultivars]). Other than in horticulture and for folk uses for food and medicine, the plants have little economic importance.

Among the approximately 135 genera there a few monotypic genera (e.g., *Lembocarpus, Titanotrichum* [plate 5.8]), as well as a few large genera with 100 species or more (e.g., *Chirita* in Southeast Asia, *Cyrtandra* on the Pacific Islands, *Besleria* and *Columnea* in the Neotropics). The flowers of Gesneriaceae usually have brightly colored corollas that attract various types of pollinators. The large epiphytic genus *Columnea*, with red corollas usually pollinated by hummingbirds in the Neotropics, has a Paleotropical counterpart in the large epiphyte genus *Aeschynanthus*, visited by sunbirds. Other flower visitors are also attracted to flower forms, such as euglossine bees, bats, and hawkmoths, each adapted to a particular flower form.

Few species are seemingly weedy as roadside species such as *Kohleria spicata*, *Nautilocalyx melittifolius*, or *Gloxinia sylvatica*, but most species in the Gesneriaceae are restricted to very specific habitats, that is, limestone cliffs or cloud forests, and usually under tree cover. When the tree cover is removed, the plants cannot survive long in full sun and low humidity.

In the Neotropics the center of distribution and high diversity for the Gesneriaceae subfamily Gesnerioideae is probably northwestern Colombia, where rainfall is extremely heavy on the western slopes of the Andean cordillera. The number of species falls off in all directions, with very few in temperate South America or the Amazon basin. None reach temperate North America north of the Tropic of Cancer, probably barred by the ecological desert of northern Mexico and the southwestern United States. This distribution contrasts with the center of distribution of the subfamily Cyrtandroideae in Southeast Asia, which extends into temperate north central China, as there is no ecological arid barrier. Other centers of distribution include eastern and southeastern Africa and the Caribbean.

Many species of Gesneriaceae are endemic to limited areas. Countries where the family is found have a few wide-ranging species and many narrow endemics. In the Neotropics, the widespread species include *Achimenes erecta, Besleria solanoides, Alloplectus ichthyoderma, Columnea sanguinea, Codonanthe crassifolia, Kohleria spicata, Gloxinia purpurascens, Sinningia incarnata,* and *Drymonia serrulata,* extending well beyond geographic borders. Narrow endemics are severely endangered in many areas, usually because of habitat destruction.

Because species of Gesneriaceae may be endemic to limited areas or restricted to specific habitats, and generally disappear when the habitats are severely disturbed, they may indicate the diversity of the flowering plants in a region. This is particularly impressive in western Ecuador just to the south of the center of distribution and diversity in northwestern Colombia. The plant biodiversity of western Ecuador is severely at risk, according to several authors (Myers 1988, 1990; Myers et al. 2000; Mittermeier et al. 1998; and others). For a relatively small area of ca. 80,000 km², western Ecuador has several unique, geographically small, and isolated forest types (Harling 1979), has a high percentage of plants known nowhere else, and has been severely deforested (Dodson and Gentry 1991; Sierra 1999; Valencia et al. 2000). There are few detailed studies of larger representative groups of plants or animals that may demonstrate the extent of extinction or endangerment apparently caused by habitat destruction in western Ecuador. Therefore, Gesneriaceae may provide a good example.

Ecuador has more than 210 species of Gesneriaceae (Skog and Kvist 1997) and 107 species (or more than half) are or were native to the elevations below 1,000 m in western Ecuador. Nearly half of the 107 species are epiphytic (and dependent on forest trees for their physical support), and 23 are endemic to western lowland Ecuador. If one looks closer at these 107 species and their former and current distributions, one sees that more than a third of the Gesneriaceae flora of western Ecuador is already extinct or endangered and that these species mainly occurred in low-elevation cloud forests and moist lowland forests, now mostly converted to agriculture (based on a study by L. P. Kvist, L. E. Skog, J. L. Clark, and R. W. Dunn, personal communication).

The status of the Gesneriaceae in western Ecuador was made plain in the recent revision of *Gasteranthus* (Skog and Kvist 2000), a genus probably indicative of the status of other Gesneriaceae as well as other flowering-plant families. The center of distribution of *Gasteranthus* is western Ecuador, where 10 of the 35 species grow in the low-elevation forests between 500 and 800 m, and another 10 species are restricted to the forests above 1,800 m. Because their distribution is limited to cloud forests, the species are vulnerable to extinction. The low-elevation cloud forests have been nearly completely destroyed, resulting in possible extinction of at least five endemic species in western Ecuador, and an additional seven species throughout the range of the genus. Due to the loss of habitat, the survival of more than ten other species of *Gasteranthus* is questionable. The recent discovery of a small low-elevation cloud forest in the Cordillera Mache-Chindul (Parker and Carr 1992) that has five species of *Gasteranthus* demonstrates the richness of the former forest cover, which was probably prevalent throughout the region and which exists now only in remnants.

Since World War II the Ecuadorian population has nearly quadrupled, with the subsequent increase in demand for farmland leading to the disappearance of more than 95% of the semideciduous and moist forests of the central and southern parts of the coastal forest (Dodson and Gentry 1991). However, the pluvial and wet forests near the Colombian border contain Gesneriaceae shared with the same forests in Colombia, and few species are endemic. Consequently, few species are endangered or extinct in this area.

Dodson and Gentry (1991) estimated that 6,300 species of flowering plants are native to western lowland Ecuador and that 20% of them are endemic to the region. The unpublished data by L. P. Kvist, L. E. Skog, J. L. Clark, and R. W. Dunn showed that similarly 20% of the Gesneriaceae in the same area are endemic, and that more than 90% (20 of 22 endemic species) are extinct or endangered. Thus, if Gesneriaceae are representative of the entire flora of flowering plants in the region, more than 1,000 endemic species may already be extinct or endangered. Also, the same study by Kvist et al. found that more than a third of all Gesneriaceae native to lowland western Ecuador may be extinct or endangered, corresponding to more than 2,000 species in the entire flora. The family may be representative of the status of the entire flora of the region, as it is well represented in the threatened cloud forest, as well as in the moist evergreen forest already largely destroyed.

A few remnants of the native forest types are in western Ecuador, and these should be identified and protected to maintain the diversity that is left. Because Gesneriaceae can be propagated easily, the endangered species that are already in cultivation might be reintroduced into protected areas. About half of the recorded species from lowland Ecuador are or were in cultivation, including three species that may be totally extinct in their native habitats (e.g., *Gasteranthus atratus*). Thus, there may be hope for preserving at least some of the lost or endangered vegetation of western Ecuador.

5.9 LITCHIS AND RAMBUTANS: FAMILY SAPINDACEAE

Pedro Acevedo-Rodríguez

THE SAPINDACEAE is a family of 140 genera and approximately 1,800 species of trees, shrubs, and lianas with mostly tropical or subtropical distribution, with a few genera extending into subtemperate zones. Their leaves are predominantly spirally arranged and variously compound, less often opposite, simple, or digitate. Compound leaves can be pinnate, bipinnate, tripinnate, trifoliolate, biternate, triternate, or a combination of these. A peculiar character of numerous Sapindaceae with pinnately compound leaves is the presence of a single terminal rudimentary leaflet or process, which distinguishes them from closely related families. Leaflets are predominantly entire, but some genera are variable, with entire, crenate, or serrate leaflets. Stipules are restricted to the climbing genera of the Paullinieae tribe. The base of petioles and the petiolules are often enlarged, with the adaxial portion very often depressed or furrowed. Indumentum in the family is quite variable, including simple, erect or curly, nonglandular trichomes and, less frequently, peltate scales or multicellular-glandular, papilliform, fasciculate, stellate trichomes.

The inflorescences in Sapindaceae are variously shaped thyrses with lateral cincinni, drepania, or dichasia, or less often thyrsoid-derived racemes. They are axillary, distal, cauliflorous, or supra-axillary, solitary or fasciculate. Inflorescences are predominantly unisexual or bisexual, but with a strong tendency for one sex to predominate. The perianth is often 5-merous, but there is considerable variation in the number of parts. Sepals are free to completely connate, and may be as few as three or as many as ten. They are usually of similar size and shape or less often dimorphic. The petals are distinct, with imbricate aestivation, inserted on the base of an extrastaminal nectary disk, commonly five, less often four or six. However, numerous genera or species completely lack a corolla. Petals are erect or reflexed with an adnate adaxial petaloid appendage, with extended inrolled basal margins, or altogether lacking appendages and inrolled margins. The presence of petaloid appendages is a diagnostic character for numerous members of the family. The extrastaminal disk is variously shaped, and in many genera, nectar is produced as a reward to pollinators. The number of stamens is often 8, but there is considerable variation (as few as 4 or as many as 74), with genera containing from 5 to 8 or from 8 to 10. The ovary is syncarpous and often three-carpellate. Numerous genera have two-carpellate ovaries in addition to the three-carpellate ones, however, and a few have unicarpellate gynoecia or as many as eight carpels. The septa are complete, with the ovary containing the same number of locules as carpels, except in *Melicoccus* and *Zollingeria*, where the septa are partially developed, resulting in a unilocular ovary. The ovules are one per carpel in subfamily Sapindoideae or two to seven or eight per carpel in subfamily Dodonaeoideae; placentation is axial, commonly in the middle or less often basal, or apical. The style is commonly terminal and simple, or sometimes two styles are present. The vast majority of Sapindaceae genera have trilocular fruits, with many of these also bearing bilocular, unilocular, or sometimes quadrilocular fruits. In addition, fruits of several genera are predominantly unilocular due to the abortion of carpels. The fruits are variable, including septifragal or loculicidal capsules, schizocarps with winged or nonwinged mericarps, a berry, or rarely a drupe; seeds are sessile, variously shaped, exalate or rarely winged, naked, with a partial to complete sarcotesta, or an arillode; the embryo usually has a radicle, which is often separated by a deep fold in the testa.

Sapindaceae are the source of numerous products, some of which are economically important either globally or locally, making an excellent case for its wise use and conservation. Among the most important products, fruits are high on the list. These include fruits such as litchi (*Litchi sinensis* Sonn.), longan (*Dimocarpus longan* Lour.), rambutan (*Nephelium lappaceum* L.), and pulasan (*N. ramboutan-ake* [Labill.] Leenh.) at the global level, and the mamoncillo or keneep (*Melicoccus bijugatus* Jacq.), *cutuplí o guaya* (*M. olivaeformis* Kunth), and pitomba (*Talisia esculenta* [Cambess.] Radlk.) at the local level. The arillodes of *Blighia sapida* Koeniguer are the source of the nutritious akee, widely consumed in Jamaica but highly toxic when eaten unripe (Rashford 2001). Numerous species of *Paullinia* L. have been reported to be useful in the preparation of medicines, caffeine-rich beverages, binding and weaving material, and arrow poisoning (Beck 1990). The seeds of *Paullinia cupana* Kunth are the source of the important Brazilian crop guarana, a source of caffeine and flavoring of soft drinks. Almost all Sapindaceae are used around the tropics for fish poisoning (Acevedo-Rodríguez 1990). The wood of some species of *Euphorianthus* Radlk., *Harpullia* Roxb., and *Schleichera* Willd. are used in the construction of houses. Numerous genera are grown as ornamentals, for example, *Acer* L., *Aesculus* L., *Arfeuillea* Radlk., *Allophylus* L., *Cardiospermum* L., *Filicium* Hook. f., *Harpullia, Koelreuteria* Laxm., *Sapindus* L., and *Xanthoceras* Bunge. Minor products include oils from *Pappea* Eckl. & Zeyh. and *Schleichera*, and arrow poison from *Paullinia pinnata* L.

As pointed out before, Sapindaceae are predominantly tropical or subtropical with areas of marked diversity. At the generic level, tropical Africa is the most diverse with 27 genera, followed by Australia with 24, Madagascar with 23, and the Americas with 22. At the species level the Americas contain the largest centers of diversity. For example, *Serjania, Paullinia,* and *Talisia* Aubl. are the most species-rich genera of Sapindaceae in northern-middle South America, *Serjania* with 230 species, *Paullinia* with about 200 species, and *Talisia* with 59 species (Acevedo-Rodríguez 2003). The Americas are followed by Australia with 59 endemic species of *Dodonaea* Mill. (West 1984).

Sapindaceae taxonomy still relies on the monographic studies of Radlkofer (1890, 1931–1934), with the exception of a few changes (Scholz 1964; Muller and Leenhouts 1976) and recent monographs (e.g., Acevedo-Rodríguez 1993a, 1993b, 2003; Adema 1991; van Welzen 1989). Conservation studies in this family are handicapped by the limited or fragmentary monographic studies, and the most we can hope for is to produce a list of species based on distribution and abundance records. We can adopt the working hypothesis that species known from one or few collections are at least rare or endangered if their known distributions are threatened by habitat destruction.

The current classification of Sapindaceae is at its best obsolete and does not reflect some of the knowledge that has been gained in the last decades. The current subfamily classification will probably stand the test of molecular classification, but the tribal classification with all probability will drastically change as new natural groups are recognized. Because of this, no attempt will be made to identify the diversity and conservation status of Sapindaceae at subfamily or tribal levels. Here I will assess the vulnerability at the generic level and at the species level only when the data are considered reliable.

There are 18 genera (13% of the family) with 20 or more species. These have widespread distributions, indicating that their vulnerability level is low. This group includes very species-rich genera such as *Allophylus* with 250 species, *Serjania* with 230 species, *Paullinia* with 200 species, and *Acer* with 126 species. Some of these genera contain taxa that are considered vulnerable by the IUCN Red List (Hilton-Taylor 2000), due to their restricted distributions. For example, *Allophylus* has 10 species that are listed as vulnerable or as insufficiently known; *Cupaniopsis* Radlk. has 15 species out of 60 that are considered vulnerable, threatened, or extinct; and *Chytranthus* Hook. f. has 1 species out of 20 that is considered vulnerable.

Seventy-nine genera (56% of the family) contain from 2 to 16 species, most of which have distributions that are widespread or moderate, but which also include a few genera restricted to a continent or a large island like Madagascar. Some of the genera include species that are listed by the IUCN Red List (Hilton-Taylor 2000) as vulnerable. Examples of this are *Atalaya* Blume, *Camptolepis* Radlk., *Dimocarpus* Lour., and *Diplokeleba* N. E. Br. In addition, the genus *Lophostigma* Radlk. (plate 5.9), known from a few collections from South America, should be added to the list (Acevedo-Rodríguez 1993a).

The remaining 43 genera (31% of the family) of Sapindaceae are monospecific; most have narrow distributions. Among these, some are considered vulnerable or insufficiently known while others are common. The following genera have their only species listed as vulnerable by the IUCN Red List (Hilton-Taylor 2000): *Athyana* Radlk., *Euchorium* Ekman & Radlk., *Eurycorymbus* Hand.-Mazz., *Gloeocarpus* Radlk., *Gongrospermum* Radlk., *Haplocoelopsis* Davis, and *Sinoradlkofera* F. G. Meyer. To this list, the genera *Bizonula* Pellegr., *Blighiopsis* van der Veken, *Chonopetalum* Radlk., *Handeliodendron* Rehder, *Pentascyphus* Radlk., *Pseudopancovia* Pellegr., *Scyphonychium* Radlk., and *Tsingya* Capuron should be added, as they are rare or known from a single collection.

A few monospecific genera are not considered vulnerable or threatened because they are common, widespread, or cultivated. These include *Amesiodendron* Hu, *Arfeuillea* Radlk., *Beguea* Capuron, *Blomia* Miranda, *Cubilia* Blume, *Euphorianthus* Radlk., *Hippobromus* Eckl. & Zeyh., *Hypelate* P. Browne, *Magonia* A. St.-Hil., *Schleichera* Willd., *Tristira* Radlk., and *Ungnadia* Endl. Some genera, although local or narrowly endemic, need further evaluation of their conservation status. These include *Bridgesia* Cambess., *Castanospora* F. V. Mueller, *Delavaya* Franch., *Dipteronia* Oliv., *Erythrophysopsis* Verdc., *Eurycorymbus* Hand.-Mazz., *Gloeocarpus* Radlk., *Gongrospermum* Radlk., *Hornea* Bak., *Loxodiscus* Hook. f., *Otonephelium* Radlk., *Pavieasia* Pierre, *Phyllotrichum* Lecomte, *Sisyrolepis* Radlk., *Smelophyllum* Radlk., *Stocksia* Benth., and *Tripterodendron* Radlk.

Elucidation of the conservation status of species requires a solid taxonomic knowledge that can be attained only through monographic studies. Species concepts depend on the knowledge level that we have of the organisms under study. The greater the knowledge, the better chance our hypotheses will stand the test of time. Anybody who has been involved in the monographic studies of a group knows very well the difficulties of this multidisciplinary endeavor. Expeditions are increasingly expensive, and the logistics for exploring remote areas are complicated. In the end, it will not be possible to gather all the necessary pieces of the puzzle, since habitat destruction is usually faster than current exploration efforts. As a result, monographs are written based on data that are less than ideal. It is not uncommon for monographs to contain species known only from one or a handful of collections. These species are thus interpreted as extremely rare or in some cases as extinct.

LITERATURE CITED

Acevedo-Rodríguez, P. 1990. The occurrence of piscicides and stupefactants in the plant kingdom. Advances in Economic Botany 8:1–23.

———. 1993a. A revision of *Lophostigma* (Sapindaceae). Systematic Botany 18:379–388.

———. 1993b. Systematics of *Serjania* (Sapindaceae). Part 1, A revision of *Serjania* sect. *Platycoccus*. Memoirs of the New York Botanical Garden 67:1–93.

———. 2003. Melicocceae (Sapindaceae): *Talisia* and *Melicoccus*. Flora Neotropica Monograph 87:1–179.

Adema, F. 1991. *Cupaniopsis* Radlk. (Sapindaceae): a monograph. Leiden Botanical Series 15:1–190.

Alongi, D. M. 1998. Coastal Ecosystem Processes. CRC Marine Science Series. CRC Press, Boca Raton, Florida.

Anderson, D. M., Cembella, A. D., and Hallegraeff, G. M. 1996. Physiological Ecology of Harmful Blooms. NATO ASI Series G, Ecological Sciences, vol. 41. Springer-Verlag, Berlin.

Ando, H. 1972. Distribution and speciation in the genus *Hypnum* in the circum-Pacific region. Journal of the Hattori Botanical Laboratory 35:68–98.

Arnell, S. 1962. *Gymnomitrion laceratum* in Peru. Bryologist 65:261.

Axelrod, D. I. 1985. Rise of the grassland biome, central North America. Botanical Review 51: 163–201.

Baldwin, B. G., Wessa, B. L., and Panero, J. L. 2002. Nuclear rDNA evidence for major lineages of helenioid Heliantheae (Compositae). Systematic Botany 27:161–198.

Barker, R. M. 1986. A taxonomic revision of Australian Acanthaceae. Journal of the Adelaide Botanic Gardens 9:1–286.

Beck, H. T. 1990. A survey of the useful species of *Paullinia* L. (Sapindaceae). Advances in Economic Botany 8:41–56.

Bentham, G. 1881. Notes on Gramineae. Botanical Journal of the Linnean Society 19:14–134.

Bentham, G., and Hooker, J. D. 1873. Genera Plantarum. Vol. 2. Lovell Reeve and Co., London.

Bowen, B. 1999. Preserving genes, species, or ecosystems? Healing the fractured foundations of conservation policy. Molecular Ecology 8 (12 supplement 1): S5–S10.

Bramwell, D. 2002. How many plant species are there? Plant Talk 28:32–34.

Bremekamp, C. E. B. 1965. Delimitation and subdivision of the Acanthaceae. Bulletin of the Botanical Survey of India 7:21–30.

Bremer, K. 1994. Asteraceae: Cladistics and Classification. Timber Press, Portland, Oregon.

Bremer, K., Jansen, R. K., Karis, P. O., Kallersjo, M., Keeley, S. C., Kim, K.-J., Michaels, H. J., Palmer, J. D., and Wallace, R. S. 1992. A review of the phylogeny and classification of the Asteraceae. Nordic Journal of Botany 12:141–146.

Brenan, J. P. M. 1952. Notes on African Commelinaceae. Kew Bulletin 7:179–208.

Brown, R. 1814. Gramineae. Pp. 580–583 in Flinders, M., A Voyage to Terra Australis. London.

Burrows, G. E., and Tyrl, R. J. 2001. Toxic Plants of North America. Iowa State University Press, Ames.

Cederwall, H., and Elmer, R. 1990. Biological effects of eutrophication in the Baltic Sea, particularly in the coastal zone. Ambio 19:109–112.

Cronk, Q. C. B. 2000. The Endemic Flora of St. Helena. Anthony Nelson, Oswestry, Shropshire, United Kingdom.

Cronquist, A. 1981. An Integrated System of Classification of Flowering Plants. Columbia University Press, New York.

Daniel, T. F. 1993. Mexican Acanthaceae: diversity and distribution. Pp. 541–558 in Ramamoorthy, T. P., Bye, R., Lot, A., and Fa, J. E., eds., Biological Diversity of Mexico: Origins and Distribution. Oxford University Press, New York.

Dettki, H., Klintberg, P., and Esseen, P. A. 2000. Are epiphytic lichens in young forests limited by local dispersal? Ecoscience 7:317–325.

Dobson, A. P., Rodriguez, J. P., Roberts, W. M., and Wilcox, D. S. 1997. Geographic distribution of endangered species in the United States. Science 275:550–553.

Dodson, C. H., and Gentry, A. H. 1991. Biological extinction in western Ecuador. Annals of the Missouri Botanical Garden 78:273–295.

Ellstrand, N., and Elam, D. R. 1993. Population genetic consequences of small population size: implications for plant conservation. Annual Review of Ecology and Systematics 24: 217–242.

Faden, R. B. 1974. Commelinaceae. Pp. 653–667 in Agnew, A. D. Q., Upland Kenya Wild Flowers. Oxford University Press, London.

———. 1983. Isolating mechanisms among five sympatric species of *Aneilema* R. Br. (Commelinaceae) in Kenya. Bothalia 14:997–1002.

———. 1994a. Commelinaceae. Pp. 303–309 in Agnew, A. D. Q., and Agnew, S., eds., Upland Kenya Wild Flowers, 2nd edition. East Africa Natural History Society, Nairobi.

———. 1994b. New species of *Commelina* (Commelinaceae) from the flora of tropical East Africa. Novon 4:224–235.

———. 1998. Commelinaceae. Pp. 109–128 in Kubitzki, K., ed., The Families and Genera of Vascular Plants, vol. 4, Flowering Plants: Monocotyledons: Alismatanae and Commelinanae (except Gramineae). Springer-Verlag, Berlin.

Faden, R. B., and Inman, K. E. 1996. Leaf anatomy of the African genera of Commelinaceae: *Anthericopsis* and *Murdannia*. Pp. 464–471 in van der Maesen, L. J. G., van der Burgt, X. M., and van Medenbach de Rooy, J. M., eds., The Biodiversity of African Plants, Proceedings of the 14th AETFAT Congress, 22–27 August 1994, Wageningen, the Netherlands. Kluwer Academic, Dordrecht.

Faden, R. B., and Suda, Y. 1980. Cytotaxonomy of Commelinaceae: chromosome numbers of some African and Asiatic species. Botanical Journal of the Linnean Society 81:301–325.

Faegri, K., and van der Pijl, L. 1979. The Principles of Pollination Ecology. 3rd edition. Pergamon Press, Oxford.

Faust, M. A. 1990. Morphologic details of six benthic species of *Prorocentrum* (Pyrrophyta) from a mangrove island, Twin Cays, Belize, including two new species. Journal of Phycology 26:548–558.

———. 1993. *Prorocentrum belizeanum, Prorocentrum elegans,* and *Prorocentrum caribbaeum*: three new benthic species (Dinophyceae) from mangrove island, Twin Cays, Belize. Journal of Phycology 29:100–109.

———. 1996. Dinoflagellates in mangrove ecosystem, Twin Cays, Belize. Nova Hedwigia 112: 447–460.

Faust, M. A., and Gulledge, R. A. 1995. Population structure of phytoplankton and zooplankton associated with floating detritus in a mangrove island, Twin Cays, Belize. Journal of Experimental Marine Biology and Ecology 197:159–175.

Fenchel, T. 1969. The ecology of marine microbenthos. Part 4, Structure and function of the benthic ecosystem, its chemical and physical factors, and the microfauna communities with special reference to the ciliate protozoa. Ophelia 6:1–182.

Fensome, R. A., Riding, J. B., and Taylor, F. J. R. 1996. Dinoflagellates. Pp. 107–169 in Jansonius, J., and McGregor, D. C., eds., Palynology: Principles and Applications, vol. 1. American Association of Stratigraphic Palynologists Foundation, Salt Lake City.

Funk, V. A., Bayer, R., Keeley, S., Chan, R., Watson, L., Gemeinholzer, B., Schilling, E.,

Panero, J., Baldwin, B., Garcia-Jacas, N. T., Susanna, A., and Jansen, R. K. In press. Everywhere but Antarctica: using a supertree to understand the diversity and distribution of the Compositae. In Friis, I., and Balslev, H., eds., Plant Diversity and Complexity Patterns: Local, Regional, and Global Dimensions. Royal Danish Academy of Sciences and Letters, Copenhagen.

Gauslaa, Y. 1997. Population structure of the epiphytic lichen *Usnea longissima* in a boreal *Picea abies* canopy. Lichenologist 29:455–469.

Geiser, L., and Reynolds, R. 2002. Using lichens as indicators of air quality on federal lands. Workshop report. U.S. Department of Agriculture, Pacific Northwest Region R6-NR-AG-TP-01–02.

Govaerts, R. 2001. How many species of seed plants are there? Taxon 50:1085–1090.

GPWG (Grass Phylogeny Working Group). 2001. Phylogeny and subfamilial classification of the grasses (Poaceae). Annals of the Missouri Botanical Garden 88:373–457.

Hansen, H. V. 1992. Studies in the Calyceraceae with a discussion of its relationship to Compositae. Nordic Journal of Botany 12:63–75.

Harling, G. 1979. The vegetation types of Ecuador: a brief survey. Pp. 165–174 in Larsen, K., and Holm-Nielsen, L. B., eds., Tropical Botany. Academic Press, London.

Hawkes, C. V., and Menges, E. S. 1996. The relationship between open space and fire for species in a xeric Florida shrubland. Bulletin of the Torrey Botanical Club 123:81–92.

Hedenas, H., and Ericson, L. 2000. Epiphytic macrolichens as conservation indicators: successional sequence in *Populus tremula* stands. Biological Conservation 93:43–53.

Hegnauer, R. 1977. The chemistry of the Compositae. Pp. 283–355 in Heywood, V. H., Harborne, J. B., and Turner, B. L., eds., The Biology and Chemistry of the Compositae. Academic Press, London.

Heinken, T. 1999. Dispersal patterns of terricolous lichens by thallus fragments. Lichenologist 31:603–612.

Heinrich, M., Robles, M., West, J. E., Ortiz de Montellano, B. R., and Rodriguez, E. 1998. Ethnopharmacology of Mexican Asteraceae (Compositae). Annual Review of Pharmacology and Toxicology 38:539–565.

Heywood, V. H., Harborne, J. B., and Turner, B. L. 1977. The Biology and Chemistry of the Compositae. Academic Press, London.

Hillebrand, H., and Sommer, U. 2000. Diversity of benthic microalgae in response to colonization time and eutrophication. Aquatic Botany 67:221–236.

Hilton-Taylor, C., comp. 2000. 2000 IUCN Red List of Threatened Species. IUCN, Gland, Switzerland, and Cambridge.

Hind, D. J. N., ed. 1996. Compositae. Vols. 1–2. Royal Botanic Gardens, Kew, London.

Hitchcock, A. S. 1935. Manual of Grasses of the United States. U.S. Department of Agriculture Miscellaneous Publication 200. U.S. Government Printing Office, Washington, DC.

Hitchcock, A. S., and Chase, A. 1951. Manual of Grasses of the United States. 2nd edition. U.S. Department of Agriculture Miscellaneous Publication 200. U.S. Government Printing Office, Washingon, DC.

James, P. W. 1973. The effect of air pollutants other than hydrogen fluoride and sulphur dioxide on lichens. Pp. 143–175 in Ferry, B. W., Baddeley, M. S., and Hawksworth, D. L., eds., Air Pollution and Lichens. University of Toronto Press, Toronto.

Jansen, R. K., and Palmer, J. D. 1987. A chloroplast DNA inversion marks an ancient evolutionary split in the sunflower family (Asteraceae). Proceedings of the National Academy of Sciences USA 84:5818–5822.

Jex-Blake, M. 1948. Some Wild Flowers of Kenya. Highway Press, Nairobi.

Keeley, S. C., and Jansen, R. K. 1991. Evidence from chloroplast DNA for the recognition of a new tribe, the Tarchonantheae, and the tribal placement of *Pluchea* (Asteraceae). Systematic Botany 16:173–181.

Kubitzki, K., ed. In press. The Compositae. In The Families and Genera of Vascular Plants. Springer-Verlag, Berlin.

Lebrun, J.-P., Audru, J., Gaston, A., and Mosnier, M. 1972. Catalogue des Plantes Vasculaires du Tchad Méridional. Institut d'Élevage Médecine Vétérinaire des Pays Tropicaux, Fort-Lamy, Chad.

Lee, K., and Geissman, T. 1970. Sesquiterpene lactones of *Artemisia*: constituents of *Artemisia ludoviciana* ssp. *mexicana*. Phytochemistry 9:403–408.

Leppik, E. E. 1977. The evolution of capitulum types of the Compositae in the light of insect-flower interaction. Pp. 61–89 in Heywood, V. H., Harborne, J. B., and Turner, B. L., eds., The Biology and Chemistry of the Compositae. Academic Press, London.

Lewis, W. H., Vinay, P., and Zenger, V. E. 1983. Airborne and Allergenic Pollen of North America. Johns Hopkins University Press, Baltimore.

Lindau, G. 1895. Acanthaceae. Pp. 274–354 in Engler, H. G. A., and Prantl, K. A. E., eds., Die Natürlichen Pflanzenfamilien IV (3b). Wilhelm Engelmann, Leipzig.

Long, R. L. 1970. The genera of Acanthaceae in the southeastern United States. Journal of the Arnold Arboretum 51:257–309.

Mabry, T. J., and Bohlmann, F. 1977. Summary of the chemistry of the Compositae. Pp. 1097–1104 in Heywood, V. H., Harborne, J. B., and Turner, B. L., eds., The Biology and Chemistry of the Compositae. Academic Press, London.

Menges, E. S., and Hawkes, C. V. 1998. Interactive effects of fire and microhabitat on plants of Florida scrub. Ecological Applications 8:935–946.

Mittermeier, R. A., Myers, N., Thomsen, J. B., da Fonseca, G. A. B., and Olivieri, S. 1998. Biodiversity hotspots and major tropical wilderness areas: approaches to setting conservation priorities. Conservation Biology 12:516–520.

Morton, S. L., and Bomber, J. W. 1994. Maximizing okadaic acid content from *Prorocentrum hoffmannianum* Faust. Journal of Applied Phycology 6:41–44.

Morton, S. L., Moeller, P. D. R., Young, K. A., and Lanoue, B. 1998. Okadaic acid production from the marine dinoflagellate *Prorocentrum belizeanum* isolated from the Belizean coral reef ecosystems. Toxicon 36:201–206.

Muller, J., and Leenhouts, P. W. 1976. A general survey of pollen types in Sapindaceae in relation to taxonomy. Pp. 407–445 in Ferguson, I. K., and Muller, J., eds., The Evolutionary Significance of the Exine, Linnean Society Symposium Series 1. Academic Press, London.

Myers, N. 1988. Threatened biotas: "hot spots" in tropical forests. Environmentalist 8:187–208.

———. 1990. The biodiversity challenge: expanded hot-spots analysis. Environmentalist 10: 243–256.

Myers, N., Mittermeier, R. A., Mittermeier, C. G., da Fonseca, G. A. B., and Kent, J. 2000. Biodiversity hotspots for conservation priorities. Nature 403:843–845.

Nimis, P. L., and Purvis, O. W. 2002. Monitoring lichens as indicators of air pollution: an introduction. In Nimis, P. L., Scheidegger, C., and Wolseley, P. A., eds., Monitoring with Lichens—Monitoring Lichens. Kluwer Academic, Dordrecht, Netherlands.

Panero, J. L., and Funk, V. A. 2002. Toward a phylogenetic subfamilial classification for the Compositae (Asteraceae). Proceedings of the Biological Society of Washington 115:909–922.

Parker, T. A. III, and Carr, J. L., eds. 1992. Status of forest remnants in the Cordillera de la Costa and adjacent areas of southwestern Ecuador. RAP Working Papers 2. Conservation International, Washington, DC.

Peroni, P. A., and Abrahamson, W. G. 1985. A rapid method for determining losses of native vegetation. Natural Areas Journal 5:20–24.

Peterson, P. M. 2003. Poaceae (Gramineae). Encyclopedia of Life Sciences. Macmillan Publishers, Nature Publishing Group. http://www.els.net.

Primack, R. B., ed. 2000. A Primer of Conservation Biology. 2nd edition. Sinauer Associates, Sunderland, Massachussetts.

Radlkofer, L. 1890. Ueber die gliederung der familie der Sapindaceen. Sitzungsberichte der mathematisch-physikalischen Classe der K. b. Akademie der Wissenschaften zu München 20:105–379.

———. 1931–1934. Sapindaceae. Pp. 1–1539 in Engler, A., ed., Das Pflanzenreich IV, 165 (Heft 98a–h). Wilhelm Engelmann, Leipzig.

Rashford, J. 2001. Those that do not smile will kill me: the ethnobotany of the ackee in Jamaica. Economic Botany 55:190–211.

Robinson, H. 1964. New taxa and new records of bryophytes from Mexico and Central America. Bryologist 67:446–458.

———. 1981. A revision of the tribal and subtribal limits of the Heliantheae (Asteraceae). Smithsonian Contributions to Botany 51:1–102.

———. 1994. Notes on the tribes Eremothamneae, Gundelieae, and Moquinieae, with comparisons of their pollen. Taxon 43:33–44.

Robinson, H., Allen, B., and Magill, R. E. 2001. The synonymy of the African moss genus *Quathlamba* with the American genus *Neosharpiella*. Evansia 18:133–136.

Rosso, A. L., McCune, B., and Rambo, T. R. 2000. Ecology and conservation of a rare old growth associated lichen in a silvacultural landscape. Bryologist 103:117–127.

Ruetzler, K., and Feller, C. 1996. Caribbean mangrove swamps. Scientific American 274:94–99.

Ruetzler, K., and Macintyre, I. G., eds. 1982. The Atlantic Barrier Reef Ecosystem at Carrie Bow Cay, Belize. Part 1, Structure and Communities. Smithsonian Contributions to the Marine Sciences 12. Smithsonian Institution Press, Washington, DC.

Scholz, H. 1964. Sapindaceae. Pp. 282–284 in Melchior, H. A., ed., Engler's Syllabus der Pflanzenfamilien, vol. 2. Gebrüder Borntraeger, Berlin.

Scotland, R. W., and Wortley, A. H. 2003. How many species of seed plants are there? Taxon 52:101–104.

Sharp, A. J., Crum, H., and Eckel, P. M., eds. 1994. The Moss Flora of Mexico. Memoirs of the New York Botanical Garden, vol. 69. New York Botanical Garden, New York.

Sierra, R., ed. 1999. Propuesta Preliminary de un Sistema de Clasificación de Vegetation para el Ecuador Continental. Proyecto INEFAN/GEF-BIRF y EcoCiencia, Quito.

Sillett, S. C., and Goslin, M. N. 1999. Distribution of epiphytic macrolichens in relation to remnant trees in a multiple-age Douglas-fir forest. Canadian Journal of Forest Research 29:1204–1215.

Sillett, S. C., McCune, B., Pick, J. E., Ranbo, T. R., and A. Ruchty. 2000. Dispersal limitations of epiphytic lichens result in species dependent on old-growth forests. Ecological Applications 10:789–799.

Simpson, B. B., and Conner-Orgorzaly, M. 1986. Economic Botany: Plants in Our World. McGraw-Hill, New York.

Skog, L. E., and Kvist, L. P. 1997. The Gesneriaceae of Ecuador. Pp. 12–23 in Valencia, R., and Balslev, H., eds., Estudios Sobre Diversidad y Ecologia de Plantas: Memorias del II Congresso Ecuatoriano de Botánica. Pontificia Universidad Católica del Ecuador, Quito.

———. 2000. Revision of Gasteranthus (Gesneriaceae). Systematic Botany Monographs 59: 1–118.

Smayda, T. J., and Shimizu, Y. 1993. Toxic Phytoplankton Blooms in the Sea. Elsevier Scientific, Amsterdam.

Soreng, R. J., Davidse, G., Peterson, P. M., Zuloaga, F. O., Judziewicz, E. J., and Filgueiras, T. S. 2000. Catalogue of New World Grasses: Suprageneric Classification. Missouri Botanical Garden. http://mobot.mobot.org/W3T/Search/nwgc.html.

Sournia, A. Chrétiennot-Dinet, M. J., and Ricard, M. 1991. Marine phytoplankton: how many species in the world ocean? Journal of Plankton Research 13:1093–1099.

Steidinger, K. A. and Tangen, K. 1996. Dinoflagellates. Pp. 387–598 in Thomas, C., ed., Identifying Marine Diatoms and Dinoflagellates. Academic Press, San Diego.

Tindall, D. R., and Morton, S. L. 1998. Community dynamics and physiology of epiphytic/benthic dinoflagellates associated with ciguatera. Pp. 293–313 in Anderson, D. M., Cembella, A. D., and Hallegraeff, G. M., eds., Physiological Ecology of Harmful Algal Bloom. Springer-Verlag, Berlin.

U.S. Fish and Wildlife Service. 1999. South Florida multi-species recovery plan. U.S. Fish and Wildlife Service, Atlanta.

Valencia, R., Pitman, N., León-Yánez, S., and Jørgensen, P. M., eds. 2000. Libro Rojo de Las Plantas Endémicas del Ecuador 2000. Herbario QCA, Pontificia Universidad Católica del Ecuador, Quito.

van Welzen, P. C. 1989. Guioa Cav. (Sapindaceae): taxonomy, phylogeny, and historical biogeography. Leiden Botanical Series 12:1–314.

Walter, K. S., and Gillett, H. J. 1997. 1997 IUCN Red List of Threatened Plants. IUCN, Gland, Switzerland, and Cambridge.

Watson, L., and Dallwitz, M. J. 1992. The Grass Genera of the World. CAB International, Cambridge.

West, J. G. 1984. A revision of Dodonaea Miller (Sapindaceae) in Australia. Brunonia 7:1–194.

Will-Wolf, S., Hawksworth, D. L., McCune, B., Sipman, H. J. M., and Rosentreter, R. 2003. Assessing the biodiversity of lichenized fungi. Pp. 340–380 in Mueller, G. M., Bills, G., and Foster, M. S., eds., Biodiversity of Fungi: Inventory and Monitoring Methods. Academic Press, Oxford.

Yahr, R. 1997. Recolonization and reintroduction of Cladonia perforata Evans, an endangered lichen at Eglin Air Force Base, Florida. Final report to Florida Division of Forestry, Tallahassee.

————. 2000. Ecology and post-fire recovery of *Cladonia perforata*, an endangered Florida-scrub lichen. Forest, Snow, and Landscape Research 75:339–356.

————. 2001. In the wake of Hurricane Opal: experimental restoration of the endangered lichen *Cladonia perforata* at Eglin Air Force Base. Final report to the U.S. Fish and Wildlife Service, Panama City, Florida.

Zoller, S., Frey, B., and Scheidegger, C. 2000. Juvenile development and diaspore survival in the threatened epiphytic lichen species *Sticta fuliginosa, Leptogium saturninum,* and *Menegazzia terebrata*: conclusions for in situ conservation. Plant Biology 2:496–503.

Zoller, S., Lutzoni, F., and Scheidegger, C. 1999. Genetic variation within and among populations of the threatened lichen *Lobaria pulmonaria* in Switzerland and implications for its conservation. Molecular Ecology 8:2049–2059.

PART III

CONTEMPORARY CAUSES OF PLANT EXTINCTION

CHAPTER 6

HABITAT FRAGMENTATION AND DEGRADATION

All habitats on the earth, whether they are terrestrial or aquatic, montane or lowland, desert or monsoonal, temperate or tropical, are constantly subjected to alteration and change. Often these changes are slight and have little long-term effect on the biota; in some cases the effects are severe and result in population declines and even extinction, depending upon the magnitude of the alteration. These forces of change can be the result of natural or human-induced processes or in some cases both acting in parallel. Over the last millennium the increase in plant extinction has been overwhelmingly the result of severe alteration of natural habitats by human activities. These activities include forest clearing, logging, agricultural expansion, cattle ranching, suburban sprawl, dam construction, commercial fishing, and the introduction of invasive species. This chapter explores five critical habitats as examples in both terrestrial and marine environments that have undergone extensive modification and describes the effects that habitat degradation has had and is having on the ecology and natural history of plant species.

6.1 FOREST FRAGMENTS AND TROPICAL PLANT REPRODUCTION IN AMAZONIAN BRAZIL

Emilio M. Bruna and W. John Kress

DEFORESTATION RESULTS in a mosaic landscape composed of forest fragments embedded in habitat with varying intensities of human use (Laurance and Bierregaard 1997; Harrison and Bruna 1999). It has been hypothesized that pollination and sexual reproduction in plants may be particularly prone to disruption in these fragmented landscapes, in part because the abundance of both plants and their pollinators can be reduced in forest fragments (Rathcke and Jules 1993). The results of studies testing this hypothesis have been idiosyncratic, however, with some plant species showing augmented pollination and fruit production in fragments while others have suffered the predicted declines (Aizen and Feinsinger 1994; Dick 2002). Furthermore, the long-term consequences of these changes in pollination remain unclear, since few studies have attempted to translate the observed changes in reproduction to changes in population genetic structure and population dynamics (Schemske et al. 1994). This lack of long-term inference is somewhat disconcerting, given the suggestion that fragmentation-related declines in reproduction could result in reduced genetic diversity and the eventual extinction of plants from forest fragments (Bond 1995).

One study that has addressed these issues and will be discussed in detail here focused on the consequences of forest fragmentation for the Amazonian understory herb *Heliconia acuminata* (Heliconiaceae; Bruna 1999, 2001, 2003; Bruna and Kress 2002). These studies were conducted at the Biological Dynamics of Forest Fragments Project (BDFFP), a collaborative research site administered by the Smithsonian Tropical Research Institute and Brazil's National Institute for Amazonian Research (Bierregaard et al. 2002). Located outside the city of Manaus, the BDFFP is home to experimentally isolated forest fragments of different sizes as well as large expanses of continuous forest, all of which contain populations of *H. acuminata*. In 1997 a series of thirteen 5,000-m² permanent plots in the BDFFP's fragments and continuous forest reserves were established to monitor the abundance, growth, survivorship, and reproduction of over 5,000 *H. acuminata* individuals (fig. 6.1, plate 6.1). Observational data have been integrated with the results of experimental manipulations and genetic analyses into simple mathematical models, with the goal of predicting how changes in ecological interactions, such as pollination and sexual reproduction, influence the long-term demography of tropical plants.

Heliconia acuminata has proven to be an excellent model system with which to investigate the consequences of fragmentation on tropical plant-pollinator interactions. Like most tropical plants this species is self-incompatible (Bawa 1990; E. M. Bruna and W. J. Kress, unpublished data), requiring the transfer of pollen by animal vectors in order to produce seeds. The suite of pollinators visiting *H. acuminata*'s flowers is specialized and limited, also a defining feature of tropical plants (Bawa 1990). In the case of *H. acuminata* the specialized pollinators are hermit hummingbirds, one of the most important and diverse groups of floral visitors in the Neotropics (Feinsinger 1987). Finally, the fruits and seeds of *Heliconia* are dispersed by birds (Kress 1985), which have frequently been shown to have reduced densities in fragmented landscapes (Bierregaard and Stouffer 1997; Sieving and Karr 1997). Since understory herbs with similar natural history are a diverse and abundant component of all tropical forests (Gentry and Emmons 1987), quantifying the effects of fragmentation on *H. acuminata* has helped to predict the consequences of landscape alterations for many other plants in these highly threatened habitats.

REPRODUCTIVE BIOLOGY OF *HELICONIA ACUMINATA*

The genus *Heliconia* is composed of over 250 species (Berry and Kress 1991), which are often divided into two groups based on the foraging strategy of the hummingbirds that pollinate them (Linhart 1973; Stiles 1975). Forest understory *Heliconia*, such as *H. acuminata*, are usually found at low density, produce few flowers and

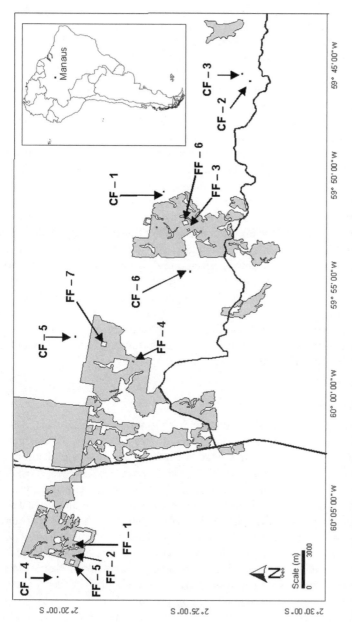

Figure 6.1 Map of the Biological Dynamics of Forest Fragments Project, with arrows indicating the location of the 13 demographic plots in which *Helico-nia acuminata* is being studied (after Bruna and Kress 2002). White areas are continuous forest, stippled areas represent pasture and clearings surround-ing fragments, and dark lines represent roads. CF, continuous forest plots; FF, forest fragment plots.

little nectar, and are visited by hermit hummingbirds that "trapline" from one flowering plant to the next rather than establishing territories to defend a group of plants (Kress 1985). In the BDFFP study sites the long-tailed hermit (*Phaethornis superciliosus*) and the straight-billed hermit (*P. bourcieri*) are *H. acuminata*'s primary pollinators (Stouffer and Bierregaard 1995; Bruna 2001). Observations of focal plants for six-hour time periods in 1997 and 1998 indicated that these hummingbirds visit *H. acuminata* flowers 0–1.8 times per hour (mean = 0.45 ± 0.49 SD visits/hour, $n = 30$ plants observed), with 26% of open flowers receiving no visits during the observation periods.

H. acuminata in the BDFFP reserves begins flowering in late January and continues through April. The probability that a plant will flower increases dramatically with plant size (Bruna and Kress 2002), and the number of reproductive plants in continuous forest demographic plots can vary considerably. For example, in 1999 the number ranged from 4 to 74 flowering plants/plot, with a mean number of 26 flowering plants/plot (Bruna and Kress 2002). These reproductive plants can have from 1 to 6 inflorescences ($n = 64$ plants), although most have only 1 (81%) or 2 (16%). Each of the 20–25 flowers produced by a plant (mean = 22.63 ± 1.36 SE, range 4–62) is open to pollinators for only one day, after which the style and perianth drop off. Only about 35% of flowers develop into fruits (Bruna and Kress 2002), probably because of low pollinator visitation rates and high levels of floral predation by the hispine beetle *Cephaloleia nigriceps* (E. M. Bruna, personal observation). The fruits produce a maximum of three seeds each (mean number of seeds/fruit = 1.9 ± 0.03 SE, $n = 873$ fruits). Although these are dispersed at the start of the dry season, they remain dormant for seven to eight months until the onset of the following rainy season in January (Bruna 1999).

REPRODUCTION IN FOREST FRAGMENTS

The patterns of *H. acuminata* reproduction in continuous forest have served as the baseline with which to compare reproduction in forest fragments. Three principal patterns emerged from this study. First, the number of flowering plants in the demographic plots in small fragments (1 ha) can be extremely low, with most plots containing only three to four flowering plants (Bruna and Kress 2002; fig. 6.2). However, this is not because plants in fragments are less likely to flower. Instead it is because populations in these fragments are smaller than those in continuous forest, and hence they contain fewer of the large plants capable of flowering. Second, the density of flowering plants in medium-sized fragments (10 ha) and continuous forest is much more variable than in small fragments. While some plots in these locations have an extremely low density of flowering plants, others have densities that are extremely high. Finally, the proportion of flowers developing into fruits is virtually identical in all locations. This observation indicates that hummingbird visits to plants in fragments are sufficiently common to ensure levels of pollination comparable to that in continuous forest.

What makes *H. acuminata* pollination insensitive to forest fragmentation? Re-

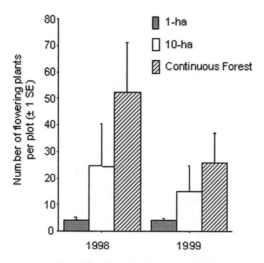

Figure 6.2 The mean number of flowering *Heliconia acuminata* in demographic plots. Plots are located in 1-ha fragments, 10-ha fragments, and continuous forest. Each plot is 5,000 m².

sults from a 12-year study documenting the abundance of *Phaethornis superciliosus* and *P. bourcieri* in the BDFFP reserves suggest that these birds move easily through the secondary growth surrounding the fragments (Stouffer and Bierregaard 1995, 1996). In addition the more common of the two species (*P. superciliosus*) actually increases in abundance in fragments during the rainy season (Stouffer and Bierregaard 1996), which is when *H. acuminata* is flowering. Although hummingbirds are rarely species-specific (Feinsinger 1987), other plant species in the understory of these fragments rarely flower during the rainy season (Gentry and Emmons 1987; E. M. Bruna and W. J. Kress, personal observation). *Heliconia acuminata* is therefore the primary source of nectar for hummingbirds moving through fragments, and flowering plants in fragments appear to receive frequent visits.

DEMOGRAPHIC CONSEQUENCES OF REDUCED POLLINATION

One of the most surprising conclusions of these studies was that even the most extreme reductions in pollination in forest fragments did not lead to the decline of *H. acuminata* populations (Bruna 2003). This lack of population decline is because the "demographic value" of seeds and seedlings is extremely low—few actually survive to reproduce and replace themselves (Horvitz and Schemske 1995). However, large *H. acuminata,* which are most likely to flower and fruit (Bruna and Kress 2002) and have extremely high survivorship (Bruna 2003), are demographically vital to populations in fragments. In computer simulations of *H. acuminata* population growth, the probability of extinction in fragmented areas increases dramatically with decreasing adult survivorship (E. M. Bruna, unpublished data). This pattern, which is frequently observed in studies of long-lived organisms (Caswell 2001), suggests plants in tropical forests may not be at immediate risk of extinction due to pollinator failure, as many previous studies have suggested (Rathcke and Jules 1993). Instead the fragmentation-related mortality of mature individuals, such as that observed in large tropical trees (Laurance et al. 2000), may be the primary factor leading to population declines.

GENETIC CONSEQUENCES OF FRAGMENTATION

Although pollination does not appear to have been *disrupted* in forest fragments, the frequent movements of hummingbirds through fragments and reduced germination of seeds in disturbed areas may have substantial consequences for gene flow and population genetic structure. Much is known about the genetics of tropical forest trees (e.g., Chase et al. 1995; Hamrick and Murawski 1991), and the few studies of tropical herbs (Roesel et al. 1996) suggest that levels of gene diversity can be high in both the upper and lower levels of tropical forests. Unfortunately, few data currently exist on the effects of fragmentation on tropical plant species in any stratum of the forest (Aldrich and Hamrick 1998; Hamilton 1999). Future studies in both continuous forest and fragments at BDFFP may provide some answers on the genetic future of plants in these fragmented habitats.

CONCLUSION

The study described here on *Heliconia*, hummingbirds, and forest fragmentation highlights the importance of a comprehensive demographic approach that simultaneously investigates several aspects of a species' natural history to evaluate the consequences of human-driven habitat modifications. While it appears the pollination of tropical plants visited by hummingbirds is resistant to disruption in fragmented habitats, other aspects of reproduction such as seedling recruitment may be substantially altered (Bruna 1999). This alteration could influence population genetic structure and, as a result, the potential to respond to future environmental changes. Finally, the results underscore the importance of small to medium-sized fragments of tropical rain forest for plant conservation efforts, because they can serve as an ecological resource for foraging hummingbirds, a source of propagules for plant populations, and important reservoirs of genetic diversity.

6.2 HABITAT ALTERATION IN THE CARIBBEAN: NATURAL AND HUMAN-INDUCED

W. John Kress and Carol Horvitz

ISLAND ECOSYSTEMS have long been recognized as unique biological laboratories by evolutionary biologists. Characterized by different degrees of isolation from species-rich continental floras and faunas, by equable maritime climates, and by reduced biological diversity, islands often contain unique biotas that experience different selective pressures from those that dominate mainland ecosystems. For example, the increased occurrence of dioecism and interspecific hybridization in plants is a well-documented island trend. Islands have also served as the focal regions of influential studies in systematics and biogeography (e.g., Wagner and Funk 1995).

The Caribbean island system consists of the Greater and Lesser Antilles and the Bahamas. Southern Florida, bounded on water by three sides and frost to the north, in a sense constitutes one of the Caribbean islands and is included here as well. This island system has played an important role in our understanding of island ecology and evolution. Classic examples of species–island area relationships (MacArthur and Wilson 1967) and adaptive radiations (Losos et al. 1998) come from this system. With its complex geologic history and proximity to North, Central, and northern South America, the Caribbean islands with a unique flora and fauna contain a fascinating mélange of interacting species. Added to this mix are two kinds of storms—physical storms in the form of hurricanes that regularly originate or pass through the Caribbean and the "storm" of human population growth and development. The former storms, along with other types of natural disturbance such as volcanic activity, have played an important role in the evolution of the biota. The latter "storm" now threatens the continued existence and ecological integrity of both terrestrial and marine Caribbean ecosystems. As summarized in Raffaele et al. 1998, both research and conservation have languished in many parts of the Caribbean to the point that biological crises and loss of biological diversity are nearly inevitable.

NATURAL PROCESSES INFLUENCING CARIBBEAN BIODIVERSITY

A recent study of global biodiversity demonstrated that the Caribbean region is one of 25 biodiversity hot spots worldwide (Myers et al. 2000). About half of the native species of plants (~13,000 species) in the Caribbean are single-island endemics. Terrestrial ecosystems include lowland and montane tropical forest, evergreen thicket, savanna, cactus and thorn scrub, mangrove, and riverine communities. Marine ecosystems in the region include the extraordinary barrier coral reefs of Belize and Andros Island, which are the second and third largest reefs on earth and smaller only than the Great Barrier Reef of Australia (Campbell 1978).

The geologic history of the Caribbean Plate has had a marked influence on the diversity of both marine and terrestrial ecosystems. A crucial geologic event was the Miocene collision of the North American and South American plates, resulting in the formation of the Isthmus of Panama (Collins et al. 1996). This event had profound biotic and oceanographic consequences, splitting the tropical American ocean in two parts and initiating the Great American Biotic Interchange between North and South America (Webb 1997). Volcanic activity, especially in the Lesser Antilles, has also had a profound effect on the evolution and ecology of the biota in the Caribbean (Howard 1977). Over 30 active or potentially active volcanoes are present in the Lesser Antilles, and some have recently erupted in Guadeloupe, Martinique, St. Vincent, and Montserrat. These geologic features are causally linked with modern environments as well as biotic diversity and distribution.

The complex geologic history of the Caribbean coupled with habitat diversity has resulted in the evolution of a remarkable flora and fauna, both terrestrial and marine. Investigations of some core organisms in this region have resulted in the formulation of widely cited theoretical models of evolutionary divergence and distribution, such as Rosen's model of vicariance biogeography (1976) and Ricklefs's concept of taxon cycles (Ricklefs and Cox 1972; Ricklefs and Bermingham 1999). Yet we have in-depth information on the evolutionary relationships and biogeographic patterns of only a handful of taxa from this region. Knowledge about the patterns of evolutionary divergence and distribution of plants is lacking. The extensive studies that have been carried out in Hawaii on biotic evolution and diversification in a "hot spot archipelago" (for summary, see Wagner and Funk 1995) are examples of the type of investigation that needs to be focused in the Bahamian and Antillean regions of the Caribbean.

RESPONSE OF ISLAND ECOSYSTEMS TO NATURAL PHYSICAL DISTURBANCE

The ecosystems of the Caribbean have developed under a high-frequency disturbance regime by virtue of being in the direct path of the Atlantic hurricane track (plate 6.2). Many of the plant species of these ecosystems are adapted to frequent canopy removal, soil disturbance, and landslides (Boucher 1990). These adaptations include rapid resprouting, recruitment of seedlings into clearings, sapling release from shade, or simply resistance to high winds (Walker et al. 1991).

Following hurricane landfall, large amounts of nutrients are transferred from the canopy to the litter layer. In some cases the canopy is simply regenerated by resprouting of the previously existing trees (Koptur et al. 1995). In these cases, the high soil nutrients sustain a rapid regeneration of the canopy driven by high light levels. In other ecosystems such as some mangrove forests, however, all overstory trees may be killed, and recruitment consists largely of other species (Smith et al. 1994). How long it might take and what the process is for the return, if ever, of an overstory of the original species is not at all clear. Despite the dominance of the disturbance regime by hurricanes, we know only how a handful of these ecosystems and individual species respond to hurricane damage. The large increases in soil nutrients, disturbed soil, unoccupied space, and light following hurricane passage make these ecosystems particularly susceptible to invasive exotic species (Horvitz et al. 1998; see below). For example, in pine forests of Florida and the Bahamas, massive secondary die-offs have occurred from insect outbreaks following hurricane passage. After Hurricane Andrew, mortality in some forest fragments was near 100% (Oberbauer et al. 1997).

Fire plays an important role in structuring some Caribbean ecosystems (Wade and Hofstetter 1980). Forest fragmentation and clearing have altered and often reduced natural fire regimes in some areas, while human-caused fires have increased in some areas under naturally low-frequency fire regimes (Snyder et al. 1990). Fire intensity interacts with hurricane disturbance; time since the last hurricane influences the fuel load available for combustion. Forest removal and fire have also been important in the loss of soil from many of these ecosystems, increasing their susceptibility to landslides on the steep topography found on some of the islands.

Global warming presents a looming threat to the Caribbean with largely unknown consequences. Some climate models predict potential increases in hurricane frequency and intensity with global warming. The increases in coral diseases noted in the Caribbean have been linked to increases in ocean temperature (Richardson 1998). Another unknown and potentially serious threat to low-lying Caribbean islands is sea-level rise due to climate warming; significant changes in the vegetation on the Florida Keys resulting from sea-level rise have already been detected (Ross et al. 1994). Such threats may result in the reduction of upland pine forests and expansion of mangrove zones.

ECOLOGICAL AND EVOLUTIONARY IMPACTS OF NONNATIVE ORGANISMS

Subtropical islands, particularly those that have suffered from habitat alteration and have served as ports of introduction, may be especially prone to invasion by nonnative organisms (Simberloff 1997). The rate and distance of movement by humans of very large numbers of species over a global spatial scale have been accelerating since the age of exploration (1500s), culminating in a globally recognized crisis only during the last 20 years (Drake et al. 1989). For example, over 25,000 species and cultivars of plants have been introduced into Florida. Of this number nearly 1,000 have become naturalized (Gordon 1998), and about 100 are currently considered invasive or harmful to native systems (Florida Exotic Pest Plant Council 2001).

Nonnative species are likely to have a large effect in disturbance-prone ecosystems (Hobbs and Huenneke 1992). Fires, floods, and hurricanes characterize Caribbean ecosystems, but only recently are we beginning to gather data on how nonnatives interact with native systems in the context of natural disturbances. For example, a study of posthurricane regeneration of tropical hardwood forests found that each native source of regeneration (e.g., seed banks, seedling banks, resprouting, etc.) is threatened by a particular suite of nonnatives, and a review of studies of other subtropical forests indicates a similar diversity of roles of nonnatives (Horvitz et al. 1998). A field experiment in the posthurricane forests of Florida showed that invasive vines from tropical Asia, Africa, and South America in diverse plant families (e.g., Rubiaceae, Oleaceae, and Dioscoreaceae) behave differently from native vines and alter the spatio-temporal distribution of recruitment sites (Horvitz and Koop 2001), shrouding the forests and rapidly creating very dense shade. The Caribbean island system, diverse in biotic, abiotic, and historical contexts but sharing the same climate zone, offers unprecedented opportunities to examine the invasiveness of nonnative species and their consequences.

The impact of newly assembled biotic interactions on native, long-evolved plant-animal interactions and the extent to which altered interactions influence the success of invasions is relatively unknown. For example, both native resident birds (e.g., gray catbirds) and native migrant birds (e.g., robins) are very attracted to abundant fleshy fruits of nonnative shrubs (Brazilian pepper, *Schinus terebinthifolius*, from Brazil and shoebutton ardisia, *Ardisia elliptica*, from Southeast Asia) in southern Florida (C. C. Horvitz and T. Fleming, unpublished data). Whether this newly assembled mutualism has altered population size or foraging traits of birds is unknown.

Native shrubs may also be negatively impacted by competition with nonnative shrubs for dispersal and pollination services of these birds. The African tulip tree (*Spathodea campanulata*) is a serious invasive in Puerto Rico, but it is not as successful in Florida (C. C. Horvitz, personal observation). In the species' natural habitats in Africa the large, orange, open-bell flowers are likely to be pollinated by hornbills. Although the pollination ecology of the tulip tree in its introduced regions is unstudied, the existence of nectarivorous passerine birds in Puerto Rico and their absence in Florida may be the cause of the differential invasiveness in these two regions.

CONSERVATION OF THE CARIBBEAN BIOTA

As stated by Bacon (1985), "There are no truly natural environments left in the Caribbean, only those which have survived 200–300 years of human impact." Like many tropical islands under European influence since the 1500s, the Caribbean islands have been ruthlessly exploited for timber, firewood, agricultural produce, and minerals, resulting in a conservation crisis of major proportions. The islands, with their spectacular land- and seascapes as well as diverse flora and fauna, face the tragic loss of a significant portion of their biological diversity (Davis et al. 1997).

Unfortunately, both research and conservation efforts have languished in many parts of the Caribbean, so that biological degradation and the loss of biodiversity are now almost inevitable (Raffaele et al. 1998).

Topography, climate, edaphic features, and historical events combine to determine the unique characteristics of the biological communities of individual islands in the Caribbean. The high rate of endemism and species diversity in the Caribbean is under clear threat of disruption and extinction due to the introduction of non-native invasive species coupled with severe habitat alteration due to human activities and natural disturbance. Determining which biological and physical characteristics have resulted in the local disruption of natural habitats for particular islands will permit extrapolation of data to other areas throughout the Caribbean.

In general, expectations are low for permanent or even moderate habitat protection in these islands. The principal reason for the lack of action is the understandable, but regrettable, reluctance of local governments to make long-term conservation commitments. Early efforts for the conservation of natural fauna and vegetation on a formal basis began largely with individuals and small groups with an amateur leaning toward natural history. Effective nongovernmental organizations with concern for the preservation of native plants and animals are still few in number. Historically, conservation has focused on establishing forest reserves for the rain forest habitats, but this focus has had the consequence of diverting attention from other vegetation types that equally deserve conservation merit and protection. It is necessary to take a broad view of the Caribbean region as a whole to determine areas that have rare or limited physiographical characteristics.

In conclusion, a plethora of biological investigations, including analysis of genetic variation among populations of endangered plants, and studies in pollination and dispersal biology, as well as conservation projects at the landscape and habitat levels are needed to effectively understand and conserve the unique biota of the Caribbean islands.

6.3 HABITAT LOSS: THE EXTREME CASE OF MADAGASCAR

Dieter C. Wasshausen and Werner Rauh

MADAGASCAR, THE FOURTH-LARGEST island in the world, is located 400 km west of Mozambique and the African continent. The island, called by the French La Grande Îsle, is approximately 580,000 km²; 1,500 km separate Tanjon' i Bobaomby (Cap d'Ambre) and Tanjon' i Vohimena (Cap Sainte-Marie), the northernmost and southernmost points respectively on the island, and almost 600 km separate the east and west coasts at the broadest part of the island. The Tropic of Capricorn passes through the south in the vicinity of Toliara (Tuléar).

In spite of its close proximity to Africa—the island is separated from it only by the 400-km-wide Mozambique Channel—Madagascar has (in the biological sense) relatively few affinities to Africa; no big animals such as elephants, giraffes, or lions inhabit the island, nor do poisonous snakes. It is estimated that Madagascar has been separated from mainland Africa for about 270 million years, and this isolation has allowed the island to develop its own flora and fauna, resulting in an unusually high percentage of endemic plants and animals. Presently botanists recognize 10,000 to 12,000 species of vascular plants occurring there, of which about 80% are endemic. Within this number are many unusual life-forms and relictual taxa, giving Madagascar the distinction of being called a "museum of living fossils." The fauna of Madagascar is also rich in endemic groups such as the lemurs, chameleons, and turtles (e.g., *Geochelone radiata*). The number of plant species that are presently known represent only a fraction of the species that once covered the wide plains of the island, for nearly 80% of the native vegetation has been destroyed within the last 1,000 years. Many plants and animals have disappeared as a result of the recent activity by humans, leading to the vast destruction of the island's biomes. The very large flightless Malagasy elephant bird, *Aepyornis maximus*, for example, has been extinct for several centuries.

According to Perrier de la Bâthie (1921) and Humbert (1955), Madagascar can be divided into two distinct phytogeographic regions, the eastern region (la région du vent) and the western region (la région sous le vent). The eastern region includes the entire eastern part of the island, which is characterized by high annual rainfall and is covered with a dense rain forest extending from the coastal plain up to 800 m elevation. This rain forest exhibits the typical stratification of most tropical rain forests: namely, an upper stratum composed of trees 25 to 30 m tall mixed with lianas; a middle stratum composed of palms, tree ferns, and the endemic Strelitzi-aceae, *Ravenala madagascariensis*; and a lower stratum rich in ferns, dwarf palms, shrubs, and herbaceous plants such as those found in the families Rubiaceae, Acanthaceae, Gesneriaceae, and Balsaminaceae. Epiphytes such as ferns and orchids make up a distinct stratum. The orchid family is especially rich in species (ca. 1,000), many of which are endemic to the island. One of the most remarkable species of orchids is the star of Madagascar, *Angraecum sesquipedale*, which has a flower with a 30-cm-long spur. This orchid flower is pollinated by a moth, *Xanthopan morgani predicta*, which has evolved with a proboscis more than 30 cm long.

In many parts of the island, especially along the coastal plain and at the lower elevations of the mountain range, the evergreen rain forest has been destroyed by the shifting cultivation practice of the indigenous people. Farmers of the eastern region employ this practice, known locally as *tavy* cultivation. They cut down the forest and clear the land by burning the dried wood. The soil is temporarily enriched with nutrients through this process, and the farmers can cultivate corn, dry rice, manioc (cassava), or *batatas* (sweet potatoes). The land can usually produce only one or, at most, two crops; afterward a new tavy must be started in a different location. Rains seriously erode the vacated land, and the soils become thoroughly leached of nutrients. This leads to the formation of a degraded secondary vegetation known as *savoka*. Savoka vegetation is primarily made up of the traveler's tree (*Ravenala madagascariensis*), a bamboo (*Ochlandra capitata*), and an arborescent Clusiaceae (*Harungana madagascariensis*). Meanwhile, the last remaining patches of primary forest in the highlands are being invaded and destroyed by introduced species such as *Solanum auriculatum, Rubus rosifolius,* and *R. mollucanus,* the latter a rampant weed.

The first inhabitants of Madagascar, the Paleo-Indonesians, arrived by boats probably about AD 500. Further waves of neo-Indonesian immigrants followed, probably in the 9th–13th centuries. The immigrants introduced zebu cattle to the island. These animals still play a significant role in the rituals of the Malagasy people and are one of the major factors in the destruction of the native vegetation of the Central Plateau.

Madagascar today has over 10 million zebu cattle, 2 million goats and sheep, and about 10 million inhabitants. In order to maintain these vast herds of zebus, the grasslands that cover the Central Plateau are burned over each year just before the start of the rainy season. The cattle cannot feed on the old, hard grass shoots. When the rains come, the burned grass tufts rapidly develop fresh, new, tender leaves, excellent fodder for the cattle. When these large areas of natural vegetation, both montane and xerophytic forests, are set on fire, very few species of plants and insects in these habitats tend to survive. Those plants that do are some grass species and a few species of palms—*Hyphaene schatan, Bismarckia nobilis, Borassus madagascariensis,* and *Chrysalidocarpus decipiens*—that are resistant to fire. These species have become the typical secondary vegetation in the western region of the Central Plateau (plate 6.3).

In the dry, semiarid regions of the southern and southwestern portions of the island, the Didiereaceae-*Euphorbia* scrub habitat is dominant. This vegetation type is most impressive, for it is rich not only in interesting endemic plants but also in rare, endemic animal species. One of the most unusual plant groups to be found here is the Didiereaceae. It is restricted in its distribution to the semidesert region of Madagascar with an annual rainfall of 30–300 mm and a dry season of up to 11 months. The Didiereaceae range is a narrow coastal strip about 50 km wide from Morondava in the north to west of Taolanaro (Fort-Dauphin) in the south. This xerophytic spiny scrub habitat cannot be set on fire. Fire cannot spread in the sparse leaf litter, and no grasses are present. Here goats tend to eat the green leaves and shoots of the Didiereaceae in spite of the sharp spines.

This same region was once home to the legendary "rok bird" or "elephant bird," *Aepyornis maximus,* which stood 3–4 m tall and whose eggs were the largest known in the world, 60 cm long, up to 30 cm in diameter, and with a 7–8-liter capacity. Occasionally, these eggs are found intact in the coastal sand dunes after being washed out by the heavy cyclonic rains in the region. This bird became extinct about 500 years ago.

This semiarid region is also the home of the well-known Madagascar star-turtle, *Geochelone radiata*. Until recently, the turtle was *fady*, or taboo, for the Mahafaly and Antandroy tribes and therefore not eaten. Now the animals are gathered up by the truckload and are shipped to Asian markets where they are sold as a culinary delicacy. Also in this southwestern thorn-shrub region resides the sifaka lemur, *Propithecus verrauxii*. This clown among the 26 species of lemur is not taboo with the local tribes and is being hunted for food. Presently these animals are found only in the protected nature reserves.

In order to maintain the large herds of zebu cattle in this arid region, the cactus *Opuntia*, especially the nearly spineless variety, *O. ficus-indica* var. *anacantha*, was introduced. Here the pads are burned before being fed to the cattle. There is another introduced species, *O. monacantha*, with long, golden-yellow spines. This species is often planted as an impenetrable hedge around fields and houses. *Opuntia* is rather easily propagated, and each fallen pad can root and grow into a new plant. In some parts of the southern scrub region *Opuntia* growth has invaded the natural vegetation and has begun to crowd out and displace the native species in a relatively short period of time.

As can be seen from the aforementioned examples, human activities have drastically altered the vegetation in most parts of Madagascar. The French botanist H. Humbert (1927, 7) wrote, "From the day when man entered Madagascar, the forests which once covered the entire island were gradually replaced by prairie, savanna and *savoka*, types of different vegetation which today cover nine-tenths of the surface of the island." These same sentiments were also voiced by numerous other French biologists, such as Perrier de la Bâthie, Paulian, Millot, Morat, Bosser, and others (Rauh 1995), who had lived and worked on the island for many years. These French and German naturalists felt that humans were morally obligated to protect and to conserve with all humanly possible efforts the last remains of this "sanctuaire de la nature."

These sentiments were integral to the establishment, in December 1927, of the first ten Réserves Naturelle Intégrales de Madagascar (RNI). Currently the island has 2 national parks, 12 integral nature reserves (RNI), and 21 special reserves (réserves spéciales) covering more than 1 million hectares, or 1.8% of the island's surface. At the 2003 IUCN Fifth World Congress on Protected Areas (WCPA, also known as the World Parks Congress), Madagascar's president, Marc Ravalomanana, announced his government's commitment to more than triple the size of its network of areas under protection, to 6 million hectares over the next five years. Under the plan, the government will expand its terrestrial coverage from 1.5 million hectares to 5 million hectares and wetland and marine wetlands from 200,000 hectares to 1 million hectares. More than two-thirds of the country's remaining forest will be placed under formal protection.

The nature reserves protect specific ecosystems and habitats, and each reserve protects a different ecosystem, landscape, flora, and fauna. For example, the Tsingy de Namoroko is located on the vast Jurassic limestone plateau south of Soalala in Mahajanga Province. It protects a deciduous forest whose understory is covered with succulent and xerophytic species including *Pachypodium ambongense, P. rutenbergianum, Euphorbia viguieri,* and *Uncarina sakalava.* Similarly, the Réserve d'Andohahela, which is northwest of Taolanaro, is unique because it exhibits the vegetation of the southeasternmost rain forest of the Madrare Valley as well as the Didiereaceae-*Euphorbia* scrub and transition woodlands of the southwest. Over a distance of only a few kilometers, one finds three contrasting vegetation types: "primary and secondary evergreen rain forests, forests adapted to semi-arid conditions with *Euphorbia* and *Alluaudia* species, and a transition zone with *Neodypsis decaryi,* the triangular palm, endemic to the region" (Andriamampianina 1984, 222). Although access to these nature reserves is possible only with permission from the Direction des Eaux et Forêts, animals such as zebus, sheep, and mohair goats (in the south) can graze in these reserves without any restrictions.

Although much work has recently been done to protect the natural environment of Madagascar by establishing natural parks, nature reserves, and special reserves and by enacting conservation laws, the reserves presently are not sufficiently protected against ongoing destruction. There are no guards present in the reserves to restrict human encroachment, and zebu cattle and goats graze at will within the reserves.

6.4 DEGRADATION OF ALGAE IN CORAL REEFS

Walter H. Adey

CORAL REEFS, the hallmarks of tropical seas, are as much accepted as ecosystems at the pinnacle of marine biodiversity as are rain forests in terrestrial environments. Small et al. (1998) estimate that the pantropical biodiversity of shallow-water coral reefs exceeds 3.5 million species, which is three and a half times greater than the previous estimate (Reaka-Kudla 1996). The algal portion of reef total biodiversity is over 700,000 species (Small et al. 1998). Yet the most recent global tally of known algal species (for all environments and regions) is less than 50,000 species (John 1994; Norton et al. 1996). While Norton et al. (1996) indicate that much higher levels are likely, their summary for tropical marine environments suggests that less than 5,000 species are known today. The implication of that analysis is that reef environments are relatively depauperate of algal species (as compared to temperate regions), but the explanations offered by those authors for a "biodiversity dip" of algae in the tropics are ecologically very unconvincing.

Coral reefs are defined here as shallow-water (less than 30 m) carbonate frameworks that have been constructed by living organisms during the Recent (last

10,000 years). Framework construction has been achieved primarily by stony corals and coralline red algae (Corallinales, Rhodophyta), with several additional invertebrates playing a less significant role. Coral-reef communities typically show extraordinary levels of primary productivity. As discussed by Adey (1998), in reviewing the collated data of Crossland et al. (1991), a mean pantropical gross primary productivity (GPP) of 7 ± 0.6 SD grams carbon/m^2/day is widely accepted. Unlike tropical rain forests, reef systems have dense animal populations, their algal production rapidly recycles, and large biomass buildups of algae do not occur. Thus, net primary productivity (NPP), at 0.7 gC/m^2/day, is low compared to the highest regional levels for tropical rain forests of about 4 gC/m^2/day. It has been demonstrated, however, that the principal productive element of typical coral reefs, algal turfs, have a community NPP that is close to that of rain forests (Adey and Goertemiller 1987), and the algal turf community provides most of the surface cover of pantropical coral reefs (Adey 1998). Higher plants, such as seagrasses, play a very limited role in providing these extraordinary levels of photosynthesis. Species of the many phyla or divisions of algae that occur abundantly in reefs are responsible for high levels of primary productivity.

Odum and Odum (1955) undertook the first, intensive study of both reef metabolism and the internal distribution of biomass in coral reefs. This demonstration, that algae dominated the Eniwetok reef flat (Marshall Islands), has withstood the test of time. Their classic pyramid of 85% algae, 13.8% herbivores, and 1.1% carnivores was generally accepted for decades. The tacit assumption has tended to be that since larger algae are not conspicuous on coral reefs (unlike in kelp beds or rocky intertidal communities), the primary productivity of reefs was primarily due to the zooxanthellae (dinoflagellates) that reside as symbionts within stony coral tissue. Odum and Odum (1955) did find that zooxanthellae dominated over macroalgae (6% of standing crop as macroalgae; 20% as symbiotic zooxanthellae). However, 74% of the total algal standing crop was in the form of algal turfs, borers, and crusts.

Considerable debate over the biomass and productivity of algal turfs has ensued over the past several decades (e.g., Larkum 1983; Carpenter 1985; Carpenter et al. 1991; Klumpp and McKinnon 1992; Johnson et al. 1995). As reviewed by Adey (1998), however, the extensive algal mats or turfs of most healthy reefs (i.e., reefs that show considerable continuing calcification and continued building) are the principal components of primary productivity (fig. 6.3). In addition McConnaughey et al. (2000) and Small and Adey (2001) demonstrated that carbon dioxide removal by both symbiotic zooxanthellae and free-living algae was responsible, in large measure, for the very high rates of calcification achieved by the major group of reef builders, the acroporids.

It has been known for over a decade that reef degradation is often accompanied by a phase shift from stony corals, coralline algae, and algal turfs to macroalgae (mostly relatively unpalatable macroalgae; Szmant 2001). Some studies have shown a direct relationship between nutrient increase and coral and coralline-algae reduction (Littler and Littler 1999), and this is backed up in microcosm studies (e.g.,

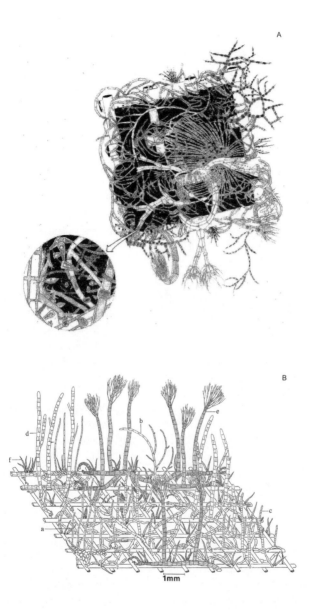

Figure 6.3 Typical algal turf cultured on plastic screens in a microcosm coral-reef environment. A, The plastic strands of the screen are 330 μm in diameter. Algae present on this screen are the diatoms *Licmorphora* sp. and *Navicula* sp.; the cyanobacteria *Anacystis dimidiata, Calothrix crustacea,* and *Oscillatoria submembranacea*; the greens *Cladophora fascicularis, Smithsoniella earleae,* and *Derbesia vaucheriaeformis*; the browns *Ectocarpus rhodochortonoides* and *Sphacelaria tribuloides*; and the reds *Asterocytis ramose, Ceramium corniculatum, Polysiphonia havanensis,* and *Herposiphonia secunda* (adapted from Adey and Hackney 1989). B, Similar screen from a different coral-reef microcosm, with the "diatom cloud" omitted for clarity. The algae present here are species of the greens *Pilinia* (a) and *Cladophora* (b), the browns *Giffordia* (c) and *Sphacelaria* (d), the red *Herposiphonia* (e), and the cyanobacteria *Calothrix* (f).

Adey and Loveland 1998). It has not been possible, however, to demonstrate a direct and consistent correlation between nutrient increase and macroalgal increase (Encore Group 2001). Algal turfs are not nutrient limited (Adey and Goertemiller 1987; Adey 1998; Encore Group 2001). However, nutrient increase directly reduces stony-coral reproduction and calcification rates (Adey et al. 2000) and changes algal turf composition to a more cyanobacteria-dominated and less productive algal turf community. Excessive fishing pressure disrupts coral-reef food webs (Roberts 1995) and can lead to both direct and indirect alteration of algal community structure. High seawater temperatures, related in part to global warming, can cause coral bleaching due to loss of zooxanthellae, and bleaching often leads to excessive coral death (Hoegh-Guldberg 1999). When stony-coral die-back rate consistently exceeds new growth (by calcification), and the new carbonate surface is not dominated by a heavily grazed and very biodiverse algal turf, then reef degradation is assured.

Significant tropical storms can break up and reduce coral framework, disrupting community structure and killing many organisms. This would almost certainly lead to local nutrient increases, as dead organisms decompose, as well as the reduction of critical mid- to higher-trophic-level predators; thus, the strong association of severe tropical storms with reef degradation and sometimes phase shifts has a rational base (Hughes 1994). Reefs that are affected by humans through extensive fishing, sediment or nutrient release, and direct surface destruction can be driven to shift phases directly or brought to the point where a relatively mild tropical storm will produce the phase shift.

In summary, healthy, actively building coral reefs represent a precariously stable state of high productivity and calcification produced by algal turfs and stony corals working in synergy in an extremely low-nutrient, heavily grazed environment (Adey and Steneck 1985; Small and Adey 2001). Human or natural factors that significantly reduce grazing, increase nutrients, or reduce physical structure will tend to shift such a reef from its stable state to a macroalgae-dominated state (Hughes and Connell 1999).

MACROALGAE IN PRISTINE CORAL REEFS

Red, brown, and green algae larger than a few centimeters in length in their mature state are common, even if not typically abundant, on coral reefs. Wynne's checklist of the tropical and subtropical benthic marine algae of the western tropical Atlantic (1998) gives 1,058 species (253 green, 150 brown, and 655 red), the majority of which occur on coral reefs. In their exquisitely illustrated manual of Caribbean reef plants, Littler and Littler (2000) describe and illustrate over 500 of these species. Species of genera such as (reds) *Amphiroa, Galaxaura, Hypnea, Coelothrix, Spyridia, Dasya, Acanthophora, Chondria,* and *Laurencia*; (browns) *Sargassum, Turbinaria, Colpomenia, Dictyopteris, Dictyota, Padina, Lobophora,* and *Stypopodium*; and (greens) *Dictyosphaeria, Valonia, Caulerpa,* and *Halimeda* can easily be found on a well-developed Caribbean reef, even if the biomass is not normally high. While many of

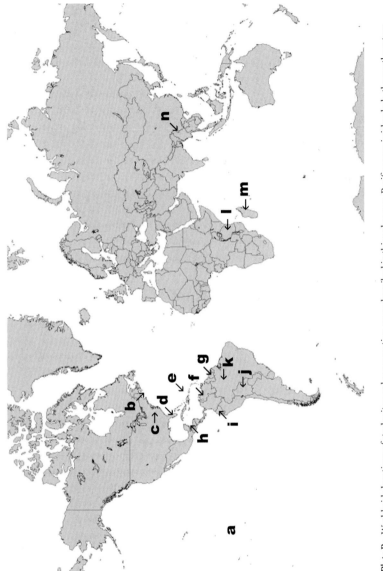

Plate P.1 Worldwide locations of study sites and conservation areas described in this volume: a, Pacific oceanic islands; b, the northwestern North Atlantic; c, Washington, DC; d, Florida; e, the Caribbean islands; f, the Andes of Venezuela; g, the Guiana Shield; h, Twin Cays, Belize; i, the Andes of Ecuador; j, Bolivia; k, Manaus, Brazil; l, Nairobi, Kenya; m, Madagascar; n, the Gaoligong Mountains, China and Myanmar.

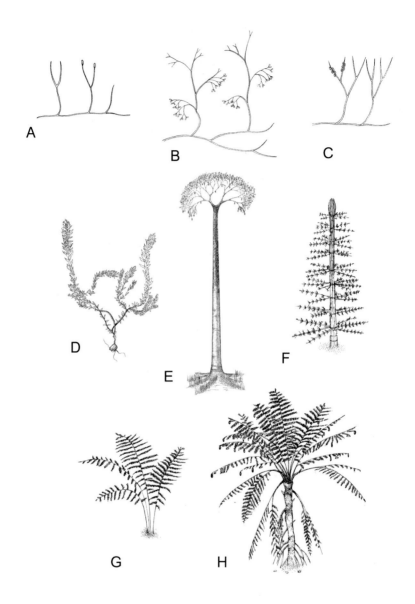

Plate 1.1 Reconstructions of vascular plants representative of major body-plan groups. A, Rhyniophyte, with naked, dichotomizing aerial axes, some terminated in sporangia. B, Trimerophyte, with unequally forked main upright axes, and paired sporangia terminating subordinate branches. C, Zosterophyll, with naked, dichotomizing aerial axes and laterally borne sporangia. D, Extant lycopsid, *Selaginella*. E, Extinct isoetalean lycopsid, *Lepidophloios*. F, Extinct sphenopsid, *Calamites*. G, Generalized small fern body plan, with fronds as the major organs of the plant. H, Extinct seed plant, *Medullosa*. Illustration by Mary E. Parrish.

GLOBAL BIODIVERSITY: SPECIES NUMBERS OF VASCULAR PLANTS

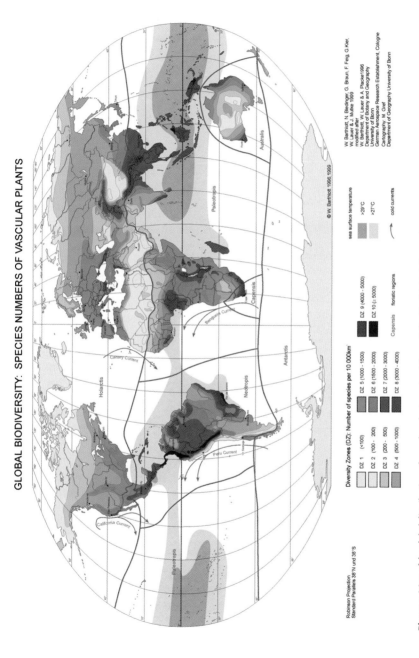

Plate 2.1 Map of the global distribution of vascular-plant species richness (after Barthlott et al. 1999a).

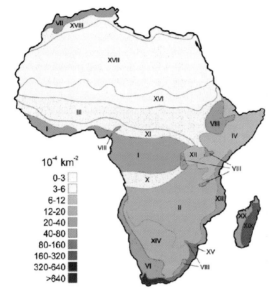

Plate 2.2 Endemism richness of African seed plants per 10,000 km². Note that only mean values for the 20 regions as delineated by White (1983) are displayed and that no further differentiation within these regions is made (modified after Kier and Barthlott 2001).

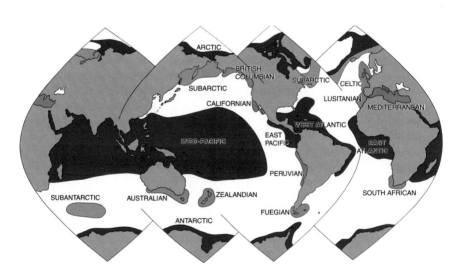

Plate 2.3 Map of the thermogeographic regions defined by figure 2.2.

Plate 3.1
Macginitiea gracilis (Platanaceae).
Photograph by Chip Clark.

Plate 3.2
Kokia cookei (Malvaceae).
Photograph by John K. Obata.

Plate 4.1
Elfin forest–páramo transition at
2,600 m in the Ramal de Guaramacal, Venezuela. Photograph by
Basil Stergios.

Plate 4.2
Kaieteur Falls, Guyana,
in the Guiana Shield.
Photograph by Carol Kelloff.

Plate 4.3
Hatiheu Valley on the
north coast of Nuku Hiva,
Marquesas Islands.
Photograph by K. R. Wood.

Plate 4.4
The Gaoligong Mountains at
3,000 m in Yunnan, China.
Photograph by Bruce
Bartholomew.

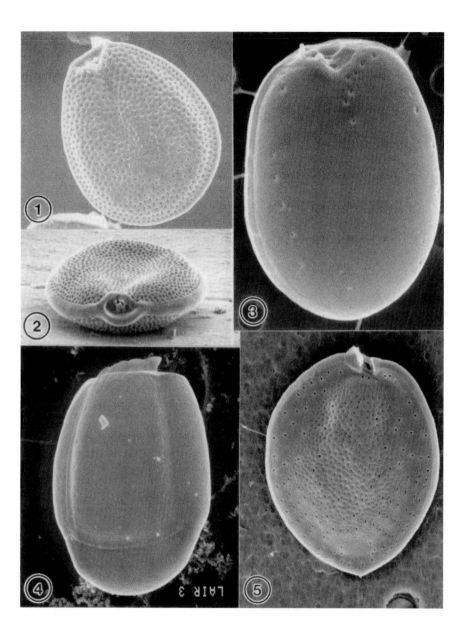

Plate 5.1 The morphology of five photosynthetic benthic dinoflagellate species is illustrated in SEM photographs. 1, *Prorocentrum hoffmannianum,* and 2, *Prorocentrum belizeanum,* two toxin-producing species. Both species are known to cause ciguatera. 3, *Prorocentrum elegans* often forms red tide in shallow warm mangrove waters. 4, *Plagodinium belizeanum* is less than 20 μm long. 5, A newly discovered *Prorocentrum* species has a heart-shaped form and rugose surface laced with pores. Photographs by Maria A. Faust.

Plate 5.2 *Cladonia perforata* (Cladoniaceae). Photograph by Rebecca Yahr.

Plate 5.3 *Leucobryum seemannii* (Leucobryaceae). Photograph by Warren L. Wagner.

Plate 5.4 *Perotis indica* (Poaceae). Photograph by Robert J. Soreng.

Plate 5.5 *Commelina nairobiensis* (Commelinaceae). Photograph by Robert B. Faden.

Plate 5.6 *Pachystachys spicata* (Acanthaceae).
Photograph by Dieter Wasshausen.

Plate 5.7 *Gazania krebsiana* (Asteraceae).
Photograph by Vicki Funk.

Plate 5.8 *Titanotrichum oldhamii* (Gesneriaceae).
Photograph by Richard Dunn.

Plate 5.9 *Lophostigma plumosum* (Sapindaceae).
Photograph by Pedro Acevedo-Rodríguez.

Plate 6.1 A 1-ha forest fragment at the Biological Dynamics of Forest Fragments Project outside Manaus, Brazil. Photograph by W. John Kress.

Plate 6.2 The aftermath of Hurricane Hugo (1989) in a forest in Puerto Rico. Photograph by Pedro Acevedo-Rodríguez.

Plate 6.3 The Tampoketsan d'Ankazobe near the Ikopa River, Madagascar, with small patches of forest near large erosion scars known as *lavaka*. Photograph by Laurence Dorr.

Plate 6.4 A coralline–sea urchin barren off the Ramea Islands on the south coast of Newfoundland, at a depth of 3–5 m. Photograph by Walter Adey.

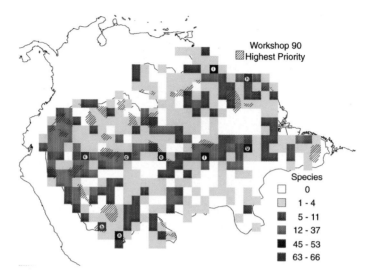

Plate 10.1 Geographic distribution in Amazonia of species diversity for all taxa combined. The number of species is indicated for each of the 472 one-degree grid cells. The Workshop 90 high-priority conservation areas are also shown. Grid cells with the highest number of species are indicated by lowercase letters (for localities see table 10.2; after Kress et al. 1998).

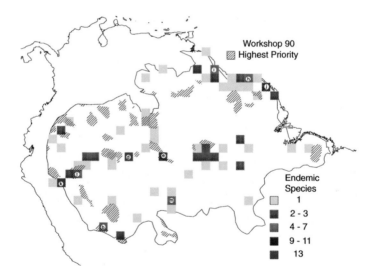

Plate 10.2 Geographic distribution in Amazonia of endemic species diversity for all taxa combined. The number of endemic species is indicated for each of the 472 one-degree grid cells. The Workshop 90 high-priority conservation areas are also shown. Grid cells with the highest concentration of endemic species are indicated by lowercase letters (for localities see table 10.2; after Kress et al. 1998).

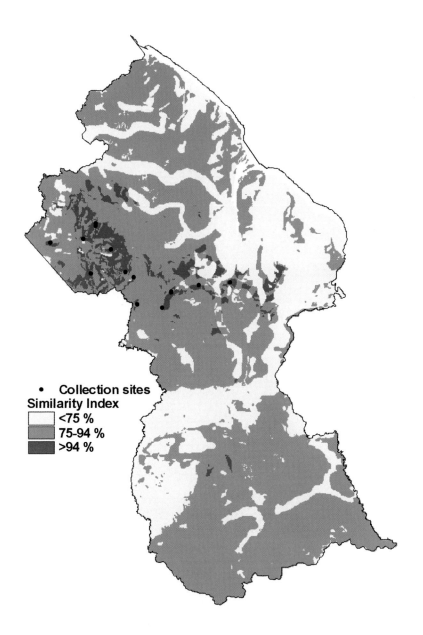

Plate 10.3 A map of Guyana illustrating the known distribution of a plant species (Melastomataceae, *Leandra purpurea* Gleason: black dots) and the distribution of the same species (shaded gray areas) based on DOMAIN modeling of the known localities with four abiotic factors. Three levels of similarity are shown; only a similarity index of 95% or more was used in the analysis (after Funk and Richardson 2002).

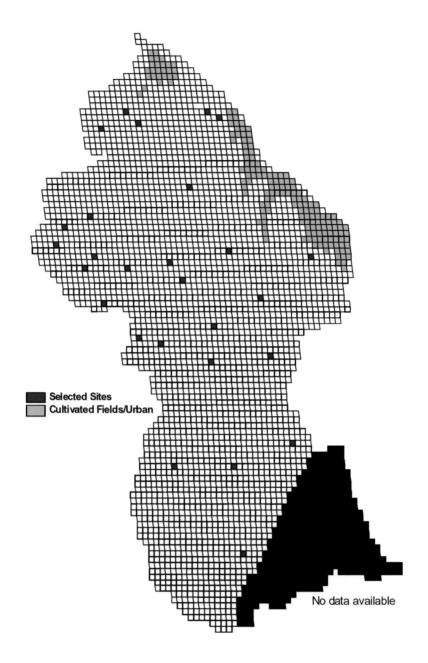

Plate 10.4 Map of Guyana with a grid showing 27 of the 33 cells that would capture each species at least three times (after Funk and Richardson 2002).

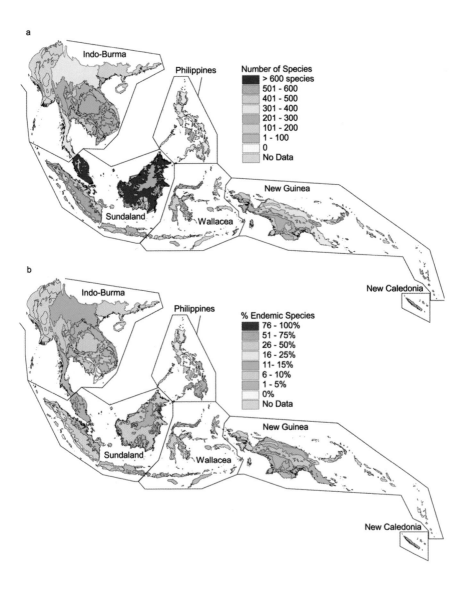

a

Indo-Burma

Philippines

Number of Species
■ > 600 species
▨ 501 - 600
▨ 401 - 500
▨ 301 - 400
▨ 201 - 300
▨ 101 - 200
▨ 1 - 100
□ 0
▨ No Data

New Guinea

Sundaland

Wallacea

b

New Caledonia

Indo-Burma

Philippines

% Endemic Species
■ 76 - 100%
▨ 51 - 75%
▨ 26 - 50%
▨ 16 - 25%
▨ 11 - 15%
▨ 6 - 10%
▨ 1 - 5%
□ 0%
▨ No Data

New Guinea

Sundaland

Wallacea

New Caledonia

Plate 10.5 Species diversity in 84 Indo-Pacific ecoregions in seven plant families (Bignoniaceae, Dipterocarpaceae, Ericaceae, Euphorbiaceae, Fagaceae, Leguminosae, and Rosaceae). a, The tabulated number of species per ecoregion. b, The tabulated percentage of endemic species per ecoregion. Endemic species are found in one and only one hot spot or ecoregion. Several ecoregions consist of multiple subunits; hence, some small distinct areas may appear to be species-rich on this map but are part of a larger ecoregion. Ecoregion names can be obtained from Wikramanayake et al. 2002. The boundaries of five hot spots and the New Guinea major wilderness area (Myers et al. 2000) are also shown. The northwestern boundary of the Indo-Burma hot spot extends into the foothills of the Himalayas (not shown) (after Krupnick and Kress 2003).

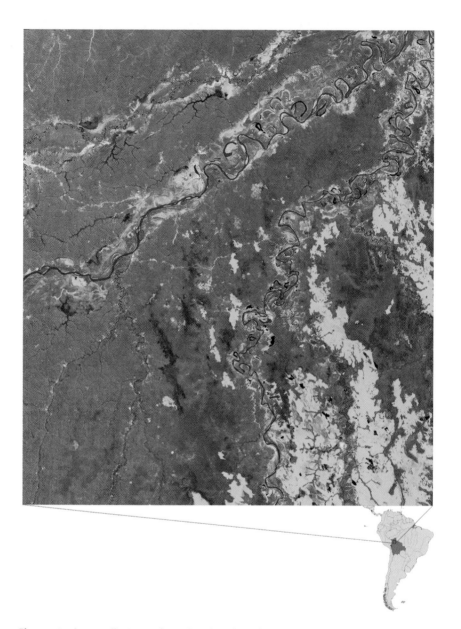

Plate 11.1 Landsat-7 satellite image of central Pando, Bolivia. The Orthon, Madre de Dios, and Beni rivers run northeast toward the Amazon. Between the rivers, upland forests with abundant Brazil-nut trees look coppery brown. In the southeast (lower right) corner of the image, long fingers of open pampas habitat (in light blue) reach up from the Department of Beni. These extensions of the southern pampas barely reach into Pando: north of the Beni River, they break up into isolated, puzzle-shaped pieces of open habitat surrounded by the extensive matrix of tall, western-Amazonian upland forests. Wine-colored areas adjacent to the open pampas are older pampas now revegetated by shrubs and trees (from NASA, Path 001, Row 068, 13 August 2000).

these algae are at least partly defended from grazing by tough tissues or distasteful compounds, they are nevertheless grazed by fish and sea urchins. Thus, on pristine reefs, they are likely to occur on sites of additional protection (e.g., next to fire coral [*Millepora*]), at the very tips of a coral branch where the potential grazer would be more subject to fish predation, or in local zones of high wave turbulence. With a more careful search, species of the more structurally or colorfully esoteric genera (reds) *Liagora, Asparagopsis, Ochtodes, Botryocladia, Chrysymenia, Dictyurus,* and *Martensia*; (browns) *Hydrochathrus*; and (greens) *Anadyomene, Microdictyon, Struvea,* and *Acetabularia* can usually be located as well. Given increased wave action on shallow exposed flats or the edges of coralline-built algal ridges, where potential grazing is even further reduced, *Gracilaria, Sargassum,* and *Turbinaria* not only appear but can locally form quite large standing crops (Adey et al. 1977; Connor and Adey 1977; Adey 1978). The same general picture can be seen of Indo-Pacific reefs, based on studies in the Australian Great Barrier Reef (Morrissey 1980; Cribb 1983, 1984).

While calcified, encrusting corallines (Corallinales, Rhodophyta), particularly of the genera *Neogoniolithon, Porolithon, Paragoniolithon, Hydrolithon, Lithophyllum,* and *Mesophyllum,* can be abundant elements of healthy coral reefs unencumbered by fine sediments, it is only in very high wave energy that they build reef frameworks (Adey 1978). In that situation, usually on the most exposed parts of forereefs, *Porolithon* and *Lithophyllum* species can raise reef structure well into and even above the intertidal zone.

ALGAL TURFS ON PRISTINE CORAL REEFS

Typically, 30 to 60% of the surface of structurally well developed and strongly calcifying coral reefs are dominated by dead coral or coralline algal carbonate overgrown by a mat or turf of mostly filamentous algae less than a few centimeters in height (reviewed by Adey 1998). Reef algal turfs are rich in species of cyanophytes and diatoms (Small et al. 1998), as well as small types of the major macroalgal groups (reds) *Centroceras, Champia, Coelothrix, Gelidiella, Jania, Wurdemannia, Ceramium, Polysiphonia, Herposiphonia, Lophosiphonia, Dasya, Callithamnion, Griffithsia,* and *Crouania*; (browns) *Sphacelaria, Ectocarpus,* and *Giffordia*; and (greens) *Enteromorpha.* Although algal turfs are rarely differentiated from macroalgae in floras, and indeed there is not a sharp morphological boundary between the two groups, Price and Scott (1992) produced a volume specifically directed at the algal-turf components of the Australian Great Barrier Reef flora. Algal turfs are sometimes referred to as epilithic algal communities, or EAC (e.g., Encore Group 2001), although this is rather a misnomer as most macroalgae are also epilithic.

Many more species of macroalgae remain undescribed in coral reefs across the tropical band of the world ocean. Littler and Littler (2000) proposed twice as many species as have already appeared in the literature. As discussed in chapter 2 of this volume, however, theoretical considerations suggest that at least five times as many species remain undescribed. The discrepancy probably lies with algal turfs, most of which require a microscope for careful evaluation and are therefore often omitted by phycologists and ecologists (fig. 6.3). More specifically, the cyanophytes and diatoms, which form significant portions of algal turfs, represent a vast and very much under-studied sector of the coral-reef flora.

In the Small et al. (1998) analysis of reef biodiversity based on microcosm analogs, 55% of the algal species were cyanophytes and benthic diatoms. The estimated minimum number of benthic diatoms in pantropical coral reefs, based on that analysis, is over 230,000 species. Thus, while the ca. 500,000 pantropical theoretically expected species cited in chapter 2 may seem inordinately large, the only direct value for coral-reef biodiversity currently available (Small et al. 1998) suggests that even this number is an underestimate. It is significant that some diatom specialists think that the worldwide biodiversity of diatoms may reach 10 million species (John 1994).

Rarely are cyanophytes and diatoms included in floras. The Littler and Littler (2000) inclusion of cyanophytes for the Caribbean flora, but only where they develop "macro-colonies," is quite exceptional in the literature, but still omits many species. The diatoms are almost universally ignored (however, see Small et al. 1998). The suggestion by Norton et al. (1996) that tropical algal biodiversity is not as high as that of rocky-shore temperate regions very likely results from a failure of phycologists to investigate intensively the algal-turf dominant plant subcommunity of coral reefs.

THE HEALTH OF CORAL REEFS AND ITS RELATIONSHIP TO ALGAE

In the past decade, numerous reviews of the status of coral-reef health have been written. The recent reviews by Hoegh-Guldberg (1999) and Adey et al. (2000) paint a very pessimistic picture of an ecocide in progress, though the latter authors suggest a "market-based" rationale for reef conservation. Bleaching, or ejections of the dinoflagellate zooxanthellae from stony corals, thought to be at least partly due to global-scale warming (Hoegh-Guldberg 1999; Fitt et al. 2001), is a significant issue. Littler and Littler (2000) and Wilkinson (1999) have a more positive view of the potential for conservation-produced turnaround. They also point out, however, that on a pantropical scale, overfishing and eutrophication are at least equal in importance to global warming. Since all of these factors result in a "phase shift" to macrolagal domination of coral reefs, it would seem that algae are the "problem" in coral-reef degradation. It is likely, however, that on reefs where a phase shift has occurred and macroalgal domination has resulted, even macroalgal biodiversity has decreased to a relatively few tough and chemically defended species (Szmant 2001). Since the primary plant biodiversity of coral-reef ecosystems very likely resides in the algal turfs, there is little question that the pantropical degradation of coral-reef ecosystems now under way is having a devastating effect on a plant biodiversity that is very much hidden from view by a lack of specialists able to provide definition.

SUMMARY

A very broad literature indicates that a pantropical-scale rapid degradation of coral reefs is under way and is likely to be complete before the end of the 21st century. The primary plant (algal) biodiversity in actively growing coral reefs lies within the algal turf subcommunity. It seems likely that this community harbors the "missing" plant biodiversity component of coral reefs that is equivalent to the higher plants in rain forests. Unless very substantial and rapid improvements are made in the global change process, much of the plant biodiversity of coral reefs is likely to vanish before it can be described. Most of this "unknown" biodiversity is likely concentrated in the vast, under-studied Indo-Pacific.

6.5 ALTERATION OF KELP COMMUNITIES IN THE NORTHWESTERN NORTH ATLANTIC

Walter H. Adey and James N. Norris

SINCE THE 16TH CENTURY, human activities have significantly disturbed subtidal communities along the entire northwestern North Atlantic coast, well north into Labrador. This disturbance has been primarily in the form of intensive fish extraction, especially that of the cod (*Gadus morhua*), but it has also included considerable hunting of seabirds and seals (Ryan 1994; Gaskell 2000). In the latter half of the 20th century, many scientists and environmentalists have begun to ask whether the ragged kelp community that we see today represents the general late Holocene state of this shore. It seems likely that European long-distance fishing in the 16th and 17th centuries, with immigration and even more intense fishing in the 18th and 19th centuries, followed by an overwhelming increase of fishing efficiency in the 20th century, has significantly altered the basic state of this ecosystem.

NORTHWESTERN NORTH ATLANTIC KELP COMMUNITIES

That kelp is an important element of the rocky subtidal of the northwest Atlantic is attested by the huge berms of mostly kelp drift cast ashore by winter storms. They generally occupy the upper subtidal rocky western shores of the North Atlantic from Long Island Sound to Labrador (Lüning 1990; Mathieson et al. 1991; fig. 6.4). Kelp communities are structurally dominated by three to five species of *Laminaria*

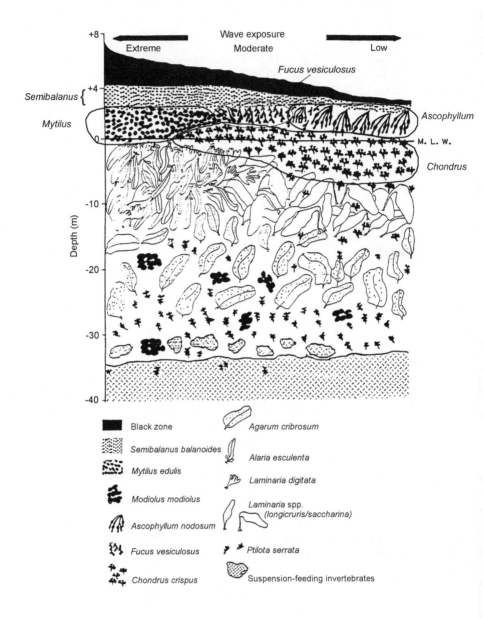

Figure 6.4 Generalized relationship of algal communities on northwestern North Atlantic shores, including the subtidal kelp community in relation to shore exposure (Mathieson et al. 1991). M.L.W., mean low water.

(mostly *L. digitata*, *L. longicruris*, and *L. saccharina*) and one species each of *Agarum* (*A. cribrosum*), *Alaria* (*A. esculenta*), and *Saccorhiza* (*S. dematodea*). In their well-developed state, the canopy-building kelps can achieve as much as several meters in height off the bottom. *Laminaria longicruris*, sometimes quite abundant, can reach 6–8 m in length. Northwest Atlantic kelp beds also contain a rich understory of leafy and filamentous red algae (including *Chondrus crispus*, *Phycodrys rubens*, *Callophyllis cristata*, *Ceramium rubrum*, and *Palmaria palmata*), brown algae (principally *Desmarestia* spp. and in the north *Chordaria*), and green algae (*Monostroma* spp., *Derbesia marina*, and *Chaetomorpha* spp.). A well-developed crust story of five to eight species of red, crustose corallines (principally *Clathromorphum circumscriptum*, *Lithothamnium glaciale*, *Phymatolithon* spp., and *Lithophyllum crouanii*) is ubiquitous. A host of invertebrates and fish occupy the rock surfaces and crevices around and beneath the sometimes dense algal cover.

Occurring subtidally at a depth of up to 20 m on most of the shore, the northwest Atlantic kelp community (at least in the latter half of the 20th century) has been quite patchy in its distribution. *Kelp forests* commonly refers to the giant kelp (*Macrocystis pyrifera*) communities along the northeastern Pacific coast from Alaska to Baja California (Foster 1985, 1990); it is rare to see the term *kelp forest* used in the scientific or environmental literature for the northwest Atlantic. The terms *kelp bed* and *kelp zone*, however, are widely used, reflecting the sometimes continuous and widespread nature of the community. The difference in terminology might also reflect the smaller sizes of *Laminaria* (3–8 m long) and related kelps as compared to the much larger *Macrocystis* (10–70 m long) and *Nereocystis* (to 40 m long) kelps of the West Coast. Nevertheless, several species form canopies (e.g., *Alaria esculenta* inshore and *Laminaria longicruris* farther offshore) floating at the surface of the water at low tide in the manner of the larger West Coast species. Ecologically, the northwest Atlantic kelp community can be extremely dense and structurally well developed, thus often warranting the terms *kelp bed* or *kelp forest*.

As noted by many researchers of the past several decades, the northwest Atlantic kelp community has a common alternate phase, a coralline–sea urchin "barren" where large numbers of the purple sea urchin (*Strongylocentrotus drobachiensis*) graze so intensely that the kelps are reduced to little more than a stubble of ragged stipes. Not infrequently, where sea urchins are particularly dense, little in the way of macroalgae, except the rock-covering crust of coralline (reviewed by Mathieson et al. 1991), can be seen (plate 6.4). This coralline–sea urchin barren can be extensive, sometimes dominating dozens of kilometers of coastline. While some researchers express the view that the coralline–sea urchin barren phase is a relatively recent phenomenon (e.g., Mathieson et al. 1991), large areas of this phase occurred in the northern Gulf of St. Lawrence and in more scattered areas in Newfoundland in the early 1960s (plate 6.4; Adey 1966). While some researchers have suggested that this might be due to ice scour (reviewed by Mathieson et al. 1991), extensive coralline barrens occur on the south coast of Newfoundland where winter sea ice is not present. According to the predator/sea urchin control model developed by Steneck and Carlton (2000), the major change inshore in the Gulf of Maine probably began in the 1940s to the 1950s. Extensive field research in the Gulf of Maine in the early 1960s found coralline–sea urchin barrens to be common in the large bays (Penobscot, Blue Hill, and Frenchman bays). However, kelp beds with a more re-duced coralline algal cover still remained abundant on the outer coast where wave action apparently inhibited sea urchin feeding (Adey 1963).

In general, grazing sea urchins have less effect on exposed shores, where they tend to be restricted to deeper water by intensive wave action. Also, except on the most protected of shorelines, a small infralittoral and uppermost sublittoral band of *Alaria esculenta* and *Laminaria digitata* can receive some wave protection from sea urchin grazing and build significant biomass. The deeper kelp *Agarum cribro-sum* is at least partially protected by secondary compounds. Smaller than some *Laminaria* species (up to 1 m), scattered individuals of *A. cribrosum* are often able to form a ragged lower edge to what is otherwise primarily a crustose coralline sur-face with scattered protected small foliose red algae such as *Phycodrys rubens* and *Callophyllis cristata* (reviewed by Mathieson et al. 1991).

On some coralline–sea urchin barrens, the sea urchins become so dense that they are subject to disease epidemics that cause population crashes (e.g., Miller 1985; Pringle 1986). The result of a sea urchin crash can be a short-term resurrection of the kelp community. For example, an intensive sea urchin fishery in the Gulf of Maine beginning in 1987 has apparently resulted in a local resurgence of the kelp community there (Steneck and Carlton 2000). The more complex, but similar, *Laminaria, Agarum,* and *Alaria* kelp communities of the western Aleutian Islands also have an alternate sea urchin barren phase. In that region, however, the flip-flop between the coralline–sea urchin barren and kelp forests is controlled by a sea urchin predator, the sea otter, *Enhydra lutris* (Estes and Duggins 1995).

SEA URCHIN FISHERIES AND KELP

Sea urchin roe is an important element of Asian cuisine, particularly Japanese, and in recent decades a purple sea urchin (*Strongylocentrotus drobachiensis*) fishery has developed along parts of the northwest Atlantic coastline to supply this market. The activities of this fishery in the 1990s in the Gulf of Maine became very intense and for much of its existence have been unregulated. Steneck and Carlton (2000) have documented the return of the kelp forest in areas where the sea urchins have been heavily harvested; more important, they have developed an intriguing link between intensive fishing, which results in the reduction of large predatory fish, an increase in sea urchins, and a decrease in kelp.

In some areas, such as the Alaskan coast of North America, the relationship be-tween sea urchins, kelp, and the predators of sea urchins, such as sea otters, is well known (Estes and Duggins 1995). While there is no resident sea otter population on the east coast of North America, there is some evidence that the ecological equiva-lent of such an animal, the extinct sea mink, *Mustela macrodon* (Mead et al. 2000), may have been responsible for some measure of control over sea urchin popula-tions. The island-based *M. macrodon* was considerably larger than mainland-based minks and was apparently driven to extinction by human overexploitation for furs in the late 19th century.

Whether or not the loss of the sea mink was a significant factor in sea urchin increase, the relationship with the reduction of large predatory fish, especially cod fishes, is strongly suspected. Even though the primary food of the cod, especially in their offshore bank cycle, is smaller finfish, the reduction of cod populations from the 1500s through the 1900s was extremely large. The younger fish in pre-European times, when entering the shore population for the first or second year in enormous numbers, would have likely been significant predators of sea urchins, especially younger individuals (less than a centimeter in diameter) (e.g., Mathieson et al. 1991). While many invertebrates in benthic communities have predator protection mechanisms, such as spines and crevice lodgment in the case of sea urchins, small sea urchins cannot both feed on algal turfs and mats growing on exposed rock surfaces and be protected from significant finfish predation.

COD FISHERIES

The northwest Atlantic cod fishery, centered on Newfoundland and the Grand Banks, is one of the largest ever developed by humans. From back in the "historical mists" of the 16th century up through the middle of the 20th century, it can be estimated that over 80 million metric tons of cod were taken (Taggert et al. 1994) from Newfoundland and Labrador waters alone (if the Gulf of St. Lawrence, Nova Scotia, and the Gulf of Maine are included, the extraction probably exceeded 200 million metric tons). Starting with low levels in the mid-1600s, the rate of extraction rose more or less steadily to 200,000 tons per year by 1800 and then to 400,000 tons by 1950. From there, the catch rate spiked to over 1 million tons per year about 1970 and then crashed dramatically to less than 100,000 tons per year just before the fishery closure in 1992. The cod, increasingly taken in the western Atlantic, was a significant source of protein for the modern development of both pre- and postindustrial Europe, and played a very significant role in European politics and in the development of the Atlantic North American states (Innis 1940). Numerous descriptions of the abundance of codfish from the 16th and 17th centuries, even when allowance is made for the social or political basis of such descriptions, describe a density that was like that of the passenger pigeon and bison on land, and not easily comprehended by today's human societies (e.g., Kurlansky 1997). Indeed, when Newfoundlanders, the archetypical western Atlantic codfishers, referred to "fish," they meant cod—other species of finfish were referred to by their common names.

Although the total cod population is divided more or less separately into north and south geographic races (and subraces) along the coast, most have a pattern of spring spawning on offshore banks (Grand Banks, Georges Bank, etc.), where winter temperatures are higher than inshore. Spawning success of the cod was keyed to the prevailing long-term winter temperature means on the banks. Spring and summer, however, brought a migration inshore along defined paths where the voracious appetite of the individual fish and their large numbers would have consumed enormous masses of invertebrates as well as smaller fish, especially capelin (e.g., for northeastern Newfoundland, see Rose 1993). To avoid being eaten by their elders, the younger-year classes undoubtedly kept close to the bottom, within the kelp and lower-story communities. Here they intensively preyed on invertebrates, and it is highly likely, given the abundance of young cod and intense feeding pressure, that feeding sea urchins, especially the very small young, even with their spiny defenses, were largely swept clear of the bottom. While there were certainly sea urchin refuges from predation, it is not likely that intertidal pools and very shallow waters provided respite. Sea gulls routinely remove sea urchins in the intertidal and infratidal. Diving and dabbling ducks are an additional source of predation near shore. All of these organisms, however, were subject to significant human predation during the 18th and 19th centuries.

The early cod fisheries were simply hook-and-line (jigger) fisheries, on the banks in the winter and inshore in summer, and because of the very high fish densities this was sufficient to provide considerable production. Baiting was used later, and the cod trap (basically a weir system) came into being in the 19th century. This technique of the inshore fishery was designed to crop a portion of the stock as it migrated along the shore during the summer. Occasionally, disruption to the fishery would occur when the expected numbers of fish would not show up for a season. This could occur locally, due to variations in migration route, or more generally due to a reduction in spawning recruitment success following unfavorable environmental conditions. Finally, bottom dragging with otter trawls, and variations thereof, were developed. At each stage of technological advance (and along with human population increase came a higher fishery demand), the total abundance of fish in the wild slowly declined; the effective effort, or efficiency of catch, however, mostly increased. Since other fish predators (especially seals and seabirds) were also being simultaneously removed by human hunting, the total harvest generally increased. Finally, the most economically efficient method for catching cod (trawling or dragging) probably has population-damaging effects that range far beyond simple reduction of adult population, namely destruction of bottom habitat (Watling and Norse 1998).

The impending crash of the cod population in northern areas became clear in the 1970s. The Canadian-Newfoundland cod fishery, except for experimental fishing, was closed in 1992 in an attempt to avoid extermination of what was at one time one of the most abundant macrospecies on earth. In ten years of closed fishery, the stock has not rebounded as expected, and the closure process, originally considered a temporary measure, has continued. As usual, there are considerable arguments over the reason for the crash; however, many regard it as unlikely that accompanying environmental changes were a significant factor, placing the blame entirely on long-term overfishing (e.g., Hutchins and Myers 1994) or bottom destruction (Watling and Norse 1998). If the hypothesis is correct, that the extremely abundant young Atlantic cod foraging in the inshore benthic community provided the primary element of sea urchin control, the required density to achieve effective sea urchin control over large stretches of coast was passed sometime in the early 20th century. During the 1960s, yearly young recruits to the cod stocks decreased roughly

70% (Hutchins and Myers 1994); as noted above, by the early 1960s, coralline–sea urchin bottoms had largely replaced kelp beds in moderately protected large bays in this region (such as outer Random Sound in Trinity Bay and White Bear Arm in southern Labrador).

STATUS OF THE NORTHWEST ATLANTIC KELP FOREST

Just as on land (e.g., in the mixed deciduous hardwood forest of North America), the northwest Atlantic kelp forest today cannot be treated as an ecosystem independent of human activity. Primarily because of fishing (though deforestation, runoff, and siltation as well as nutrification are certainly significant factors locally), it is likely that this community can be understood only when humans are considered a top predator species. In addition, there is considerable indication that human disturbance processes influenced these ecosystems well before the great overfishing of the 20th century. For example, prehistorically one of the most abundant large predator species of the region was the flightless great auk (*Alca pinguinis*). It was driven to regional extinction in the early 19th century (Gaskell 2000), almost certainly from the relentless commercial overharvest of these birds and their eggs, when these birds were, in dense populations, carrying out their short breeding season on remote rocky islets. While in the early years the great auks were taken primarily for food, the final regional extinction was driven primarily by harvest for feathers (total extinction occurred in Iceland in the 1860s). Likewise the considerable harvest of the principal two abundant seal populations that have pups on offshore sea ice (i.e., harp seals, *Phoca groenlandica*, and hood seals, *Cystophora cristata*) especially in the first half of the 19th century, undoubtedly significantly affected the regional biota (Ryan 1994). The seal fishery was directed at the reproducing populations (on sea ice, rather than on rocky islets for the great auk). Extinctions were forestalled by the combination of declining oil prices in the late 19th century, due to the commercial development of mineral oil, and the extreme physical danger of the fishery. Seal harvest yields continued to fall, even though fishery pressure and capability increased. Almost certainly, the seals were being subjected to overharvesting.

It is difficult not to conclude that the continual increase of cod harvests in the first half of the 19th century occurred in part because the primary pre-European cod predators, seals, were also being heavily harvested. This may have been at least partly true earlier in the 18th century because of the human harvest of the great auk. When the far more efficient harvesting method of trawling began on the banks in the later 19th century, a population crash was inevitable (though another three-quarters of a century was needed to apply the coup de grace). Although crashes of dense sea urchin populations related to disease have been documented (e.g., Mathieson et al. 1991), this is not likely a primary, pre-European pattern. Disease tends to be a significant factor in wild animal populations when population limitations are released by a lack of predation pressure. Sea urchins were likely controlled in pre-European times either directly or indirectly by high predator populations.

THE ATLANTIC SUBARCTIC BIOGEOGRAPHIC REGION

Adey and Steneck (2001) have shown (using crustose red algal coralline cover) that the coast from Cape Cod to northern Nova Scotia is a transition zone between the Atlantic boreal zone and the Atlantic Subarctic zone with a roughly 40:60 species mix. Boreal algal introductions from Europe (e.g., species of *Fucus, Codium,* and *Grateloupia*) are known to have occurred in historical times, and these introductions are likely to increase in the future, perhaps even independent of human activities. The core Atlantic Subarctic zone occupies much of the island of Newfoundland, southern Labrador, and the northern Gulf of St. Lawrence. The Arctic biota extends northward from northern Labrador and sends relatively few species southward into the Atlantic Subarctic, primarily in deeper water. In the northwest Atlantic there was no significant human use of kelps or the kelp community that might have provided some insight as to its nature prior to the 19th century. In the core Subarctic region, given a long-term suspension of cod fishing and a subsequent long-term reduction of coastal sea urchin populations, it may be possible, however, to glimpse a view of what northwest Atlantic kelp communities were before the fisheries.

SUMMARY

The northwest Atlantic kelp forest has a basic macroalgal upper- and midlevel canopy and an understory that is similar in species composition and diversity to kelp forests in the northeast Pacific. This kelp community, with some endemics, represents a relatively small Atlantic extension of the much larger North Pacific Subarctic.

Five centuries of extensive fisheries exploitation of larger predators (especially seabirds, seals, and cod) in the northwestern North Atlantic have reduced some of these predators to extinction and others to economic extinction. Recently, a linkage between the reduction of larger fish predators and sea urchin population explosions has been demonstrated. When populations of sea urchins, *Strongylocentrotus drobachiensis,* are restrained primarily by disease (rather than by predators), the kelp communities become generally degraded, occurring only as fleeting patches in time and space as sea urchin disease alternately and patchily explodes and dies back.

An indefinite moratorium has been placed on codfish extraction in Atlantic Canada. Thus, this core Atlantic Subarctic region provides a possible conservation area to witness the rebirth of northwest Atlantic kelp communities. Long-term study of these communities could be carried out without the additional confusing elements of boreal introductions, which are continuing in the New England/Nova Scotia transition regions to the south.

LITERATURE CITED

Adey, W. 1963. *Phymatolithon, Clathromorphum,* and *Pseudolithophyllum* in the Gulf of Maine. Ph.D. diss., University of Michigan, Ann Arbor.

———. 1966. The distribution of saxicolous crustose corallines in the northwestern North Atlantic. Journal of Phycology 2:49–54.

———. 1978. Algal ridges of the Caribbean Sea and West Indies. Phycologia 17:361–367.

———. 1998. Coral reefs: algal structured and mediated ecosystems in shallow, turbulent, alkaline waters. Journal of Phycology 34:393–406.

Adey, W., Adey, P., Burke, R., and Kaufman, L. 1977. Holocene reef systems of eastern Martinique. Atoll Research Bulletin. 281:1–40.

Adey, W., and Goertemiller, T. 1987. Coral reef algal turfs: master producers in nutrient poor seas. Phycologia 26:374–386.

Adey, W., and Hackney, J. 1989. Harvest production of coral reef algal turfs. Pp. 1–135 in Adey, W., ed., The Biology, Ecology, and Mariculture of *Mithrax spinosissimus* based on Cultured Algal Turfs. Mariculture Institute, Los Angeles.

Adey, W., and Loveland, K. 1998. Dynamic Aquaria: Building Living Ecosystems. 2nd edition. Academic Press, San Diego.

Adey, W., McConnaughey, T., Small, A., and Spoon, D. 2000. Coral reefs: endangered, biodiverse genetic resources. Pp. 33–42 in Sheppard, C., ed., Seas at the Millennium. Elsevier, Amsterdam.

Adey, W., and Steneck, R. 1985. Highly productive eastern Caribbean reefs: synergistic effects of biological, chemical, physical, and geological factors. NOAA Symposium Series for Undersea Research 3:163–187.

———. 2001. Thermogeography over time creates biogeographic regions: a temperature/space/time-integrated model and an abundance-weighted test for benthic marine algae. Journal of Phycology 37:677–698.

Aizen, M. A., and Feinsinger, P. 1994. Forest fragmentation, pollination, and plant reproduction in a chaco dry forest, Argentina. Ecology 75:330–351.

Aldrich, P. R., and Hamrick, J. L. 1998. Reproductive dominance of pasture trees in a fragmented tropical forest mosaic. Science 281:103–105.

Andriamampianina, J. 1984. Nature reserve and nature conservation in Madagascar. Pp. 219–227 in Jolly, A., Oberlé, P., and Albignac, R., eds., Key Environments: Madagascar. Pergamon Press, Oxford.

Bacon, P. R. 1985. Environmental impact assessment. Caribbean Conservation Association News 4:10–12.

Bawa, K. S. 1990. Plant-pollinator interactions in tropical rain forests. Annual Review of Ecology and Systematics 21:399–422.

Berry, F., and Kress, W. J. 1991. *Heliconia*: An Identification Guide. Smithsonian Institution Press, Washington, DC.

Bierregaard, R. O. Jr., Gascon, C., Lovejoy, T. E., and Mesquita, R., eds. 2002. Lessons from Amazonia: The Ecology and Conservation of a Fragmented Forest. Yale University Press, New Haven.

Bierregaard, R. O. Jr., and Stouffer, P. C. 1997. Understory birds and dynamic habitat mosaics in Amazonian rainforests. Pp. 138–155 in Laurance, W. F., and Bierregaard, R. O. Jr., eds., Tropical Forest Remnants: Ecology, Management, and Conservation of Fragmented Communities. University of Chicago Press, Chicago.

Bond, W. J. 1995. Assessing the risk of plant extinction due to pollinator and disperser failure. Pp. 131–146 in Lawton, J. H., and May, R. M., eds., Extinction Rates. Oxford University Press, New York.

Boucher, D. H. 1990. Growing back after hurricanes: catastrophes may be critical to rainforest dynamics. BioScience 40:163–166.

Bruna, E. M. 1999. Seed germination in rainforest fragments. Nature 402:139.

————. 2001. Effect of habitat fragmentation on the reproduction and population dynamics of an Amazonian understory herb (*Heliconia acuminata*, Heliconiaceae). Ph.D. diss., University of California, Davis.

————. 2003. Are plants in rain forest fragments recruitment limited? Test with an Amazonian herb. Ecology 84:932–947.

Bruna, E. M., and Kress, W. J. 2002. Habitat fragmentation and the demographic structure of an Amazonian understory herb (*Heliconia acuminata*). Conservation Biology 16:1256–1266.

Campbell, D. G. 1978. The Ephemeral Islands: A Natural History of the Bahamas. Macmillan, London.

Carpenter, R. 1985. Relationships between primary production and irradiance in coral reef algal communities. Limnology and Oceanography 30:784–793.

Carpenter, R., Hackney, J., and Adey, W. 1991. Measurements of primary productivity and nitrogenase activity of coral reef algae in a chamber incorporating oscillatory flow. Limnology and Oceanography 36:40–49.

Caswell, H. 2001. Matrix population models: construction, analysis, and interpretation. Sinauer Associates, Sunderland, Massachusetts.

Chase, M. R., Boshier, D. H., and Bawa, K. S. 1995. Population genetics of *Cordia alliodora* (Boraginaceae), a Neotropical tree. Part 1, Genetic variation in natural populations. American Journal of Botany 82:468–475.

Collins, L. S., Budd, A. F., and Coates, A. G. 1996. Earliest evolution associated with closure of the Tropical American Seaway. Proceedings of the National Academy of Science USA 93:6069–6072.

Connor, J., and Adey, W. 1977. The benthic algal composition, standing crop, and productivity of a Caribbean algal ridge. Atoll Research Bulletin 211:1–15.

Cribb, A. 1983. Marine Algae of the Southern Great Barrier Reef: Rhodophyta. Australian Coral Reef Society, Brisbane.

————. 1984. Algal vegetation of the Capricornia Section, Great Barrier Reef Marine Park. Pp. 79–86 in Ward, T., and Saenger, P., eds., Royal Society of Queensland Symposium: The Capricornia Section of the Great Barrier Reef, Past, Present, and Future. Royal Society of Queensland, Brisbane.

Crossland, C., Hatcher, B., and Smith, S. 1991. Role of coral reefs in global ocean production. Coral Reefs 10:55–64.

Davis, S. D., Heywood, V. H., Herrera-MacBryde, O., Villa-Lobos, J., and Hamilton, A. C. 1997. Centres of Plant Diversity. Vol. 3, The Americas. IUCN Publications Unit, Cambridge.

Dick, C. W. 2002. Habitat change, African honeybees, and fecundity in the Amazonian tree *Dinizia excelsa* (Fabaceae). Pp. 146–157 in Bierregaard, R. O. Jr., Gascon, C., Lovejoy, T. E., and Mesquita, R., eds., Lessons from Amazonia: The Ecology and Conservation of a Fragmented Forest. Yale University Press, New Haven.

Drake, J. A., Mooney, H. A., and di Castri, F. 1989. Biological Invasions: A Global Perspective. Cambridge University Press, Chichester, United Kingdom.

Encore Group. 2001. The effect of nutrient enrichment on coral reefs: synthesis of result and conclusions. Marine Pollution Bulletin 42:91–120.

Estes, J. A., and Duggins, D. O. 1995. Sea otters and kelp forests in Alaska: generality and variation in a community ecological paradigm. Ecological Monographs 65:75–100.

Feinsinger, P. 1987. Approaches to nectarivore-plant interactions in the New World. Revista Chilena de Historia Natural 60:258–319.

Fitt, W., Brown, B., Warner, M., and Dunne, R. 2001. Coral bleaching: interaction and thermal tolerance limits and thermal thresholds in tropical corals. Coral Reefs 20:51–65.

Florida Exotic Pest Plant Council. 2001. List of Florida's Invasive Species. Florida Exotic Pest Plant Council. http://fleppc.org/01list.htm.

Foster, M. S. 1985. The Ecology of Giant Kelp Forests in California: A Community Profile. U.S. Fish and Wildlife Service, Biological Report 85 (7.2). U.S. Fish and Wildlife Service, Washington, DC.

———. 1990. Organization of macroalgal assemblages in the northeast Pacific: the assumption of homogeneity and the illusion of generality. Hydrobiologia 192:21–33.

Gaskell, J. 2000. Who Killed the Great Auk? Oxford University Press, Oxford.

Gentry, A. H., and Emmons, L. H. 1987. Geographical variation in fertility, phenology, and composition of the understory of Neotropical forests. Biotropica 19:216–217.

Gordon, D. R. 1998. Effects of invasive non-indigenous plant species on ecosystem processes: lessons from Florida. Ecological Applications 8:975–989.

Hamilton, M. B. 1999. Tropical tree gene flow and seed dispersal. Nature 401:129–130.

Hamrick, J. L., and Murawski, D. A. 1991. Levels of allozyme diversity in populations of uncommon Neotropical tree species. Journal of Tropical Ecology 7:395–399.

Harrison, S., and Bruna, E. 1999. Habitat fragmentation and large-scale conservation: what do we know for sure? Ecography 22:225–232.

Hobbs, R. J., and Huenneke, L. F. 1992. Disturbance, diversity, and invasion: implications for conservation. Conservation Biology 6:324–337.

Hoegh-Guldberg, O. 1999. Climate change, coral bleaching, and the future of the world's coral reefs. Marine and Freshwater Research 50:839–866.

Horvitz, C. C., and Koop, A. 2001. Removal of non-native vines and post-hurricane recruitment in tropical hardwood forests of Florida. Biotropica 33:268–281.

Horvitz, C. C., Pascarella, J. B., McMann, S., Freedman, A., and Hofstetter, R. H. 1998. Functional roles of invasive non-indigenous plants in hurricane-affected subtropical hardwood forests. Ecological Applications 8:947–974.

Horvitz, C. C., and Schemske, D. W. 1995. Spatiotemporal variation in demographic transitions of a tropical understory herb: projection matrix analysis. Ecological Monographs 65:155–192.

Howard, R. A. 1977. Conservation and the endangered species of plants in the Caribbean islands. Pp. 105–114 in Prance, G. T., and Elias, T. S., eds., Extinction Is Forever. New York Botanical Garden, New York.

Hughes, T. 1994. Catastrophes, phase shifts, and large-scale degradation of a Caribbean coral reef. Science 265:1547–1551.

Hughes, T., and Connell, J. 1999. Multiple stressors on coral reefs: a long-term perspective. Limnology and Oceanography 44:932–940.

Humbert, H. 1927. La Destruction d'une Flore Insulaire par le Feu: Principaux Aspects de la Végétation à Madagascar: Dominants Photographiques et Notices. Mémoires de l'Académie Malgache 5:1–80.

———. 1955. Les territoires phytogéographiques de Madagascar: leur cartographie. Colloque sur les Régions Écologiques du Globe, Année Biologique 31:195–204.

Hutchins, J., and Myers, R. 1994. What can be learned from the collapse of a renewable resource? Atlantic cod, *Gadus morhua*, of Newfoundland and Labrador. Canadian Journal of Fisheries and Aquatic Sciences 51:2126–2146.

Innis, H. 1940. The Cod Fisheries: The History of an International Economy. University of Toronto Press, Toronto.

John, D. 1994. Biodiversity and conservation: an algal perspective. Phycologist 38:3–15.

Johnson, C., Klumpp, D., Fied, J., and Bradbury, R. 1995. Carbon flux on coral reefs: effects of large shifts in community structure. Marine Ecology Progress Series 126:123–143.

Klumpp, D., and McKinnon, A. 1992. Community structure, biomass, and productivity of epilithic algal communities on the Great Barrier Reef: dynamics at different spatial scales. Marine Ecology Progress Series 86:77–89.

Koptur, S., Oberbauer, S., and Whelan, K. R. T. 1995. A comparison of damage and short-term recovery from Hurricane Andrew in four upland forest types of the Everglades. Report to Everglades National Park. National Park Service, Homestead, Florida.

Kress, W. J. 1985. Pollination and reproductive biology of *Heliconia*. Pp. 267–271 in D'Arcy, W. G., and Correa, M. D., eds., The Botany and Natural History of Panama. Missouri Botanical Garden, St. Louis.

Kurlansky, M. 1997. Cod: A Biography of the Fish That Changed the World. Walker and Co., New York.

Larkum, A. 1983. The primary productivities of plant communities on coral reefs. Pp. 221–230 in Barnes, D., ed., Perspectives on Coral Reefs. Australian Institute of Marine Science, Townsville, Australia.

Laurance, W. F., and Bierregaard, R. O. Jr. 1997. Tropical Forest Remnants: Ecology, Management, and Conservation of Fragmented Communities. University of Chicago Press, Chicago.

Laurance, W. F., Delamonica, P., Laurance, S. G., Vasconcelos, H. L., and Lovejoy, T. E. 2000. Rainforest fragmentation kills big trees. Nature 404:836.

Linhart, Y. B. 1973. Ecological and behavioral determinants of pollen dispersal in hummingbird pollinated *Heliconia*. American Naturalist 107:511–523.

Littler, D., and Littler, M. 2000. Caribbean Reef Plants. Offshore Graphics, Washington, DC.

Littler, M., and Littler, D. 1999. Bottom up and top down interactions determine dominances of benthic producers on tropical reefs. 16th International Botanical Congress Abstract 4773. International Botanical Congress, St. Louis.

Losos, J. B., Jackman, T. R., Larson, A., de Queiroz, K., and Rodriguez-Schettino, L. 1998. Contingency and determinism in replicated adaptive radiations of island lizards. Science 279:2115–2118.

Lüning, K. 1990. Seaweeds and Their Environment, Biogeography, and Ecophysiology. John Wiley, New York.

MacArthur, R. H., and Wilson, E. O. 1967. The Theory of Island Biogeography. Princeton University Press, Princeton, New Jersey.

Mathieson, A. C., Penniman C. A., and Harris, L. G. 1991. Northwest Atlantic rocky shore ecology. Pp. 109–191 in Mathieson, A. C., and Harris, P. H., eds., Intertidal and Littoral Ecosystems. Ecosystems of the World 24. Elsevier, Amsterdam.

McConnaughey, T., Adey, W., and Small, A. 2000. Optimizing calcification on the coral reef. Limnology and Oceanography 45:1667–1671.

Mead, J. I., Speiss, A. E., and Sobolik, K. D. 2000. Skeleton of extinct North American sea mink (*Mustela macrodon*). Quaternary Research 53:247–262.

Miller, R. J. 1985. Succession in sea urchin and seaweed abundance in Nova Scotia, Canada. Marine Biology 84:275–286.

Morrissey, J. 1980. Community structure and zonation of macroalgae and hermatypic corals on a fringing reef flat of Magnetic Island (Queensland, Australia). Aquatic Botany 8:91–139.

Myers, N., Mittermeier, R. A., Mittermeier, C. G., da Fonseca, G. A. B., and Kent, J. 2000. Biodiversity hotspots for conservation priorities. Nature 403:853–858.

Norton, T. A., Melkanion, M., and Andersen, R. 1996. Algal biodiversity. Phycologia 35:308–326.

Oberbauer, S. F., Kariyawasam, P., and Burch, J. L. 1997. Comparative analysis of growth, nutrition, carbon isotope ratios, and hurricane-related mortality of slash pines in south Florida. Florida Scientist 60:210–222.

Odum, H., and Odum, E. 1955. Trophic structure and productivity of a windward coral reef community on Eniwetok atoll. Ecological Monographs 35:291–320.

Perrier de la Bâthie, H. 1921. La Végétation Malgache. Annales du Musée Colonial de Marseille, series 3, 9:1–268.

Price, I., and Scott, F. 1992. The Turf Algae of the Great Barrier Reef. Part 1. Rhodophyta. Botany Department, James Cook University, Townsville, Australia.

Pringle, J. D. 1986. A review of urchin macro-algal associations with a new synthesis for nearshore, eastern Canadian waters. Monografias Biologica 4:191–218.

Raffaele, H., Wiley, J., Garrido, O., Keith, A., and Raffaele, J. 1998. A Guide to the Birds of the West Indies. Princeton University Press, Princeton, New Jersey.

Rathcke, B. J., and Jules, E. S. 1993. Habitat fragmentation and plant-pollinator interactions. BioScience 65:273–277.

Rauh, W. 1995. Succulents and Xerophytic Plants of Madagascar. Vol. 1. Strawberry Press, Mill Valley, California.

Reaka-Kudla, M. 1996. The global biodiversity of coral reefs. Pp. 83–108 in Reaka-Kudla, M., Wilson, D., and Wilson, E. O., eds., Biodiversity 2: Understanding and Protecting Our Natural Resources. Joseph Henry/National Academic Press, Washington, DC.

Richardson, L. L. 1998. Coral diseases: what is really known? Trends in Ecology and Evolution 13:438–443.

Ricklefs, R. E., and Bermingham, E. 1999. Taxon cycles in the Lesser Antillean avifauna. Ostrich 70:49–59.

Ricklefs, R. E., and Cox, G. C. 1972. Taxon cycles in the West Indian avifauna. American Naturalist 106:195–219.

Roberts, C. 1995. Effects of fishing on the ecosystem structure of coral reefs. Conservation Biology 9:65–91.

Roesel, C. S., Kress, W. J., and Bowditch, B. M. 1996. Low levels of genetic variation in *Phenakospermum guyannense* (Strelitziaceae), a widespread bat-pollinated Amazonian herb. Plant Systematics and Evolution 199:1–15.

Rose, G. 1993. Cod spawning on a migration highway in the northwest Atlantic. Nature 366: 458–461.

Rosen, D. E. 1976. A vicariance model of Caribbean biogeography. Systematic Zoology 24: 431–464.

Ross, M. S., O'Brien, J., and Sternberg, L. 1994. Sea-level rise and the reduction in pine forests in the Florida Keys. Ecological Applications 4:144–156.

Ryan, S. 1994. The Ice Hunters: A History of Newfoundland Sealing to 1914. Breakwater, St. Johns, Newfoundland.

Schemske, D. W., Husband, B. C., Ruckelshaus, M. H., Goodwillie, C., Parker, I. M., and Bishop, J. G. 1994. Evaluating approaches to the conservation of rare and endangered plants. Ecology 75:584–606.

Sieving, K. E., and Karr, J. R. 1997. Avian extinction and persistence mechanisms in lowland Panama. Pp. 156–170 in Laurance, W. F., and Bierregaard, R. O. Jr., eds., Tropical Forest Remnants: Ecology, Management, and Conservation of Fragmented Communities. University of Chicago Press, Chicago.

Simberloff, D. 1997. Strangers in Paradise: Impact and Management of Non-indigenous Species in Florida. Island Press, Washington, DC.

Small, A., and Adey, W. 2001. Reef corals, zooxanthellae, and free-living algae: a microcosm that demonstrates synergy between calcification and primary production. Ecological Engineering 16:443–457.

Small, A., Adey, W., and Spoon, D. 1998. Are current estimates of coral reef biodiversity too low? The view through the window of a microcosm. Atoll Research Bulletin 458:1–20.

Smith, T. J. III, Robblee, M. B., Wanless, H. R., and Doyle, T. W. 1994. Mangroves, hurricanes, and lightning strikes: assessment of Hurricane Andrew suggests an interaction across two differing scales of disturbance. BioScience 44:256–262.

Snyder, J. R., Herndon, A., and Robertson, W. B. Jr. 1990. South Florida Rockland. Pp. 230–277 in Myers, R. L., and Ewel, J. J., eds., Ecosystems of Florida. University of Central Florida Press, Orlando.

Steneck, R. S., and Carlton, J. T. 2000. Human alterations of marine communities: students beware! Pp. 445–468 in Bertness, M. D., Gaines, S. D., and Hay, M. E., eds., Marine Community Ecology. Sinauer, Sunderland, Massachusetts.

Stiles, F. G. 1975. Ecology, flowering phenology, and hummingbird pollination of some Costa Rican *Heliconia* species. Ecology 56:285–301.

Stouffer, P. C., and Bierregaard, R. O. Jr. 1995. Effects of forest fragmentation on understory hummingbirds in Amazonian Brazil. Conservation Biology 9:1085–1094.

————. 1996. Forest fragmentation and seasonal patterns of hummingbird abundance in Amazonian Brazil. Ararajuba 4:9–14.

Szmant, A. 2001. Why are coral reefs world-wide becoming overgrown by algae? Algae, algae everywhere and nowhere a bite to eat! Introduction to the special issue, Coral Reef Algal Community Dynamics. Coral Reefs 19:299–302.

Taggert, C. T., Anderson, J., Bishop, C., Colburne, E., Hutchings, J., Lilly, G., Morgan, J., Murphy, E., Myers, R., Rose, G., and Sheldon, P. 1994. Overview of cod stocks, biology, and environment in the northwest Atlantic region of Newfoundland, with emphasis on northern cod. ICES Marine Science Symposium 198:140–157.

Wade, D., Ewel, J., and Hofstetter, R. 1980. Fire in South Florida Ecosystems. U.S. Department of Agriculture, Forest Service General Technical Report SE-17. Southeastern Forest Experiment Station, Asheville, North Carolina.

Wagner, W. L., and Funk, V. A. 1995. Hawaiian Biogeography. Smithsonian Institution Press, Washington, DC.

Walker, L. R., Brokaw, N. V. L., Lodge, D. J., and Waide, R., eds. 1991. Ecosystem, Plant, and Animal Responses to Hurricanes in the Caribbean. Biotropica, special issue, 23:313–521.

Watling, I., and Norse, E. 1998. Disturbance of the seabed by mobile fishing gear: a comparison to forest clearcutting. Conservation Biology 12:1180–1197.

Webb, S. D. 1997. The great American faunal interchange. Pp. 97–122 in Coates, A. G., ed., Central America: A Natural and Cultural History. Yale University Press, New Haven.

Wilkinson, C. 1999. Global and local threats to coral reef functioning and existence: review and predictions. Marine and Freshwater Research 50:867–878.

Wynne, M. 1998. A checklist of benthic marine algae of the tropical and subtropical western Atlantic: first revision. Nova Hedwigia 116:1–150.

CHAPTER 7

INVASIVE SPECIES

Jessica Poulin, Ann Sakai, Stephen Weller, and Warren L. Wagner

THE CONSERVATION PROBLEMS caused by invasive species are less widely recognized than some other conservation issues such as tropical deforestation and air pollution, even though invasions represent a serious conservation threat and have major economic and ecological ramifications. Invasive species (also referred to as invaders, exotics, weeds, and aliens) are species that colonize and flourish in areas where they have not previously occurred (Mack 1985). Pimentel et al. (2000) estimated that containment, treatment, eradication, and damage control of invasives in the United States costs $137 billion annually. The U.S. Department of Agriculture estimates that programs associated with a single species, such as the fire ant (*Solenopsis* spp.), can cost more than $1 billion per year (USDA 2000). European gypsy moths (*Lymantria dispar*) have spread throughout the Northeast and upper Midwest, defoliating approximately 3 million acres of forest annually (Simberloff 1996). These moths caused $764 million in losses from eastern forests in 1981 alone (Goetz 2000). Animals are not the only costly invaders. Estimates of the costs associated with the eradication of all exotic plants are close to $100 billion over the course of their invasion histories (U.S. Congress OTA 1993). The cost of invasive species is perhaps most clearly seen in the effects of exotics on American crop plants. When citrus canker, a disease caused by a wind-dispersed bacterial pest (*Xanthomonas axonopodis* pv. *citri* [*Xac*]), was found in Florida citrus groves, rigorous quarantine and cleaning practices cost $160 million (Eden et al. 1985). Additionally, close to 11 million citrus plants had to be destroyed to contain the infestation at unknown cost to the citrus industry.

While these economic effects are of obvious concern, the ecological effects of invaders are also of growing concern to scientists, policymakers, and the public. Invasive species are the second leading cause of species endangerment and extinction worldwide, following only habitat destruction (Wilcove et al. 1998). Invasives negatively affect native habitats and species through habitat alteration, competition, and predation, and through introduction of novel parasites and pathogens. One of the most drastic ways that an invader can alter a native habitat is through fire cycling. D'Antonio and Vitousek (1992) suggest that grass invasions create feedback loops that increase the frequency and intensity of fires in invaded habitats. Invasive grasses produce large amounts of dry biomass that increases the chance of fire, particularly in ecosystems that contain few native grasses (Mack and D'Antonio 1998). The effects of these fires may be especially devastating in communities that have not experienced fire in their ecological and evolutionary history. In California, the ice plant (*Carpobrotus edulis*), a perennial native to South Africa, spreads in dense circular mats, outcompeting less densely aggregated native species (D'Antonio 1993; Weber and D'Antonio 1999). In the Florida Keys, the native semaphore cactus (*Opuntia corallicola*) has all but disappeared because the cactus moth (*Cactoblastis cactorum*), introduced into the Lesser Antilles to control prickly pear (*Opuntia* spp.), has arrived in the Keys and eats the semaphore cactus (Johnson and Stiling 1998). The introduction of chestnut blight fungus (*Endothia parasitica*) and Dutch elm disease (*Ophiostoma ulmi*) to this country in the early part of the 20th century has all but eradicated chestnuts and elms, respectively, from American forests (Council for Agricultural Science and Technology 2002).

While these examples of economic and ecological effects are troubling, they are of special concern because cases like these are common around the world. Up to 65% of species in a country's flora are nonnative (Vitousek et al. 2000). Some geographic areas are prone to invasion. In Florida, approximately 20% of vascular plants species are nonnative, and around 52% of plant species in Hawaii are nonnative (Simberloff 1996; Vitousek et al. 2000; see Wagner et al. 2001–present). While the exact figure is unknown, the proportion of invasives in the U.S. flora as a whole is almost certainly lower than in these highly invaded areas.

Not all invaders have extensive ecological impacts on their new habitats. In fact, many people would argue that invasion, or colonization, is a natural process without serious consequence for the environment. Tectonic shifting and normal alterations in climate and biological composition have led to both natural local extinctions and successful invasions (Lodge 1993). What has changed recently is the rate of new introductions. With the boom in human population size, increasing air travel, and expanding international trade, previously formidable barriers to species movement are now easily crossed. These effects have facilitated dispersal rivaling that of even the most extreme natural processes.

Even with this acceleration in transfer of species, most nonnative species that arrive in a novel habitat do not become invasive, even if they survive in the new area. The invasion process (fig. 7.1) is lengthy and requires arriving species to survive through many stages before they can begin to spread and have ecological impacts on their new environments.

SURVIVAL IN TRANSPORT

The first phase of the invasion process is surviving transport to the new habitat. This stage includes finding a successful means of transport, as well as arriving and gaining entrance to the new habitat. Human-facilitated invasions are often the result of deliberate introductions, including introducing ornamental plants for gardens, crop species as food resources, nonindigenous plants for erosion control, sporting species to encourage hunting and fishing, and biocontrol agents. In these cases, the means of transport and successful arrival are guaranteed by human intervention. Other invaders are brought to their new homes unintentionally. Unplanned introductions to the United States include zebra mussels (*Dreissena poly-*

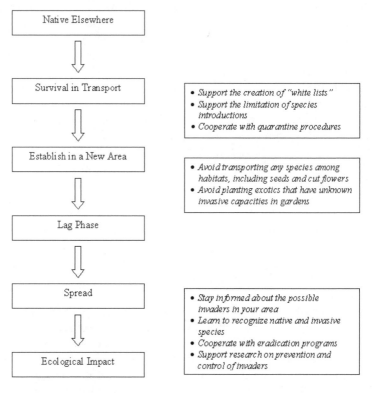

Figure 7.1 The invasion process, including actions the public can take to prevent invasion success at each stage (based on a figure from Sakai et al. 2001).

morpha) from the ballast water of ships, the fungal disease karnal bunt (*Tilletia indica*) from shipments of wheat, and white pine blister rust (*Cronartium ribicola*) from nursery stock from Europe (Goetz 2000; Vitousek et al. 2000; Council for Agricultural Science and Technology 2002).

The best chance for stopping an invasion is preventing it from occurring at all; thus, this first stage of the invasion process is critical. Unfortunately, current U.S. federal laws on the importation of particular species are generally created only after a problem species is recognized. In most cases the invader has already spread with significant ecological impact before the United States considers banning its import. This list of known problem species excluded from import is referred to as a black list. Another more effective method for preventing invasion is a white list, or "guilty until proven innocent" approach. With a white list, only species known not to have negative ecological impacts could be imported. This does not guarantee invasion-free imports—some white-list species may eventually become invasive—but it greatly reduces the scope of the problem (Simberloff 1996).

Sometimes "the cure is worse than the disease" (Civeyrel and Simberloff 1996), and the negative impacts of deliberate introductions for control of invasive species have not been anticipated. The predatory rosy wolf snail (*Euglandina rosea*) was introduced to the Hawaiian Islands as a biological control agent for the giant African snail (*Achatina fulica*), another invader. Without managing to control the giant African snail, the rosy wolf snail has caused the extinction of 30 native snail species (Civeyrel and Simberloff 1996). Many government agencies have encouraged the spread of invasives for erosion control, hunting, and other activities with little thought to their long-term ecological effects. Kudzu (*Pueraria lobata*) was cultivated as an ornamental plant and extensively planted as erosion control by the U.S. Soil Conservation Service with disastrous effects (Pappert et al. 2000). The kudzu vine can grow up to a foot a day and kills trees by blocking the sunlight, as well as quickly covering buildings, cars, and other stationary objects. Many state wildlife programs continue to actively encourage the import and release of nonnative game species to encourage tourism and the sale of hunting and fishing licenses (Simberloff 2000). Better coordination and understanding among conservation agencies, as well as those with different missions and goals, are needed to control the negative impacts of invasive species.

Better quarantine procedures, more inspection facilities and staff, and larger budgets could potentially limit the spread of unintentionally introduced species. There is now less than a 50% chance of finding a hidden plant or crop pests on a passenger plane or in cargo (Council for Agricultural Science and Technology 2002). Many commercial shipments are packed far from ports and their inspection sites, and are delivered far from inspection areas in recipient countries. This practice makes it all the easier for unintentional invaders to pass into new countries unscrutinized.

ESTABLISHING IN A NEW AREA

Many prospective invaders fail to establish in a new area because they are not suited to the new climate or are not accustomed to native predators and other biotic interactions (Lodge 1993). Competition, disease, novel parasites, pollinator limitation, and other factors may also play a role in the inability of exotic species to establish, but are more difficult to measure. The number of "failed invasions" is unknown because they are traditionally underrecorded, with an overrepresentation of easily observed species. Because of this, most estimates for the percentage of colonizers that successfully naturalize are probably too high (Lodge 1993). Williamson (1989) estimates that 10% of the colonists that survive transport become established.

Many traits that are commonly associated with "weed" species enhance establishment (Sakai et al. 2001). Such traits include high growth rates, the capacity for both vegetative and sexual reproduction, and the ability to survive in a variety of habitats (Baker 1965). Competitive ability is also important in establishing a self-sustaining population (Holway 1999; Byers 2000).

Containment of an invader is especially critical at this early stage of establishment. Human-aided spread of invasives may place the invader in the ideal habitat for establishment. Greater public awareness that the simple act of picking a fruit or wearing muddy boots may lead to the unintentional dispersal of seeds or insects into new natural areas may help limit the spread of invasive species. Similarly, more stringent laws and better public education of gardeners and horticulturists may help to limit the spread of many ornamental species that have weedy characteristics and can begin establishing in gardens, quickly spreading to more natural areas. Such repeated spread may be the key to naturalization of some alien species.

LAG PHASE

Most invasive species go through a lag period after they become naturalized, when they are present and sustained in their novel habitat, but are not spreading rapidly (Mack 1985). Estimates of the duration of lag times vary widely by species, ranging from 3 to 170 years (Crooks and Soulé 1996; Ellstrand and Schierenbeck 2000; Council for Agricultural Science and Technology 2002). The causes of lags are also variable and not always well understood. Some lags occur simply because small populations do not grow very fast, but as they slowly increase in number, their ability to spread improves. Other lags may be due to a lack of genetic mutations for traits such as disease resistance in the new invaders. Once such a mutation occurs, the population can spread quickly.

Lags may also occur because of the need for multiple introductions. The most successful invasive birds in North America, the starling and house sparrow, began to spread only after repeated introductions (Ehrlich 1989). Genetic variability may have increased through the repeated introductions of individuals, and may be necessary for the evolution of critical genetic changes needed to form a viable population in the new area. Multiple introductions might also be required simply to produce enough individuals so that the new population can grow (Sakai et al. 2001). This may indicate that planned introductions with limited genetic variability and low overall numbers of individuals may limit species with the potential for spreading.

Populations in the lag phase may also spread because of human activities. The wood-boring isopod, *Limnoria tripunctata,* was present in southern California harbors before 1900 and remained there for more than 50 years at a low population size (Crooks and Soulé 1996). Pollution controls established in the late 1960s that removed harmful chemicals from the harbor area led to a rapid spread of these isopod populations and subsequent damage to wooden structures in the area.

SPREAD AND ECOLOGICAL IMPACT

Problems associated with invasive species become more obvious as the species spread. Invaders may impact the ecology, and possibly economy, of their new habitats in many ways, including habitat destruction via disrupted nutrient cycling, altered fire cycling, erosion, and soil disturbance. Invasive organisms can outcompete natives by stealing nutrients and dominating resource use with high levels of reproduction (D'Antonio 1993; Holway 1999). Invasions can result in overgrazing or overpredation of native species, causing drastic declines in populations, and novel parasites that have not coevolved with the native species decimate populations with no natural resistance (Johnson and Stiling 1998; Byers 2000; Cabin et al. 2000).

These negative effects do not operate alone and can combine to facilitate the invasion process. In dry forests on the Hawaiian Islands, introduced ungulates (cattle and feral goats) overgraze native vegetation, causing habitat disturbance that may facilitate the invasion of alien plant species due to decreased competition from native species (Cabin et al. 2000). Once the cattle have reduced populations of native species, invasives have an easier time establishing. Introduced rats also lower native tree populations by preferentially eating native seeds and seedlings, further encouraging establishment of nonnative plant species. The combination of habitat disturbance, overpredation, and greater competitive ability has resulted in virtually no regeneration by native tree species in Hawaiian dry forests (Cabin et al. 2000).

The ecological impacts of invaders are not always easy to determine. Not only are the effects of invaders often combined, but also they can be masked by other factors. On the slopes of Diamond Head Crater on the island of Oahu, there is a single, small, shrinking population of the endemic native *Schiedea adamantis* (Caryophyllaceae). Plants of *S. adamantis* are limited to extremely rocky outcroppings. A several-year drought has undoubtedly contributed to the decline of this population, but alien invasives may also interact to limit recruitment of this highly endangered native. Most of the crater, including potential *S. adamantis* habitat, is covered with the nonnative plant *Leucaena leucocephala* (Fabaceae), once introduced for cattle feed and now widespread in drier habitats. In the absence of *L. leucocephala* native *S. adamantis* probably has some adaptations to survive even extended droughts. When present, *L. leucocephala* may limit *S. adamantis* to the most unfavorable regions of the crater, where *L. leucocephala* cannot survive, thus making *S. adamantis* highly susceptible to extinction during droughts. This native species may go extinct because of interactions between natural variation in climate and competition from an aggressive introduced competitor (S. G. Weller and A. K. Sakai, unpublished data).

CONCLUSIONS

One of the most significant factors in the spread and impact of invasive species is a lack of education about the problems they cause. The public is often uninformed, even in cases where experts know of potential or demonstrated problems caused by an invader. While the public may be resistant to the total elimination of certain invasive species such as those used in hunting, in other cases intensive education programs have been initiated and public support for eradication or containment programs has been strong.

For example, public education and eradication efforts have slowed the spread of *Miconia calvescens* (Melastomataceae), an aggressive invasive species in Hawaii with potentially devastating effects. This tree species, native to South and Central America, has attractive leaves that make it desirable to gardeners, who plant it, unknowingly starting infestations. Its seeds are also spread by birds and can be dispersed over long distances into pristine native forest. *Miconia* escaped from a botanical garden in Tahiti in the late 1930s, and 60% of the island is now covered with *Miconia* groves. This is particularly threatening to native species because the shallow root systems of *Miconia* lead to landslides that destroy areas of native vegetation and animal habitats. *Miconia* also thrives in sun or shade conditions, making it an exceptional competitor. A full 25% of Tahiti's native species are considered endangered because of *Miconia*. The Tahitian problems with *Miconia* could easily be repeated in other tropical habitats. In Hawaii, a fast response and intensive education campaign have helped to keep *Miconia* populations from more rapid expansion. People are discouraged from planting it and encouraged to remove it whenever they find it, and an extensive program for reporting *Miconia* populations has been instituted. Wanted posters are widespread on the islands, explaining the dangers of *Miconia* and how to destroy it. It remains to be seen if early identification of this problem species, along with the current levels of education and eradication efforts, will be sufficient to protect native species and habitats.

Invasive species may be a serious ecological and economic threat, especially if they begin to spread without preventive actions. Programs that could help stop the invasion process in the early stages include broad education about problems with invaders, increased public tolerance of inspection and eradication efforts, tighter regulation in introduction and movement of species, and increased research on prevention and control of invasives. Research on the invasion process itself is critical because the problems caused by invaders are interconnected, and their interactions with their invaded habitats are not easily predicted. These complexities make understanding and predicting the outcome of colonization events very difficult. The impact of an educated and dedicated public is also critical. The concept of thinking globally and acting locally is ideally suited to invasive species (Vitousek et al. 2000). By recognizing and preventing the spread of possible invaders and by supporting related legislation, the public can be a critical factor in preventing economic and ecological disasters caused by invasive species.

LITERATURE CITED

Baker, H. G. 1965. Characteristics and modes of origin of weeds. Pp. 147–169 in Baker, H. G., and Stebbins, G. L., eds., The Genetics of Colonizing Species. Academic Press, New York.
Byers, J. E. 2000. Competition between two estuarine snails: implications for invasions of exotic species. Ecology 81:1225–1239.
Cabin, R. J., Weller, S. G., Lorence, D. H., Flynn, T. W., Sakai, A. K., Sandquist, D., and Hadway, L. J. 2000. Effects of long-term ungulate exclusion and recent alien species control on the preservation and restoration of a Hawaiian tropical dry forest. Conservation Biology 14:439–453.
Civeyrel, L., and Simberloff, D. 1996. A tale of two snails: is the cure worse than the disease? Biodiversity Conservation 5:1231–1252.

Council for Agricultural Science and Technology. 2002. Invasive plant species: impacts on agricultural production, natural resources, and the environment. Issue Paper no. 20. Council for Agricultural Science and Technology, Ames, Iowa.

Crooks, J., and Soulé, M. 1996. Lag times in population explosions of invasive species: causes and implications. Pp. 39–46 in Sandlund, O. T., Schei, P. J., and Viken, A., eds., Proceedings of Norway/United Nations Conference on Alien Species. Directorate for Nature Management and Norwegian Institute for Nature Research, Trondheim, Norway.

D'Antonio, C. M. 1993. Mechanisms controlling invasion of coastal plant communities by the alien succulent *Carpobrotus edulis.* Ecology 74:83–95.

D'Antonio, C. M., and Vitousek, P. M. 1992. Biological invasions by exotic grasses, the grass/fire cycle, and global change. Annual Review of Ecology and Systematics 23:63–87.

Eden, W. G., Brush, R. R., Cox, H. C., Kingsolver, C. H., Lovitt, D. F., and Mulhern, F. J. 1985. Protecting United States Agriculture from Foreign Pests and Diseases. Blue Ribbon Panel Report. U.S. Department of Agriculture, Washington, DC.

Ehrlich, P. R. 1989. Attributes of invaders and the invading processes: vertebrates. Pp. 315–328 in Drake, J. A., Mooney, H. A., di Castri, F., Groves, R. H., Kruger, F. J., Rejmanek, M., and Williamson, M., eds., Biological Invasion: A Global Perspective. John Wiley and Sons, New York.

Ellstrand, N. C., and Schierenbeck, K. A. 2000. Hybridization as a stimulus for the evolution of invasiveness in plants? Proceedings of the National Academy of Sciences USA 97:7043–7050.

Goetz, R. 2000. Exotic invaders. Agricultures vol. 3, no. 3. Purdue University, West Lafayette, Indiana. http://www.agricultures.purdue.edu/agricultures/past/summer2000/features/feature_04.html.

Holway, D. A. 1999. Competitive mechanisms underlying the displacement of native ants by the invasive Argentine ant. Ecology 80:238–251.

Johnson, D. M., and Stiling, P. D. 1998. Distribution and dispersal of *Cactoblastis cactorum* (Lepidoptera: Pyralidae), an exotic *Opuntia*-feeding moth, in Florida. Florida Entomology 81:12–22.

Lodge, D. M. 1993. Biological invasions: lessons for ecology. Trends in Ecology and Evolution 8:133–137.

Mack, M. C., and D'Antonio, C. M. 1998. Impacts of biological invasions of disturbance regimes. Trends in Ecology and Evolution 13:195–198.

Mack, R. N. 1985. Invading plants: their potential contribution to population biology. Pp. 127–142 in White, J., ed., Studies in Plant Demography: A Festschrift for John L. Harper. Academic Press, London.

Pappert, R. A., Hamrick, J. L., and Donovan, L. A. 2000. Genetic variation in *Pueraria lobata* (Fabaceae), an introduced, clonal, invasive plant of the southeastern United States. American Journal of Botany 87:1240–1245.

Pimentel, D., Lach, L., Zuniga, R., and Morrison, D. 2000. Environmental and economic costs of non-indigenous species in the United States. BioScience 50:53–65.

Sakai, A. K., Allendorf, F. W., Holt, J. S., Lodge, D. M., Molofsky, J., With, K. A., Baughman, S., Cabin, R. J., Cohen, J. E., Ellstrand, N. C., McCauley, D. E., O'Neil, P., Parker, I. M., Thompson, J. N., and Weller, S. G. 2001. The population biology of invasive species. Annual Review of Ecology and Systematics 32:305–322.

Simberloff, D. 1996. Impacts of introduced species in the United States. Consequences 2: 13–22.

———. 2000. Non-indigenous species: a global threat to biodiversity and stability. Pp. 325–334 in Raven, P., and Williams, T., eds., Nature and Human Society: The Quest for a Sustainable World. National Academy Press, Washington, DC.

U.S. Congress. Office of Technology Assessment (OTA). 1993. Harmful Non-indigenous Species in the United States. OTA-F-565. Government Printing Office, Washington, DC.

USDA (U.S. Department of Agriculture). 2000. Pest Risk Assessment for Importation of Solid Wood Packing Materials into the United States. U.S. Department of Agriculture, Animal and Plant Health Inspection Service and U.S. Forest Service, Washington, DC.

Vitousek, P. M., D'Antonio, C. M., Loope, L. L., and Westbrooks, R. 2000. Biological invasions as global environmental change. American Scientist 84:468–478.

Wagner, W. L., Herbst, D. R., and Palmer, D. D. 2001–present. Flora of the Hawaiian Islands. Smithsonian Institution. http://ravenel.si.edu/botany/pacificislandbiodiversity/hawaiianflora/index.htm.

Weber E., and D'Antonio, C. M. 1999. Phenotypic plasticity in hybridizing *Carpobrotus* spp. (Aizoaceae) from coastal California and its role in plant invasion. Canadian Journal of Botany 77:1411–1418.

Wilcove, D. S., Rothstein, D., Dubow, J., Phillips, A., and Losos, E. 1998. Quantifying threats to imperiled species in the United States. BioScience 48:607–615.

Williamson, M. 1989. Mathematical models of invasions. Pp. 329–350 in Drake, J. A., Mooney, H. A., di Castri, F., Groves, R. H., Kruger, F. J., Rejmánek, M., and Williamson, M., eds., Biological Invasions: A Global Perspective. SCOPE 37. John Wiley and Sons, Chichester, United Kingdom.

CHAPTER 8

GLOBAL CLIMATE CHANGE:
THE SPRING TEMPERATE FLORA

Paul M. Peterson, Stanwyn G. Shetler, Mones S. Abu-Asab,
and Sylvia S. Orli

A CONSENSUS OF scientists and environmentalists has been reached over the past few years that the earth's atmosphere is warming, and that humans are the main contributors to this trend. This conclusion was reached by the United Nations' Intergovernmental Panel on Climate Change (IPCC). The IPCC report (Watson et al. 2001), the most definitive on the subject, declared that the 1990s had the warmest years on record, and most of the increase is attributed to human activities—the burning of fossil fuels and the release of carbon dioxide and greenhouse gases. The report also predicts that global average surface temperature will increase by 1.4 to 5.8°C during this century.

Spring advances in first-flowering in plants from the Washington, DC, area provide evidence for global warming. The trend of average first-flowering times per year for 100 species with 20 years' or more data shows a significant advance of 3.2 days over a 32-year period. When 10 species that exhibit later first-flowering times are excluded from the data set, the remaining 90 species show a significant advance of 5.1 days. Trends for 80 species flowering significantly earlier range from −2.8 to −44.2 days, while those for 7 significantly later-flowering species range from +2.2 to +9.7 days. Advances of first-flowering in these 90 species are directly correlated with local increase in minimum temperature (T_{MIN}). These results are consistent with other studies showing that changes in air temperature have advanced the average annual growing season by 6 days in Europe and that warmer temperatures and higher carbon dioxide levels have promoted summer plant growth, thereby advancing the season by 7 days in northern latitudes.

CLIMATE CHANGE

The effects of global warming in the 20th century, mostly caused by human activity, have been marked by a rise in the average surface temperature (about 0.6°C), a 40% thinner Arctic ice cover, and increased levels of carbon dioxide in the atmosphere (Houghton et al. 1995). The northern high latitudes have warmed by about 0.8°C just since the early 1970s (Hansen et al. 1999). These changes appear to have had a profound effect on the first-flowering times of plants in the Washington, DC, area.

Climate change is an important environmental issue affecting not only entire ecosystems but also the dominant plants of a given region (Melillo 1999). Other studies of the natural vegetation indicate that nocturnal global warming may be an important factor in plant growth (Alward et al. 1999; Chapin et al. 1995). A study of short-grass steppe vegetation over a ten-year period shows that net primary production of a C_4 grass (*Bouteloua gracilis*) was lower with increasing T_{MIN}, whereas native and exotic C_3 forbs increased in production and abundance (Alward et al. 1999). Long-term ecological research in the Alaskan Arctic indicates that climate change, when simulated by increasing the mean daily air temperature above the vegetation by 3.5°C, can alter plant species composition (Chapin et al. 1995). Changes in the periodicity in plants as related to climatic events (phenology) have been found in Europe, where leaf unfolding in spring has advanced 6 days, whereas leaf coloring has been delayed 4.8 days, thus lengthening the average annual growing season by 10.8 days (Menzel and Fabian 1999).

Flowering in angiosperms is an important phenological cycle. Plants in temperate areas, such as the mid-Atlantic region of the North America, are adapted to an annual seasonal cycle with a winter dormancy period that is sensitive to temperature and light. Flowering time is directly related to temperature. To investigate potential changes in the timing of first-flowering, Abu-Asab et al. (2001) examined first-flowering records of 100 plant species, representing 44 families of angiosperms, for 31 years of the 32-year period 1970–2001 (1984 not recorded) in the Washington, DC, area. Because this investigation suggested that global climate change has an effect on plant flowering in the capital of the United States, more details are provided in this chapter.

MEASURING FIRST-FLOWERING

From a database of first-flowering records for over 600 species, the 100 species with the greatest number (20 or more) of recorded years were selected (Abu-Asab et al. 2001). First-flowering here refers to the stage at which a mono- or diclinous-flower begins anthesis or is receptive to pollen. Washington and vicinity (35-mile radius from center of DC; latitude 38°24′ to 39°23′ N, longitude 76°24′ to 77°42′ W) is a large area, and thus many different observers (>125) were involved in the recording. In actual practice the date of first-flowering is the first observed date and not necessarily the absolute earliest (Shetler and Wiser 1987). Using average first-flowering time per year, linear trends and their significance according to the F-test were calculated, along with confidence intervals, for the 100 species, the 90 species that exhibit only an earlier flowering, and 10 species that exhibit only later flowering (fig. 8.1A–C). Trends for the majority of the species (80%) are statistically significant (table 8.1).

To investigate potential correlation between flowering trends and environmental factors during the last 30 years, linear trends and confidence intervals for local climate data were calculated. The climate data include average minimum temperature (T_{MIN}), average precipitation (fig. 8.2), and precipitation per month. In

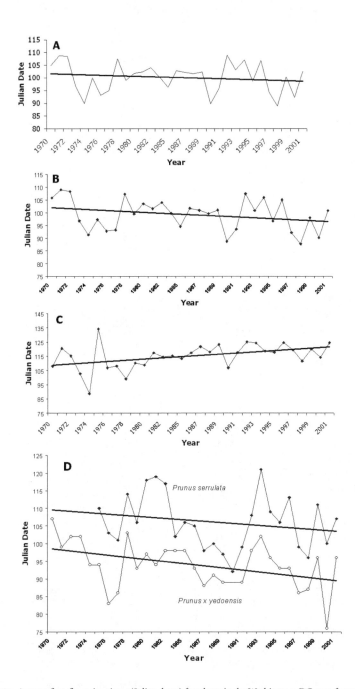

Figure 8.1 Average first-flowering times (Julian dates) for plants in the Washington, DC, area for 1970–2001. Bold line represents the trend, and regular line indicates the average. A, For all 100 plant species. B, For 80 plant species flowering significantly earlier. C, For 7 species flowering significantly later. D, Trends of the Japanese cherry trees *Prunus serrulata* (6.1 days' advance in flowering) and *Prunus × yedoensis* (9 days' advance in flowering).

Table 8.1 Eighty species with statistically significant advances in first-flowering

Species	Advance in flowering (days)	Standard deviation	Number of years on record
Acer negundo	11.67	7.59	29
Acer rubrum	19.20	14.23	31
Achillea millefolium	11.62	8.77	23
Alliaria petiolata*	11.77	7.56	28
Alnus serrulata	5.45	15.41	26
Amelanchier canadensis	6.28	6.74	22
Anemone quinquefolia	19.17	9.39	24
Anemonella thalictroides	9.18	9.02	26
Anthoxanthum odoratum*	17.84	7.71	23
Aquilegia canadensis	17.57	9.94	27
Arisaema triphyllum	10.70	6.39	27
Aronia arbutifolia	20.91	10.05	24
Asarum canadense	5.16	8.21	30
Asimina triloba	17.64	8.80	27
Barbarea vulgaris*	2.83	7.05	26
Cardamine hirsuta*	41.17	25.60	30
Cercis canadensis	9.48	9.81	28
Chelidonium majus*	17.32	7.64	22
Chionanthus virginicus	8.63	6.03	27
Chrysogonum virginianum	28.21	19.32	25
Cichorium intybus*	11.40	6.30	23
Claytonia virginica	7.54	13.07	29
Cornus florida	8.75	8.48	29
Corydalis flavula	8.99	8.28	26
Dicentra cucullaria	2.86	7.00	29
Dirca palustris	13.81	8.83	22
Duchesnea indica*	44.21	31.33	28
Erigenia bulbosa	9.17	15.27	23
Erodium cicutarium*	8.94	7.65	31
Fragaria virginiana	22.91	18.79	26
Geranium maculatum	14.71	8.54	29
Glechoma hederacea*	6.56	10.84	29
Hepatica americana	15.88	23.62	30
Ilex opaca	7.37	6.52	26
Kalmia latifolia	8.18	6.68	26
Lamium purpureum*	36.53	19.69	30
Lindera benzoin	8.28	8.57	30
Liquidambar styraciflua	12.27	7.24	21
Liriodendron tulipifera	12.15	7.93	31
Maianthemum canadense	5.07	6.43	22
Mertensia virginica	16.01	11.02	29
Nyssa sylvatica	36.38	11.07	24

Table 8.1 *(continued)*

Species	Advance in flowering (days)	Standard deviation	Number of years on record
Panax trifolius	17.19	10.17	25
Phlox subulata	7.47	14.20	27
Plantago lanceolata*	15.51	7.91	24
Poa annua*	21.13	20.16	25
Podophyllum peltatum	11.03	7.34	30
Polygonatum biflorum	9.65	7.63	22
Potentilla canadensis	9.85	7.00	26
Prunus serrulata+	6.08	7.72	26
Prunus × yedoensis+	9.02	6.63	30
Ranunculus abortivus	3.18	7.83	29
Ranunculus bulbosus*	20.82	20.95	25
Rhododendron periclymenoides	15.99	8.76	28
Robinia pseudoacacia	5.19	6.44	28
Rumex acetosella*	7.58	9.81	23
Rumex crispus*	17.56	9.69	22
Salvia lyrata	17.51	8.44	24
Sanguinaria canadensis	12.47	7.91	27
Sassafras albidum	8.69	8.30	29
Senecio aureus	14.23	13.22	29
Smilacina racemosa	4.10	4.99	28
Staphylea trifolia	7.02	6.92	25
Stellaria media*	10.24	26.50	29
Taraxacum officinale*	23.81	28.88	26
Toxicodendron radicans	5.83	6.44	23
Trifolium pratense*	7.62	6.16	25
Tussilago farfara*	5.37	14.80	28
Ulmus americana	3.65	10.12	30
Uvularia perfoliata	11.43	8.81	22
Vaccinium corymbosum	24.53	13.93	22
Vaccinium stamineum	15.40	10.97	27
Veronica hederaefolia*	6.09	13.97	25
Viburnum acerifolium	15.20	6.31	22
Viburnum prunifolium	10.17	7.87	28
Viola bicolor	8.69	9.06	26
Viola pubescens	4.41	9.80	23
Viola sororia	10.71	7.84	31
Viola striata	14.09	10.38	24
Zizia aurea	21.55	18.30	22

Note: Species marked with an asterisk are naturalized, those marked with + are cultivated, and all others are native. Nomenclature largely follows Kartesz 1994.

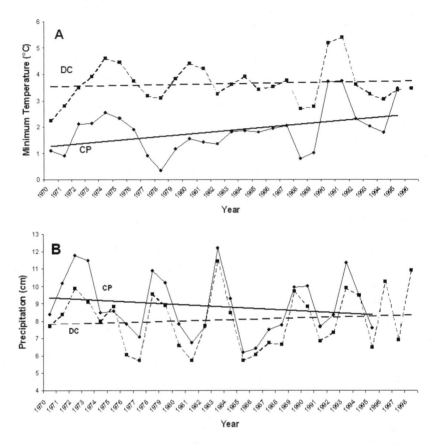

Figure 8.2 Climate data and their trends for Washington, DC (DC), and College Park, Maryland (CP), for 1970–1997. Solid lines with circles represent CP averages, and dashed lines with squares indicate DC averages. A, Average minimum temperature (T_{MIN}) for six months (December–May). B, Average precipitation for the same period. Data from National Oceanic and Atmospheric Administration, Climate Diagnostics Center.

addition, Pearson correlation coefficients were calculated to test for relationships between T_{MIN} and average precipitation versus first-flowering times.

EARLY FLOWERING

The linear trend of average first-flowering time for all 100 species shows a significant advance of 3.2 days for the 32-year period (fig. 8.1A). However, the elimination from the data set of the 10 species that exhibit the later flowering trend results in a significant advance of 5.1 days for the 32-year period (fig. 8.1B). Eighty of the 90 species exhibit a significant advance (table 8.1), and together show an average decrease of −5.8 days for the 32-year period. Ten of these 90 species have insignificant trends (an asterisk indicates naturalized species): *Capsella bursa-pastoris, Dentaria* *laciniata, Epigaea repens, Euonymus americanus, Galium aparine,* *Paulownia tomentosa, Phlox divaricata, Prunus serotina, Saxifraga virginiensis,* and *Sedum ternatum.* Significant trends for the 80 earlier flowering species range from −2.8 to −44.2 days. Of these 80 species, *Duchesnea indica* (false strawberry) shows the greatest advance in flowering time, with a trend of −44.2 days, whereas *Barbarea vulgaris* (yellow rocket) shows the smallest advance with a trend of −2.8 days. Both are naturalized species.

The following ten species exhibit later first-flowering times (an asterisk indicates naturalized species): *Acer saccharinum,* **Ajuga reptans, Houstonia caerulea,* **Lamium amplexicaule,* **Lonicera japonica,* **Melilotus officinalis, Osmorhiza claytonii,* **Solanum dulcamara, Stellaria pubera,* and *Trillium sessile.* Seven of the ten species show significant delays in first-flowering; of these, *Lonicera japonica* (Japanese honeysuckle) has a trend of +9.7 days, and *Osmorhiza claytonii* (sweet cicely) has a trend of +2.2 days.

From 1970 to 1999, the average T_{MIN} for December to May, covering the recorded season of first-flowering, shows an increase of +1.2°C at College Park, Maryland (CP), and +0.2°C at Washington, DC (fig. 8.2A). The monthly average precipitation for the six-month period at DC has increased +0.6 cm, while it has decreased −1.1 cm at CP (fig. 8.2B). After surveying the precipitation by month, it appears that both stations show a decrease in precipitation during December (−0.7 cm for DC; −1.6 cm for CP), February (−0.2 cm for DC; −1.9 cm for CP), April (−0.6 cm for DC; −1.3 cm for CP), and May (−1.1 cm for DC; −0.8 cm for CP); whereas an increase in precipitation is seen in January (+1.7 cm for DC; +0.2 cm for CP) and March (+1.8 cm for DC and CP). Average flowering time (in Julian date) per year of the 100 species shows a significant negative correlation with T_{MIN} over the studied period for both CP and DC. No significant correlation was found between first-flowering and precipitation trends.

Although the onset of flowering is cued by photoperiodism, break of dormancy is triggered by temperature (Salisbury and Ross 1992). Because the annual photoperiodic cycle presumably has not changed during the last 32 years, we must assume that other factors, such as temperature and precipitation, are causing earlier spring flowering. The analysis presented here suggests that trends toward earlier flowering are the result of a warming trend in the study area. This warming trend, +1.2°C at CP, coincides with the global warming trend (Easterling et al. 1997). However, other causal factors not investigated by this study, such as carbon dioxide, cannot be eliminated as being important in regulating spring flowering. Likewise, why ten species exhibit later first-flowering times is difficult to explain. At this time, compelling evidence exists that the rise in the average minimum temperatures (T_{MIN}) is contributing to earlier flowering in at least 80 of the 100 species investigated. The results presented here are consistent with other studies showing that changes in air temperature have advanced the average annual growing season by six days in Europe and that warmer temperatures and higher carbon dioxide levels have promoted summer plant growth, thereby advancing the season by seven days in northern latitudes (Menzel and Fabian 1999).

Possible consequences of the warming and earlier-flowering trends should be investigated. Other studies (Shetler and Wiser 1987) have shown that the earliest noncultivated plants to flower in the Washington area are predominantly naturalized introductions, such as *Stellaria media* (common chickweed), *Taraxacum officinale* (dandelion), and *Lamium amplexicaule* (henbit). These species can exploit the slightest bit of warmth during the cold season and bloom quickly. The warming trend resulting in milder winters is likely to favor increasingly the naturalization of exotic species. The introduced element already appears to constitute about 40% of the local flora (Shetler and Orli 2000), and clearly these introductions have been rising steadily over the years.

Spring ephemerals require a tight synchrony with their pollinators, and the effects of earlier flowering on this relationship should be studied. Are the insects keeping pace? Likewise, the synchrony of the flowering cycles of the ephemerals with canopy closure needs examination. Of practical consequence is the earlier flowering of wind-pollinated trees and shrubs, which starts the season for allergy sufferers earlier. Finally, there is the Cherry Blossom Festival in Washington, DC, each spring. On average the two principal species, *Prunus serrulata* (Kwanzan cherry and other varieties) and *P.* × *yedoensis* (Yoshino cherry), bloom six and nine days earlier, respectively, than they did in 1970 (fig. 8.1D). This has major tourist implications since predicting peak-flowering from year to year can be problematic.

ACKNOWLEDGMENTS

This chapter draws substantially on an earlier article by the authors (Abu-Asab et al., *Biodiversity and Conservation* 10 [2001]: 597–612). The authors thank the more than 125 people who contributed first-flowering dates through the years, especially Aaron Goldberg and the late John J. Wurdack; and Susan J. Pennington for comments that improved the manuscript.

LITERATURE CITED

Abu-Asab, M. S., Peterson, P. M., Shetler, S. G., and Orli, S. S. 2001. Earlier plant flowering in spring as a response to global warming in the Washington, DC, area. Biodiversity and Conservation 10:597–612.

Alward, R. D., Detling, J. K., and Milchunas, D. G. 1999. Grassland vegetation changes and nocturnal global warming. Science 283:229–231.

Chapin, F. S., Shaver, G., Giblin, A., Nadelhoffer, K., and Laundre, J. 1995. Responses of arctic tundra to experimental and observed changes in climate. Ecology 76:694–711.

Easterling, D. R., Horton, B., Jones, P. D., Peterson, T. C., Karl, T. R., Parker, D. E., Salinger, M. J., Razuvayev, V., Plummer, N., Jameson, P., and Folland, C. K. 1997. Maximum and minimum temperature trends for the globe. Science 277:364–367.

Hansen, J., Ruedy, R., Glascoe, J., and Sato, M. 1999. GISS analysis of surface temperature change. Journal of Geophysical Research 104:30997–31022.

Houghton, J. T., Meria Filho, L. G., Callender, B., and Harris, N., eds. 1995. The Science of Climate Change. Cambridge University Press, Cambridge.

Kartesz, J. T. 1994. A Synonymized Checklist of the United States, Canada, and Adjacent Greenland. 2nd edition. Timber Press, Portland, Oregon.

Melillo, J. M. 1999. Climate change: warm, warm on the range. Science 283:183–184.

Menzel, A., and Fabian, P. 1999. Growing season extended in Europe. Nature 397:659.

Salisbury, F. B., and Ross, C. W. 1992. Plant Physiology. 4th edition. Wadsworth Publishing, Belmont.

Shetler, S. G., and Orli, S. S. 2000. Annotated Checklist of the Vascular Plants of the Washington-Baltimore Area. Part 1, Ferns, Fern Allies, Gymnosperms, and Dicotyledons. Smithsonian Institution (National Museum of Natural History, Department of Botany), Washington, DC.

Shetler, S. G., and Wiser, S. K. 1987. First flowering dates for spring-blooming plants of the Washington, DC, area for the years 1970 to 1983. Proceedings of the Biological Society of Washington 100:993–1017.

Watson, R. T., and the Core Writing Team, eds. 2001. Climate Change 2001: Synthesis Report. Cambridge University Press, Cambridge.

CHAPTER 9

GENETIC CONSEQUENCES OF REDUCED DIVERSITY: HETEROZYGOSITY LOSS, INBREEDING DEPRESSION, AND EFFECTIVE POPULATION SIZE

Paul M. Peterson and Carrie L. McCracken

BIODIVERSITY COMPRISES the variation of forms of life. This can be the diversity of species in a habitat, or it can be the genetic diversity within a species. When managing a particular habitat, species diversity is a primary concern. When attempting to ensure the continued survival of a species, however, managing for genetic diversity is a concern. If, for any number of reasons such as habitat loss, disruption of pollinator-plant interactions, genetic drift, inbreeding, or outbreeding, the fitness (the ability of organisms to contribute to subsequent generations) of individuals in a population or a species is lowered, the continued existence of the individuals, and thus the population or species, is in jeopardy. Avise's introduction to conservation genetics (1996) highlights the issue of heterozygosity as a means of measuring within-population genetic variability. He poses four questions of paramount concern to plant conservation genetics: (1) Is genetic variation reduced in endangered species? (2) Is this a cause for ecological or evolutionary concern? (3) Should a population be managed for increased variation? (4) How serious a problem is inbreeding depression? This chapter provides some bases for evaluating these important questions. We define terms and concepts and then present examples from the literature to illustrate the possible management choices for the conservation biologist.

THE HARDY-WEINBERG MODEL

Conservation genetics has its foundations in population genetics—the study of gene frequencies in populations and the forces that change them through time and space. A group of organisms of the same species living in a particular area at the same time is known as a population. A fundamental concept of population genetics is the Hardy-Weinberg model. In this model, several assumptions are made to imagine an "ideal" population. It can be informative to test real populations against this ideal model. First, it is assumed that this "ideal" population is of diploid organisms—organisms that have two copies of each locus (a location in the genome). Second, the generations are assumed to be nonoverlapping in time. Third, there is no mutation; that is, new alleles (alternate forms of the locus) will not be created in the ideal population. Fourth, the study organism is sexually reproducing, so that each copy of the locus comes from a different parent. Fifth, the genes are not subject to natural selection; one allele is as fit as another.

In a simplified version of the model, a single locus has two alleles that can arbitrarily be named A and B. Individuals in the population can have the genotypes, or genetic constitution, of AA, AB, and BB. Individuals with the genotypes AA and BB are homozygous at this locus. A homozygous individual has two copies of the same allele at a locus. A heterozygous individual has two different alleles at a locus. In this example, individuals with genotype AB are heterozygous. The frequency of the allele A, p, can be determined as twice the number of individuals of genotype AA plus the number of individuals with the genotype AB divided by twice the population size ($p = [2 \times AA + AB]/2N$ where N is the population size). The frequency of the allele B, q, is the sum of the number of individuals of genotype AB and twice the number of individuals of genotype BB divided by twice the population size ($q = [AB + 2 \times BB]/2N$). Because all alleles in the population are either A or B, $p + q$ is equal to 1. Another assumption about this "ideal" population is that mating is random, so that the probability of any two alleles combining is based on their frequency. In this "ideal" population the expected frequency of the genotype AA is p^2 and the expected frequency of genotype BB is q^2. The expected heterozygosity (H_e), the frequency of the genotype AB, is $2pq$. This is the combination of the probability of receiving an A from the mother and receiving a B from the father, pq, and the probability of receiving a B from the mother and receiving an A from the father, pq. This expected heterozygosity is a commonly used measure of genetic diversity of a population.

SMALL POPULATIONS

When a population size drops, it violates a sixth assumption of the Hardy-Weinberg model, the assumption of a large population size. A small population is more likely to be subject to changes due to genetic drift, that is, random changes in gene frequency. The probability of an allele reaching fixation (100% frequency in a population) increases with decreasing population size. Then, at that locus, there are no more heterozygotes. A genetic bottleneck occurs in nature when there is a sharp reduction in the number of individuals of a species in a particular place and time and there is a corresponding loss of genetic diversity (Barrett and Kohn 1991). When allele frequencies change or the alleles are lost by chance in these small populations, the process is referred to as random genetic drift.

In a small population, there is an increased chance of nonrandom mating. The seventh assumption of the Hardy-Weinberg model is that the individuals in a population mate randomly. With fewer potential mates available in a small population, the probability of close relatives mating increases. Inbreeding will result in a loss of heterozygotes. To illustrate, the most extreme case of inbreeding is self-fertilization. A heterozygous individual that reproduced by self-fertilization would have only

50% heterozygous offspring. If all of these offspring reproduce by self-fertilization, only half of them, the heterozygotes, could produce heterozygous offspring. Only half of the offspring produced by these heterozygotes would be heterozygotes themselves. In this example, in each generation the proportion of heterozygotes would be halved until the number of heterozygotes would be negligible. Inbreeding (F_{IS}) can be measured by the difference in the observed heterozygosity (H_o) and the expected heterozygosity (H_e): $F_{IS} = (H_e - H_o)/H_e$. The observed heterozygosity (the number of heterozygous individuals divided by the total number of individuals in the population) is less than the expected heterozygosity in an inbreeding population.

INBREEDING DEPRESSION

Inbreeding can result in a loss of fitness. This is called inbreeding depression. Inbreeding may result in decreased fitness because there is increased expression of deleterious recessive alleles. A recessive allele is one that is expressed only when it occurs as a homozygote. When an organism is heterozygous at that locus, its fitness is not affected by the presence of the deleterious recessive allele. Because the allele is deleterious, selection will remove the homozygotes. Because it is recessive, however, it will persist in the population in the heterozygous form at a reduced frequency. Therefore, in a randomly mating population, these homozygotes will occur only rarely. Inbreeding causes an increase in the proportion of homozygotes, so these lower-fitness individuals will occur more often. In a population that is experiencing inbreeding due to decreased population size, inbreeding depression could cause the loss of even more individuals, further threatening the persistence of the population.

OUTBREEDING DEPRESSION

The eighth assumption of the Hardy-Weinberg model is that there is no migration between populations. Outbreeding is the crossing of individuals from populations that normally do not cross. It is possible that genetic exchange between previously isolated populations will result in outbreeding depression. When populations are isolated in a particular habitat, they may become locally adapted. Genes introduced from another population are likely to be less adapted to the local habitat. Hybrids of the locally adapted individuals and individuals from other habitats will be less adapted to the local environment than their locally adapted parents. Another problem that can cause outbreeding depression is the breakdown of coadapted gene complexes. Genes at different loci may evolve to work together, forming coadapted gene complexes (Brown and Feldman 1981; Schaeffer and Miller 1993). They may form independently in different populations. Crossing these separate populations may result in the recombining of genes, so that the coadapted genes are no longer present in an individual (Waser 1993).

HYBRIDIZATION

The extreme case of outbreeding is hybridization, which is the crossing of individuals of different species. The most commonly used definition of *species* is the biological species concept, that species are a set of interbreeding organisms (Ridley 1993). Hybridization may or may not result in either the loss or gain of a species. When a rare species hybridizes with a more common species, on one hand, eventually there may be no individuals left of the original rare species. All offspring of the rare species are hybrids, and the rare species is lost. Potentially the offspring may suffer from outbreeding depression. On the other hand, hybridization may be a mechanism by which new plant species form (see examples in Arnold 1993 and Wolfe et al. 1998), potentially forming a new species that is adapted to a niche intermediate to its parental species.

PLANT REPRODUCTION MECHANISMS

The common condition for plants is to reproduce sexually by outcrossing or xenogamy (cross-pollinating involving flowers on different individuals). At the other end of the spectrum is asexual reproduction, where a plant may reproduce by budding or vegetative propagation. Plants may also self-pollinate (known as autogamy), usually done by plants with cleistogamous (closed) flowers. Therefore, the historical heterozygosity level in a population can depend on these life-history characters. An autogamous plant will have lower heterozygosity. However, it is possible that because of this, there are fewer deleterious recessive alleles in an autogamous plant than in a xenogamous plant. The recessive alleles are expressed and then eliminated more often because they occur much more often as a homozygote. A high level of inbreeding is expected in these species, even when there is a large population of fit individuals.

EFFECTIVE POPULATION SIZE

The underlying objective of the conservation biologist is to manage and thus perpetuate, preferably in situ, the effective population size of a species. The effective size of a population is the number of individuals in an ideal population having the same magnitude of random genetic drift as the actual population (Hartl and Clark 1989). There are three ways to determine the genetically effective population size (Hartl and Clark 1989; Vrijenhoek 1996); the one most commonly employed by biologists is the inbreeding effective size, which measures the effect of variation in the family size (Wright 1931). In populations with only a few families contributing the most offspring to the next generation, the effective population size is less than the actual population size. Inbreeding effective size (N_{ef}) can be calculated as ($4N - 4$)/($V_2 + 2$), where $V_2 = s(1 - s)\bar{x} + s^2 V$ and $V =$ variance of family size, $\bar{x} =$ mean of family size, $s = 2/\bar{x}$ (Cook and Callow 1999). The inbreeding effective size is inversely related to the inbreeding coefficient. Therefore, in small populations (low *N* value), it is more likely that individuals share common ancestors, that is, the population is more inbred (Cook and Callow 1999).

MOLECULAR BIOLOGY AND CONSERVATION GENETICS

Advances in molecular biology have allowed scientists to measure genetic diversity on the level of the genome. The first method commonly used was allozyme electrophoresis. Proteins are the direct expression of DNA. Allozymes are different forms of an enzyme (which are proteins) specified by alternative alleles at a single gene locus, whereas isozymes may be used to designate forms of an enzyme specified by structural genes at different loci (Crawford 1989). Electrophoresis is a technique that allows a scientist to separate allozymes based on differences in size and electric charge. The different allozymes are equivalent to the alleles discussed in the earlier models. Newer methods have been developed that allow the scientist to directly assess the variation in DNA. Microsatellites are one such technique. They are short sequences of DNA that are repeated in tandem in portions of the genome that are not expressed. Microsatellites vary in the number of times that they are repeated. Another commonly used technique is AFLPs (amplified fragment length polymorphisms). With this technique random portions of the genome are amplified when the recognition sequence is present.

CASE STUDIES

Enzyme Electrophoresis
A classic allozyme study of a serpentine endemic, *Clarkia franciscana* F. H. Lewis & P. H. Raven (Onagraceae), and two more widely spread congeners, *C. amoena* (Lehm.) A. Nelson & J. F. Macbr. and *C. rubicunda* (Lindl.) F. H. Lewis & M. R. Lewis, was performed by Gottlieb (1973). Previous cytological research by Lewis and Raven (1958) suggested that *C. amoena* was an ancient species that gave rise to *C. rubicunda*, which in turn was the progenitor to *C. franciscana*, a self-pollinating annual now known from fewer than five occurrences (CNPS 2001) but initially known from a single location just south of the Golden Gate in the Presidio of San Francisco. *Clarkia franciscana* was found to lack variability for the eight enzymes investigated ($H_o = 0.00$; $F_{IS} = 1.00$), whereas the outcrossers, *C. amoena* and *C. rubicunda* ($H_o = 0.11$), showed higher levels of genetic variability. The genetic identity (*I*) or genetic similarity (alleles held in common) between *C. franciscana* and *C. rubicunda* was very low ($I = 0.28$). This result led Gottlieb to conclude that perhaps *C. franciscana* was a more ancient taxon, and its lack of genetic variation is consistent with the high degree of self-fertilization prevalent in this species. It also appears that *C. franciscana* has repeatedly passed through genetic bottlenecks, because it has duplicated loci, which are fixed for alternate alleles. These results are consistent

with the widespread belief that localized endemic species generally have less genetic variation than their widespread congeners (Hamrick and Godt 1996).

After discovery of a second population of *C. franciscana* in the Oakland Hills in East Bay Regional Park, Gottlieb and Edwards (1992) investigated 16 enzymes encoded by 31 loci for the Oakland Hills and Presidio populations to determine whether the Oakland Hills population was genetically distinct and therefore indigenous to the site, or possibly a recent introduction since seeds of the Presidio population had been cultivated. All individuals of the Oakland Hills population proved to have identical electrophoretic mobility for each of the 31 loci. Likewise, all individuals for the Presidio population exhibited identical electrophoretic mobility for each locus. However, the Oakland Hills population differed from the Presidio population at five loci (16%). This finding strongly suggests that the Oakland Hills population did not originate by seed transfer from the Presidio population and that it must be regarded as indigenous to the site. Even though the Oakland Hills and Presidio populations have very little genetic variability, *C. franciscana* has a reasonable level of interpopulational variability, and this may prove useful for future conservation management decisions (Gottlieb and Edwards 1992). Perhaps ex situ breeding experiments with individuals from the Oakland Hills and Presidio populations will be undertaken to determine effects on fitness and genetic diversity in the offspring.

Clearly hybridization through interspecific gene flow can have both beneficial and harmful consequences for the conservation of biological diversity (Cade 1983). Hybridization can lead to loss of genetic diversity through genetic assimilation of a smaller population by a more widespread species or via outbreeding depression where reduced fitness of the small population can result in extinction (Rieseberg 1991). To examine the question of hybridization in small populations, an analysis of 17 isozyme loci demonstrated that *Cercocarpus traskiae* Eastw. (Rosaceae; California's rarest tree, with only seven known individuals) had extremely low mean values for polymorphic loci (0.12), alleles per locus (1.12), and observed heterozygosity ($H_o = 0.04$) (Rieseberg et al. 1989). This species is confined to a single canyon in Wild Boar Gully on the southwest side of Santa Catalina Island. Levels of genetic variation in *Cercocarpus betuloides* var. *blancheae* (C. K. Schneid.) Little, its closest widespread congener, are only slightly greater (mean polymorphic loci of 0.18, 1.24 alleles per locus, and $H_o = 0.05$ [Rieseberg et al. 1989]). By comparing isozyme expression in *C. traskiae* with its putative relative *C. betuloides* var. *blancheae*, Rieseberg et al. (1989) concluded that five of the individuals were "pure" adult *C. traskiae* trees, two individuals were hybrid trees, and one was "pure" *C. betuloides* var. *blancheae*. Management suggestions discussed in Rieseberg et al. (1989) for the continued existence of *C. traskiae* were (1) remove the single individual of *C. betuloides* var. *blancheae* from Wild Boar Gully; (2) establish cuttings representing the five "pure" *C. traskiae* trees and transplant these to other areas on Santa Catalina Island where risk of hybridization with *C. betuloides* var. *blancheae* is minimal; and (3) remove all nonnative herbivores from Wild Boar Gully.

Microsatellites

Allozyme electrophoresis has provided a valuable tool for studying genetic variation. Since the time that allozyme electrophoresis was first introduced, however, new molecular biology techniques have been developed. One that has been used in studies of population genetics is microsatellites. They are also called simple sequence repeats (SSRs) because they are tandem repeats of very short simple sequences, usually dinucleotides or trinucleotides. Microsatellites occur in the regions of the genome that do not code for any genes, so they are not subject to the same selection as allozymes, where a mutation might cause an allele's being lost from the population. Microsatellites are extremely variable in the number of repeats present. The different alleles in a microsatellite system differ in length based on the number of repeats. There are generally more alleles at each locus for microsatellites than there are for allozymes. Therefore, there is a greater expected heterozygosity. This makes it easier to detect when the observed heterozygosity is significantly different than the expected heterozygosity.

Microsatellites were used in a study of a plant that has been observed in decline, the Mauna Loa silversword (*Argyroxiphium kauense* [J. F. Rock & Neal] O. Deg. & I. Deg.; Friar et al. 2001; see also section 11.1). The introduction of ungulates to Hawaii has resulted in the reduction of the distribution of the Mauna Loa silversword from its historical range on midelevation slopes of the Mauna Loa volcano on the island of Hawaii to its current limited range that includes only three locations and fewer than a thousand individuals. Microsatellites were used to assess the impact of this range restriction on the genetic diversity of the species. The risks of inbreeding and outbreeding depression for population reintroduction were tested. If populations have already lost significant genetic diversity and are experiencing inbreeding, there may be further decline due to loss of fitness that may be countered by augmenting populations at risk with individuals that would increase the genetic diversity of the populations. If the populations are found to be genetically distinct, however, augmenting the populations with outside individuals may cause outbreeding depression. In this study, seven microsatellite markers were used. The data did not show any significant inbreeding in any of the populations, and the genetic variation was similar to that found in a closely related species, *A. sandwicense* DC. of Hawaii and Maui. Because severe range restrictions may have occurred since the 1970s and the plant has a long generation time (20–30 years), there may not have been time for the decline in numbers to affect the genetic diversity. The island of Hawaii is less than 0.5 million years old, so this species is relatively new. Newly formed species may have low genetic variation simply as a relict of the bottleneck that occurred in the founder population of the species. Over time, new alleles will form via mutation. However, it was found that the populations are genetically distinct from one another. This is not unexpected. They are geographically distinct populations that occur in ecologically different habitats. Morphological differences had also been observed. Therefore, there is the potential for outbreeding depression in this example. It is reasonable to believe that there may be local adaptation in these different populations. Fortunately, the high diversity remaining within each population makes it seem reasonable that reintroduction projects can use germ plasm from a single site. If enough samples from the source site are used, there is no reason to expect inbreeding depression to occur.

Another plant that has been studied with the use of microsatellite markers is *Caryocar brasiliense* Cambess., an endangered tree in Brazil (Collevatti et al. 2001). The tree grows in an increasingly fragmented habitat. Ten microsatellite markers were used to test for inbreeding in four populations of *C. brasiliense*. The four populations were chosen from different sites to represent varying amounts of human disturbance. The only population found to have an inbreeding coefficient significantly different than zero was the one with the greatest amount of human disturbance. The density of reproductive individuals of *C. brasiliense* was the lowest at this site. As with the Mauna Loa silversword, the generation time is long, approximately 70 years. The impact of current human disturbance may not yet be reflected in the population genetics of the species. Seed abortion was observed in this species, however, and seed abortion has been demonstrated to increase in controlled self-pollination experiments designed to increase inbreeding. Therefore, increased inbreeding may result in decreased fitness in the form of seed abortion. Genetic diversity was found to be generally low in *C. brasiliense*. This may be accounted for by the foraging behavior of the primary pollinator, glossophagine bats. These bats tend to forage in small areas and disperse the pollen only short distances. Large birds and mammals disperse the seeds of *C. brasiliense*. The seedlings, however, are often established beneath the mother tree. Increased fragmentation of the forest may further restrict the movement of the seed-dispersing animals and the pollinating bats, as was observed in the *C. brasiliense* population with the greatest amount of human disturbance.

AFLPs

Another useful molecular biology tool that is often used by population biologists is amplified fragment length polymorphisms (AFLPs). They are useful because they are easily and quickly developed in the laboratory and the same general protocol applies for all taxa. They do, however, have the problem that they are not codominant, as the allozyme and microsatellite markers are. In a codominant marker, both alleles can be detected for the heterozygote. For allozymes, the alleles that are different sizes and charges can be distinguished from one another, and for microsatellites, the alleles of different numbers of repeats can be separated. For a codominant marker, the heterozygotes are distinct from the homozygotes. However, AFLPs are a dominant marker. The data are the presence or absence of DNA fragments. Therefore, the fragments are present if a certain sequence is present, and the fragment is absent without that sequence. A heterozygote that has a particular DNA sequence would be indistinguishable from a homozygote for the same sequence. With the AFLP technique, heterozygotes are not observed. However, because AFLPs are so variable and easy to use in the laboratory, they are useful. AFLPs can be used with microsatellites to assess the genetic variability of a population.

Mariette et al. (2001) used microsatellites and AFLPs to assess genetic variation in two groups of the maritime pine, *Pinus pinaster* Aiton. One population was on the island of Corsica and the other was in the southwest of France, Aquitaine. The microsatellites were used to estimate the inbreeding coefficient, and there was no indication of inbreeding. Then the genetic variation was assessed based on AFLPs because they can quickly provide a large amount of data for analysis. Both markers found lower genetic variation in the Corsican populations than in the Aquitaine populations. This could be a result of a small founder population that established *P. pinaster* on Corsica originally. Alternatively, this could be due to the differences in life history. The populations on Corsica are subject to frequent fires and could have suffered from multiple bottlenecks over the years. Further investigation with AFLPs, after assessment of inbreeding with microsatellites, can provide useful information for management. In this example, the two populations of *P. pinaster* were found to be genetically divergent. This information can be used to plan management strategies that avoid mixing the two populations and risking outbreeding depression.

Studies have used AFLPs without microsatellites to assess genetic variation in a population. Gaudeul et al. (2000) used AFLPs in a study of *Eryngium alpinum* L., an endangered alpine species. Within populations, genetic variation did not appear lowered by environmental impact, but a significant correlation was found for population size and genetic variation. Therefore, small populations of *E. alpinum* may not be able to maintain genetic variation. Also, a significant correlation was found between geographic distance and genetic distance among populations. In this case, therefore, crossing individuals from distant populations may introduce the risk of outbreeding depression. As the authors note, however, because AFLPs are a dominant marker, some genetic statistics such as the inbreeding coefficient cannot be derived. Inbreeding could cause the low genetic variation observed in this study. Yet there are other causes of low genetic variation. For example, it could be due to random genetic drift. In a population with low observed genetic variation, inbreeding is more likely to occur, but it is only speculation without more data from codominant markers.

CONCLUSIONS

Is genetic variation reduced in endangered species? Gitzendanner and Soltis (2000) found in their review of 36 studies comparing genetic diversity in rare and widespread congeners that there is less genetic variation in rare species than in their widespread congeners. Is this a reason for ecological or evolutionary concern? To address this, the evolutionary history and life history of the species must be considered. A newly derived species may have lowered genetic variation. Many rare species are newly derived. Also, self-pollinating plants may have lowered genetic variation. Often when there is a history of low genetic variation, deleterious alleles have been purged, and the species is better adapted to future bottlenecks. When there is a decrease in genetic variation in a species with a history of higher genetic variation, however, there is a reason to be concerned that fitness could be affected by inbreeding.

Should populations be managed for increased variation? Again this depends on evolutionary and life history. The conservative approach would be to manage for historical levels of genetic variation, to avoid decreasing variation with a corresponding increase in inbreeding, and to avoid increasing variation with its corresponding potential for outbreeding. How serious a problem is inbreeding depression? It has the potential to be a serious problem. Species that have a history of inbreeding have survived a period of increased risk during which recessive deleterious alleles were expressed and many individuals had an increased risk of mortality. For these species, inbreeding is no longer a problem. In species that normally outcross, however, there is a risk of extinction when mortality increases with increased inbreeding and expression of deleterious recessive alleles. For heavily impacted species, that is, populations or species with few individuals and low levels of genetic variation, this could be a real threat to species persistence. Where possible, maintenance of the historical genetic variation is the most conservative approach.

As Hamrick and Godt (1996, 297) point out, "the goals of any conservation program should be the insurance of the long-term survival of species and the maintenance of ecological and evolutionary processes." Therefore, it is critical to manage the in situ habitat for continued survival of a species, particularly a habitat of limited size. Protecting sites from unnatural perturbations, that is, human-mediated, is by far the preferred method of conservation. This is the best-case scenario, but as most of us realize, other types of measures may be necessary. A short-term plan for ex situ preservation of a species could be to cultivate representatives in gardens or to store seed or other propagules. Because of our overall ignorance of the population biology of rare species, studies of the ex situ cultivated populations measuring the genetic variation via molecular markers are desperately needed to understand the biology of rare species (Templeton 1991).

LITERATURE CITED

Arnold, M. L. 1993. *Iris nelsonii* (Iridaceae): origin and genetic composition of a homoploid hybrid species. American Journal of Botany 80:577–583.
Avise, J. C. 1996. Introduction: the scope of conservation genetics. Pp. 1–9 in Avise, J. C., and Hamrick, J. L., eds., Conservation Genetics: Case Histories from Nature. Chapman and Hall, New York.
Barrett, S. C., and Kohn, J. R. 1991. Genetic and evolutionary consequences of small population size in plants: implications for conservation. Pp. 3–30 in Falk, D. A., and Holsinger, K. E., eds., Genetics and Conservation of Rare Plants. Oxford University Press, New York.
Brown, A. H. D., and Feldman, M. W. 1981. Population structure of multilocus associations. Proceeding of the National Academy of Sciences USA 78:5913–5916.
Cade, T. J. 1983. Hybridization and gene exchange among birds in relation to conservation. Pp. 288–310 in Schonewald-Cox, C. M., Chambers, S. M., MacBryde, B., and Thomas, W. L., eds., Genetics and Conservation: A Reference for Managing Wild Animal and Plant Populations. Benjamin-Cummings, Menlo Park, California.

CNPS (California Native Plant Society). 2001. Inventory of Rare and Endangered Plants of California. 6th edition. Edited by D. P. Tiber. California Native Plant Society Special Publication 1. California Native Plant Society, Sacramento.

Collevatti, R. G., Grattapaglia, D., and Hay, J. D. 2001. High resolution microsatellite based analysis of the mating system allows the detection of significant biparental inbreeding in *Caryocar brasiliense,* an endangered tropical tree species. Heredity 86:60–67.

Cook, L. M., and Callow, R. S. 1999. Genetic and Evolutionary Diversity: The Sport of Nature. 2nd edition. Stanley Thornes, Cheltenham, United Kingdom.

Crawford, D. J. 1989. Enzyme electrophoresis and plant systematics. Pp. 146–164 in Soltis, D. E., and Soltis, P. S., eds., Isozymes in Plant Biology. Dioscorides Press, Portland, Oregon.

Friar, E. A., Boose, E. L., LaDoux, T., Roalson, E. H., and Robichaux, R. H. 2001. Population structure in the endangered Mauno Loa silversword, *Argyroxiphium kauense* (Asteraceae), and its bearing on reintroduction. Molecular Ecology 10:1657–1663.

Gaudeul, M., Taberlet, P., and Till-Bottraud, I. 2000. Genetic diversity in an endangered alpine plant, *Eryngium alpinum* L. (Apiaceae), inferred from amplified fragment length polymorphism markers. Molecular Ecology 9:1625–1637.

Gitzendanner, M. A., and Soltis, P. S. 2000. Patterns of genetic variation in rare and widespread plant congeners. American Journal of Botany 87:783–792.

Gottlieb, L. D. 1973. Enzyme differentiation and phylogeny in *Clarkia franciscana, C. rubicunda,* and *C. amoena.* Evolution 27:205–214.

Gottlieb, L. D., and Edwards, S. W. 1992. An electrophoretic test of the genetic independence of a newly discovered population of *Clarkia franciscana.* Madroño 39:1–7.

Hamrick, J. L., and Godt, M. J. W. 1996. Conservation genetics and endemic plant species. Pp. 281–304 in Avise, J. C., and Hamrick, J. L., eds., Conservation Genetics: Case Histories from Nature. Chapman and Hall, New York.

Hartl, D. L., and Clark, A. G. 1989. Principles of Population Genetics. 2nd edition. Sinauer Associates, Sunderland, Massachusetts.

Lewis, F. H., and Raven, P. H. 1958. Rapid evolution in *Clarkia.* Evolution 12:319–336.

Mariette, S., Chagne, D., Lezier, C., Pastuszka, P., Raffin, A., Plomion, C., and Kremer, A. 2001. Genetic diversity within and among *Pinus pinaster* populations: comparison between AFLP and microsatellite markers. Heredity 86:469–479.

Ridley, M. 1993. Evolution. Blackwell Scientific, London.

Rieseberg, L. H. 1991. Hybridization in rare plants: insights from case studies in *Cercocarpus* and *Helianthus.* Pp. 171–181 in Falk, D. A., and Holsinger, K. E., eds., Genetics and Conservation of Rare Plants. Oxford University Press, New York.

Rieseberg, L. H., Zona, S., Aberbom, L., and Martin, T. D. 1989. Hybridization in the island endemic, Catalina mahogany. Conservation Biology 3:52–58.

Schaeffer, S. W., and Miller, E. L. 1993. Estimates of linkage disequilibrium and the recombination parameter determined from segregating nucleotide sites in the alcohol dehydrogenase region of *Drosophila pseudoobscura.* Genetics 135:541–552.

Templeton, A. R. 1986. Coadaption and outbreeding depression. Pp. 105–116 in Soulé, M. E., ed., Conservation Biology: The Science of Scarcity and Diversity. Sinauer, Sunderland, Massachusetts.

————. 1991. Off-site breeding of animals and implications for plant conservation strategies. Pp. 182–194 in Falk, D. A., and Holsinger, K. E., eds., Genetics and Conservation of Rare Plants. Oxford University Press, New York.

Vrijenhoek, R. C. 1996. Conservation genetics of North American desert fishes. Pp. 367–397 in Avise, J. C., and Hamrick, J. L., eds., Conservation Genetics: Case Histories from Nature. Chapman and Hall, New York.

Waser, N. M. 1993. Population structure, optimal outbreeding, and assortative mating in angiosperms. Pp. 173–190 in Thornhill, N. W., ed., The Natural History of Inbreeding and Outbreeding: Theoretical and Empirical Perspectives. University of Chicago Press, Chicago.

Wolfe, A. D., Xiang, Q.-Y., and Kephardt, S. R. 1998. Assessing hybridization in natural populations of *Penstemon* (Scrophulariaceae) using hypervariable intersimple sequence repeat (ISSR) bands. Molecular Ecology 7:1107–1125.

Wright, S. 1931. Evolution in Mendelian populations. Genetics 28:114–138.

PART IV

THE CONSERVATION OF
PLANT DIVERSITY

ASSESSMENT,
MANAGEMENT STRATEGIES,
AND ACTION

CHAPTER 10

MAPPING BIOLOGICAL DIVERSITY

Many natural history scientists are concerned with understanding the extent and distribution of organismic diversity as well as using this knowledge for conservation purposes. It is clear that only a portion of the remaining forested areas can be preserved unless a considerable change takes place in the current social, political, and economic priorities of the world. If only a portion of these habitats can be maintained, it is imperative that, given a level of available resources and budget, areas be identified that maximize the amount of diversity from the genome level to the biome level. A number of approaches have been advocated for defining what constitutes an area or set of areas of maximum diversity, including those that use species diversity (i.e., total species richness, endemic species richness, complementary richness) and phylogenetic diversity as their basic criterion.

The underlying assumption of all conservation planning is that one must use the best available data at any given time. However, what constitute the best data? Unfortunately, a database that describes the full distribution of all biodiversity is never available. This absence of complete data requires the use of surrogates, such as vegetation maps, modeled distributions, or numerical classifications of environmental variables, to quantify biodiversity. None of these surrogates, however, is robust enough on its own to be heralded as optimal. The most convincing biodiversity surrogates come from a combination of environmental variables and species data that are always limited by the availability of information about a specific region.

The first two sections of this chapter explore the use of species data from herbarium collections and published floras to identify conservation priority regions; the third section discusses how the evolutionary history of species is used to rank geographic regions when setting conservation priorities.

10.1 HERBARIUM COLLECTIONS, FLORAS, AND CHECKLISTS

W. John Kress and Vicki A. Funk

AN ESTIMATED 3 billion specimens of organisms, including plants, animals, and microorganisms, are housed in the world's natural history collections (Edwards et al. 2000). These collections are a crucial subset of systematic data that can be used for conservation planning. Each of these specimens contains a suite of information about the organism, including name, rank, and the locality where it was collected. Taken together, these data constitute a rich library about life on earth, providing important information about the distribution of individual species, genera, and families by region, country, and habitat. They also supply crucial information on the recent history of a species as they constitute a permanent species record at a given location and at a specific time. A record attached to a specimen housed in a herbarium is considered "vouchered," which means it can be checked by experts at any time in the future for accuracy of the information and proper identification. As environmental degradation continues around the world, the analyses of biological information derived from museum collections will provide a wide array of information that will assist in preserving the earth's biodiversity.

The limitations to using collections data for conservation decision making are that they can be (1) geographically biased, favoring more easily accessed areas; (2) taxonomically incomplete, including only easy-to-study species thus giving undue weight to a few taxa; and (3) temporally biased, that is, based on a single survey in a single season (Faith and Walker 1996; Ferrier 1997; Funk et al. 1999). Several techniques have been developed to deal with these limitations. Some exclude the use of collections data altogether, for example, techniques that use abiotic surrogates of biodiversity including land classifications, vegetation maps, numerical classification of environmental variables, and ordination of environmental variables (Mackey et al. 1988; Mackey et al. 1989; Belbin 1993, 1995; Pressey and Logan 1995; Faith and Walker 1996; Wessels et al. 1999; Faith et al. 2001). The problem with these techniques is that they are not informed by the biological data, which means that the resulting conservation decisions are made without regard to what species may or may not be in those areas. As a baseline, collections data serve as the only direct evidence of species distributions.

Most conservation planners accept that a network of conservation sites needs to be complementary, where each site complements the biodiversity of other sites (Vane-Wright et al. 1991; Pressey et al. 1993; Pressey et al. 1994; Margules and Austin 1994). Incorporating the concept of "complementarity" ensures that sites are selected to maximize the representation of different species. Recently the concept of complementarity has been enhanced by the introduction of "irreplaceability" (Pressey et al. 1993; Ferrier et al. 2000). *Irreplaceability* refers to a measure of uniqueness, where the irreplaceability value of a site reflects the relative importance of that site for achieving an explicit conservation target (Ferrier et al. 2000). Although the political decision to designate a site for conservation may depend upon additional analyses of economic, political, and other potential land uses, and the methods depend on comparable lists of taxa, this complementarity-irreplaceability approach has been used successfully to select areas of high biodiversity priority in Australia, South Africa, and the United States (Pressey et al. 1993; Rebelo 1994; Pressey 1994; Lombard et al. 1997; Lombard et al. 1999; Davis et al. 1999).

In this section two examples of the use of collections data in biodiversity and conservation studies in South America are provided. In the first example, taxonomic data documented directly from museum collections are used alone to determine areas of high species diversity and endemism in the Amazon Basin. These results are then compared to field-based "ad hoc" estimates, and the limitations of these approaches for assessing biodiversity are discussed. In the second example, taxonomic data from museum specimens collected in Guyana are combined with environmental factors to predict total distributions of the organisms that are then used to identify areas for conservation.

THE AMAZON REGION

The Amazon region of tropical South America contains the largest remaining expanse of pristine forest on the planet. Yet with over 100 years of collecting data on species diversity and distribution in the Amazon Basin, until the 1990s only scattered information on the areas of greatest species concentrations was available, and hence baseline data for conservation purposes were lacking (Nelson et al. 1990; Voss and Emmons 1996). In 1990 a group of taxonomic specialists and conservationists made an attempt to identify high species-diversity regions for conservation priority areas in Amazonia based on their individual field experience with various groups of organisms (Workshop 90 1991). They outlined five levels of priority areas for conservation based on a qualitative synthesis of species diversity and endemism criteria.

In response to this initial identification of diversity areas based on qualitative assessment, scientists in the Neotropical Lowlands Research Program at the National Museum of Natural History utilized quantifiable specimen data, although recognizably incomplete, to verify these centers of biodiversity (for details of the analyses, see Kress et al. 1998; Heyer et al. 1999). Before specimen localities were mapped, the geographic area of Amazonia (as defined by Ab'Sáber 1977) was divided into 472 one-degree (latitude by longitude) grid cells. Distributional data from five main taxonomic groups found in Amazonia, that is, plants, arthropods, fishes, amphibians, and primates, were selected for study. Any species with a significant portion of its distribution within the Amazonian domain (and below 350 m) was included. In total 3,991 records of 421 species in 33 genera were included in the analysis (table 10.1). These collections, which provide a repeatable and reliable data set for analyzing the distribution of biological diversity, are housed at museums and universities distributed throughout the world and in some cases are an exhaustive record for a particular species. Although other taxa could have been selected for the analyses, these genera exemplify a cross section of both rare and common species found in Amazonia and represent one of the most extensive taxonomic samples currently available for this geographic region. A geographic information system (GIS) analysis was then used to determine the distribution in Amazonas of each species by pinpointing the exact locality of each specimen record in a specific grid cell on the map. Both total species richness (i.e., maximizing the number of species) and endemic species richness (i.e., maximizing the concentration of rare species) were calculated.

Table 10.1 Taxa used in the analyses of Amazonian biodiversity distribution

Taxon	Number of species	Number of records
Plants		
Heliconia	30	440
Phenakospermum	1	17
Talisia	35	198
Total for plants	66	655
Arthropods		
Agra	101	122
Batesiana	21	79
Deinopis	6	16
Geballusa	2	4
Gouleta	3	28
Hemiceras	108	342
Total for arthropods	241	591
Amphibians		
Leptodactylus	14	536
Total for frogs	14	536
Fishes		
Boulengerella	5	112
Caenotropus	3	55
Copeina	1	25
Copella	3	38
Cyphocharax	2	6
Lebiasina	1	15
Nannostomus	15	123
Pyrrhulina	2	17
Steindachnerina	33	419
Total for fishes	65	810
Primates		
Alouatta	2	152
Aotus	3	104
Ateles	3	43
Cebuella	1	41
Cacajao	3	50
Callicebus	2	191
Callimico	1	27
Callithrix	2	58
Cebus	2	19
Chiropotes	2	108
Lagothrix	1	74
Pithecia	5	187
Saguinus	7	236
Saimiri	1	109
Total for primates	35	1,399
Total	421	3,991

Table 10.2 Species diversity and endemism in Amazonia

Identifying locality within grid	Latitude, longitude (NW corner of grid cell)
a. Tambopata, Peru*	12°S, 70°W
b. Cocha Cashu–Manu, Peru*‡	11°S, 72°W
c. Iquitos, Peru*	3°S, 74°W
d. Along upper Rio Solemoes, Brazil*‡	3°S, 69°W
e. Tefé, Brazil*‡	3°S, 65°W
f. Manaus–Ducke Reserve–INPA, Brazil*	3°S, 60°W
g. Santarém, Brazil*	2°S, 55°W
h. Parimaribo Region, Surinam*‡	6°N, 55°W
i. Georgetown, Guyana*‡	7°N, 59°W
j. Cayenne, French Guiana‡	5°N, 53°W
k. Moyobamba, Peru‡	6°S, 77°W
l. Rio Ucayali, Peru‡	5°S, 75°W
m. Pôrto Velho, Brazil‡	8°S, 64°W

Notes: Areas with highest species diversity (43–67 species) are marked with *; areas with highest concentrations of endemism (4–13 endemic species) are marked with ‡. Identifying letters correspond to grid cells marked in plates 10.1 and 10.2.

The distribution of species across all taxa within the 472 one-degree grid cells in Amazonia comprised six categories between 0 and 66 species per grid cell (plate 10.1). Over one-fourth of the grid cells had no representative species of the groups under study, that is, no collections. Only 2.0% of the total grid cells had high species diversity (45–66 species) and 57.8% had low species diversity (1–11 species). The nine areas with highest diversity are scattered throughout the region and in general correspond to well-known and historical collection localities (e.g., Tambopata Reserve, Iquitos, Tefé, Manaus, Cayenne, etc.; table 10.2). There was no obvious species-diversity gradient between east and west or north and south.

For endemic species, 64 grid cells in Amazonia contained from 1 to 13 endemic taxa (plate 10.2). Nine areas composed the three highest categories (4–13 endemic species); the remaining 55 grid cells contained from 1 to 3 endemic species. Five of the nine grid cells with the highest endemicity corresponded to areas with the highest species diversity (45–66 species). The overall distribution of endemic species was significantly associated with the distribution of total species number.

With respect to numbers of records of species, individual grid cells ranged from 0 (129 cells) to 143 records (1 cell). Of the grid cells with records present, 64.2% had 27 or fewer records. The remaining 40 grid cells with more than 27 records had an average of 49.3 records per cell. If individual taxonomic groups are considered separately, insects are the least collected, with nearly 84% of grid cells with 0 records, and primates are the most evenly sampled, with over 53% of the grid cells with at least 1 record.

As previously noted (Williams et al. 1996), little correspondence exists between the high-priority areas of the Workshop 90 report and the actual specimen-based data (plates 10.1 and 10.2). Of the total area of Amazonia, 25.3% (120 grid cells) overlapped in part with the Workshop 90 high-priority areas. The highest species areas (45–66 species) identified by the specimen data corresponded with only 2.5% of the Workshop 90 high-priority areas, whereas 52.5% of their priority areas had low species diversity for the actual taxa studied (less than 11 species). Of the nine highest areas of endemicity identified by specimens, only four fell within the high-priority areas of Workshop 90; the two grid cells with the highest endemicity based on specimens (13 endemic species) were not identified as high-priority areas by Workshop 90.

The centers of high diversity identified with actual specimen data (plate 10.1; table 10.2) correspond to many areas that historically have been the focus of museum collectors, for example, areas around Iquitos, Manaus, Santarém, and Cayenne (Nelson et al. 1990). The nine highest species grid cells in Amazonia all corresponded to the most intensively collected areas. If total number of species is accepted as the only criterion for determining genetic diversity, one might recommend that the top nine most diverse regions identified here, and especially those five areas that overlap with high levels of endemism, be considered high-priority areas for conservation.

This study based on museum collections demonstrated that in most cases no areas in Amazonia have been thoroughly or even adequately sampled. It is clear that the perceived species diversity of any area in Amazonia is a direct function of how many collections have been made in that area and not necessarily the absolute level of diversity. In general very few localities have more than a single collection per species even in the areas with high numbers of collections. It is therefore likely that all areas will prove more diverse when additional collections are made. It is remarkable that even after a century of inventory and collecting efforts by museum scientists in this super-high-diversity region, adequate distributional data are lacking for most organisms. A coordinated plan to intelligently sample Amazonia is clearly in order.

The museum-based collection data supported the recognition of at least a subset of the high-priority areas of Workshop 90 as regions of exceptional biodiversity. However, it is noteworthy that the majority of the Workshop 90 high-priority areas lack significant collection data to verify the recommendations for conservation. Such recommendations should be treated as hypotheses of centers of diversity and not as conservation planning mandates. Areas of high diversity that are documented with collection data should be given conservation priority over areas of suspected diversity that lack supporting collection-based data. Additional biological information about species, which can be provided by systematists, field biologists, and local naturalists, also must be taken into account when identifying high-diversity regions. This investigation of biodiversity in the Amazon region demonstrated that collections information as currently available is necessary but not always sufficient for identifying priority areas for conservation. Reliable taxonomic data should be complemented with data on multiple environmental factors (as discussed in the next section) to provide a baseline upon which conservation recommendations can be formulated.

GUYANA

Many of the problems facing other tropical countries are not an issue in Guyana: it is not a large country (215,000 km²); it has a large amount of its land intact or only marginally damaged (ca. 70%); and it has a small human population that is concentrated along the coast (ca. 800,000 in 10% of its territory). In addition, although previously poorly known biologically, exploration in the last 20 years has generated a wealth of information for some organisms in some parts of the country. Most importantly, although there are few protected areas in Guyana, the government is interested in developing a national protected-area system. Based on this initiative, several preliminary studies on how existing data might be used in the development of such a system have been conducted (see Ter Steege 1998; Funk et al. 1999; Richardson and Funk 1999; Ter Steege et al. 2000), but no final decisions have yet been made.

The Biological Diversity of the Guianas Program (BDG) of the Smithsonian Institution (Hollowell et al. 2004) has had an active field program in Guyana for the last 18 years. The BDG has developed a database from historical collections made in Guyana and now housed at museums and herbaria around the world as well as from recent field collections. These collections data have been used in investigations to identify sampling gaps, to improve survey design, and to reduce collecting biases (Funk and Richardson 2002). Two aspects of conservation assessment using collections data are considered here: (1) building richness, restricted range (endemicity), and distribution maps; and (2) selecting priority biodiversity sites for possible conservation. Distributions were examined using 25,111 records representing 5,123 species of plants and animals from Guyana. Data on climatic variables and vegetation types were assembled from various sources (see Funk and Richardson 2002).

Building Distribution and Species-Richness Maps

Sites with greater species richness have generally been considered more important for conservation than the sites deemed "species-poor" (Myers 1988, 1990; Mittermeier and Werner 1990). Given that complete inventories of species are impractical, particularly in species-rich tropical areas, the utility of species richness and other species-based approaches depends on the extent to which results from limited data sets can be generalized. In this example, known locality data and potential distribution data from modeled distributional maps were used to enhance species-richness maps (Funk and Richardson 2002). Other studies have examined how well certain taxonomic groups act as indicators for other taxonomic groups (Pearson and Cassola 1992; Prendergast et al. 1993; Williams et al. 1996; Moritz et al. 2001).

Modeling of species' distributions assumes that differences in species composition and abundance at any given location can largely be explained by differences in environmental factors, such as temperature, moisture, nutrients, and evaporation (Nix 1982; Busby 1986; Margules et al. 1988; Belbin 1995). The following steps were taken to model the potential distributions of each species: (1) The digital elevation model, vegetation map, lithology map, and mean monthly rainfall of the driest month (October) were selected as the variables to model species distributions. (2) Using the selected variables, each species was modeled with the program DOMAIN (a presence-only data modeling technique; Carpenter et al. 1993; for an example, see plate 10.3). (3) A similarity map that was produced for each species showed the likelihood of the species' being present in a given area (a similarity value of 95% or greater was chosen as a conservative cutoff point). (4) The modeled distributions were then used to improve upon the species-richness map (Funk and Richardson 2002).

A map of the restricted range values, which is a measurement of endemism, was calculated in the same manner, using an index of restricted range. Species with very restricted ranges had higher scores, with the most restricted species (found only in one grid cell) scoring 1.0 on the restricted range scale. Both the known locality data and the modeled data were used to calculate restricted range values and to produce the resulting map of restricted range.

The species-richness maps of only the known locality data and of the known locality and modeled distribution data reveal differences in both the number of species in a given grid cell and the distribution of the sites with the highest species richness. In the analysis of species richness for Guyana using only the known locality data, only 0.15% of the total grid cells had high species richness, and 42% had low species richness. Using the known locality and modeled distribution data, species richness increased to 8.5%. Areas of estimated high species richness are Kaieteur Falls and the Potaro River gorge; Kurupukari and the central Essequibo; Bartica and the lower Cuyuni, Mazaruni, and Essequibo rivers; the Pakaraima Mountains including Mounts Ayanganna, Roraima, and Wokomong; the upper Cuyuni and Mazaruni rivers; the Kanuku Mountains and the Rewa River; the upper Berbice River, and a few scattered areas in the Rupununi savannas, disregarding the sites in the far southeast of the country where virtually no collecting has taken place and very little abiotic data are available. Interestingly, the overlap between areas of high species richness and areas of very restricted species was found to be quite high (71.4%). This may indicate that certain species-rich areas in Guyana are also centers of endemism and that these areas have similar biogeographic features.

Location of Priority Biodiversity Sites

One of the main drawbacks with using only species-richness or restrictedness data to select priority biodiversity sites for conservation is that they do not provide any means of ensuring that different species in an area are conserved. For instance, a grid cell might be relatively species-poor, but if it adds the most species not already represented in an existing network of conservation sites, then it may be the most important in terms of conservation (Flather et al. 1997). However, it is possible to represent many more species in a network of sites if decisions are made to use endemism-richness, species-richness, or restricted-range values and the complementarity principle to select sites (in this case an index of summed irreplaceability) (Vane-Wright et al. 1991; Pressey et al. 1993; Margules and Austin 1994).

The known and modeled data of 320 species were used to select priority sites for biodiversity conservation using a grid size of 8 km × 8 km (3,553 grid cells across the country; Funk and Richardson 2002). This grid size was chosen for demonstrative purposes only. Although the size of the grid is arbitrarily set in most conservation planning exercises, the size may influence the quantity and location of priority biodiversity sites (Flather et al. 1997; Reid 1998; R. Richardson and V. A. Funk, unpublished data). Excluded from the analyses were 157 grid cells representing urban areas and cultivated fields and 418 grid cells from the southeast corner of the country where very little information has been collected due to logistical and political problems.

Priority biodiversity sites were selected using an interactive software package, C-Plan (New South Wales National Park Service 1999), that runs as an extension in ArcView (version 3.2; ESRI 2000). Sites were selected to maximize the rate of species accumulation, using an iterative process based on estimated summed irreplaceability, defined as the sum for all species of the likelihood that a site would be required as part of a network of sites to achieve a set target: in this instance the representation of each species within at least three sites (Pressey et al. 1993; Ferrier et al. 2000). The minimum set of sites needed to satisfy the target was calculated using an interactive stepwise algorithm that selected sites based on their highest summed irreplaceability. For Guyana, in order to capture each species at least three times in a network of sites, 33 grid cells of a possible 2,978 were required (plate 10.4). These 33 sites were selected to maximize the complementarity of species between the sites and the relative irreplaceability value of each site. A few sites are in the northwest both near the Venezuelan border and the coast, and quite a number are in the Pakaraima Mountains, including the vicinity of Kaieteur, Mount Ayanganna, and the upper reaches of the Mazaruni River. In the northeast corner there are three sites, the Essequibo River between Bartica and Kurupukari has three sites, and the Berbice River has several. Below the fourth parallel there are four sites, two in the Kanuku Mountains, one near the border with Surinam, and one in the far south just north of Gunn's. No doubt additional data from southern Guyana would change the results. The rate of species accumulation for plants and animals shows that 80% of the plants species (95% of animals species) are represented in 19 grid cells; however, it requires a further 14 grid cells to capture all species at least three times.

CONCLUSIONS

Analyses of biological data derived from museum collections can provide predictive tools for identifying critical biodiversity regions for conservation. Collections data may not be the perfect surrogate of biodiversity; however, they can be used to assist in the identification of areas with high endemism and species diversity as well as the prioritization of conservation areas. Yet the acquisition of biological information is only the first step in a many-tiered process for determining conservation areas that includes social, political, and economic factors as well. In order to take this first step, it is clear that our current knowledge of the distribution and diversify of the biota in such areas as Amazonia and Guyana must be greatly expanded. A renewed and structured effort to inventory with vouchered collections the various habitats of the tropics is imperative if informed decisions on conservation priorities are to be made in the near future.

10.2 HOT SPOTS AND ECOREGIONS

Gary A. Krupnick

THE NUMBER OF SPECIES threatened with extinction due to human activities is at an all-time high (Pitman and Jørgensen 2002; see also chapters 3 and 11 in this volume). Actions by conservation biologists, politicians, and land managers are urgently needed to prevent further extinctions. In a just world, conservation biologists would be able to focus their efforts on protecting all species, habitats, and ecosystems on the planet. However, with limited funds, time, and human resources, this is not an easy option. Conservation biologists should thus focus on areas that maximize biodiversity, thereby protecting the highest proportion of species and the most evolutionarily unique species at the lowest necessary cost.

Since resources for conservation are limited, it is essential to establish priorities to maintain earth's biodiversity. One tool that is used in setting priorities is a detailed map of the total species diversity. There have been several approaches to creating maps of biodiversity. The data upon which these maps are based come from different sources ranging from detailed species distribution maps to the expert opinions of taxonomic specialists. Here a variety of approaches that have been used in mapping out conservation priorities are compared.

CREATING MAPS

Identifying the areas that maximize biodiversity is the first step in setting conservation priorities. To many people, biodiversity is simply a count of the number of species within a given area. Yet biodiversity can be defined more elaborately—a measure of spatial variation and turnover (beta diversity), or a count of species rarity and endemism. Other criteria may also be used when setting conservation priorities, including an assessment of habitats within an ecosystem, or the occurrences of rare evolutionary and ecological phenomena (e.g., unique higher-level taxonomic groups, large vertebrate migrations, globally rare habitat types).

Biodiversity can be measured using a variety of methods. For instance, every organism in a given area can be inventoried, which would provide the most accurate account of total species richness. At most geographic scales of interest, however, this is not practical. Thus, most measures of biodiversity rely upon estimates. Krupnick and Kress (2003) illustrate four hierarchical steps in measuring biodiversity. The first, and most informative, is a complete inventory of every species within a given area, region, or continent. Efforts are under way to produce this information (e.g., Brugman and Secretariat 2004; All Species Foundation 2004); but it may be many years, if not decades, until we have an accurate count of all of earth's species. The second measure of data is a registry of species held within the museums and herbaria of the world. This registry represents a sample of the known biota, and can provide detailed information about biodiversity over time and space. Harnessing the information contained in these collections is now the focus of an international effort (see GBIF Secretariat 2004), but such data are not yet available except in a few cases (e.g., Conabio 2004). The third class of data is the geographic distribution of species contained in regional and national floras and faunas. In this class, rigorous in-depth investigations of the first two classes of data are compiled by a taxonomic specialist for a given geographic area. The fourth and final level is the knowledge and experience of the world's authorities on the taxonomy of plants, animals, and microorganisms. This level of information is typically most readily available, but not necessarily repeatable or testable. All four classes of biodiversity data are found within the resources of natural history institutions.

Since it is not yet possible to provide an accurate count of every living species, surrogate measures of taxic diversity are typically used in setting conservation priorities (e.g., see Faith 2003). These surrogates tend to focus on a relatively well known group of species, such as mammals, birds, or butterflies. The presence of one taxon may indicate the presence of other taxa, that is, an indicator taxon may be used as a surrogate for the total biodiversity of a region. For instance, areas of high plant diversity have been shown to correlate well with those for some animals, especially invertebrates (e.g., see Panzer and Schwartz 1998). However, other studies show that priority areas for one taxon may differ drastically from those for another (Prendergast et al. 1993; Kress et al. 1998). Even within the plant kingdom, vascular-plant species richness has been shown to be a poor predictor of "lower"-plant species diversity, including bryophytes and lichens (Pharo et al. 2000).

Another method employed is the use of higher-taxon (e.g., families, genera) richness as a broad surrogate for species richness (Gaston and Williams 1993; Williams and Gaston 1994; Williams et al. 1996). These types of data can be collected faster and at a lower cost than species richness. Critics have argued, however, that these measures are inadequate because they use a very coarse geographic resolution, and may not provide an accurate measure of biodiversity for local and regional conservation needs (Fjeldså 2000).

HOT SPOTS, ECOREGIONS, AND CENTERS OF PLANT DIVERSITY

There have been several approaches to mapping global biodiversity for the purpose of setting conservation priorities. These include the "25 biodiversity hotspots" (Myers et al. 2000), the "terrestrial ecoregions of the world" (Ricketts et al. 1999; Olson et al. 2001; Wikramanayake et al. 2002), and the "centres of plant diversity" (WWF and IUCN 1994–1997). Each of these three approaches looks at the distribution of life beyond political boundaries. Most conservation actions occur within governmental jurisdictions, but because nature does not recognize political boundaries, it is important to realize that many species have distributions that cross county, state, and country lines.

Each approach has adopted specific criteria for its selection of sites. A hot spot is a relatively large area of land with high species endemism (at least 0.5% or 1,500 of the world's 300,000 plant species) and a high degree of threat (at least 70% of primary vegetation lost). In contrast, ecoregions are defined as a unit of land containing a geographically distinct assemblage of natural communities, species, and ecological dynamics. Centers of plant diversity (CPD) are areas with high concentrations of plant diversity. Each CPD site has at least 1,000 vascular-plant species, 10% of which are endemic (the criteria for islands differs in that the flora must contain at least 50 endemic species or at least 10% of the flora must be endemic). In most cases ecoregions are at a regional scale, finer than hot spots, and CPD sites are at a local scale, even finer than ecoregions. For instance, there are 20 ecoregions that overlap the Sundaland hot spot (see plate 10.5), and within those 20 ecoregions there are 45 CPD sites.

In each of these three approaches, international taxonomic authorities were asked to provide assessments of plant species richness and endemism. The data that they provided came from firsthand fieldwork and museum study, and typically took the form of indirect estimates (the fourth step of measuring biodiversity; see above). These results are readily available but not necessarily repeatable or testable. If this type of information continues to be used for conservation purposes, it is critical that we test its quality with verifiable data at finer scales (e.g., regional inventories, specimen records, and accumulated distribution data).

MEASURING PLANT DIVERSITY USING TAXONOMIC DATA

Considering that between 310,000 and 422,000 flowering plant species exist in the world today (Prance et al. 2000; Govaerts 2001), many of them poorly known, mapping global plant richness to 25 hot spots, 866 terrestrial ecoregions, or 234 CPD sites is an enormous task. In a few cases, an accurate count of all plant species is available within an area of concern. For instance, the California Floristic Province hot spot is known to have 4,426 plant species, a figure determined after years of detailed taxonomic research (Myers et al. 2000). In most cases, however, only a rough estimate of the number of species is available. For example, approximately 10,000 plant species are found within Brazil's Cerrado hot spot (Myers et al. 2000).

Surrogates can be used to assess species richness in areas of the world where knowledge about the extent and distribution of biodiversity is poor. Krupnick and Rubis (2002), for instance, mapped 525 species of the Dipterocarpaceae to 110 ecoregions of the Indo-Pacific. Dipterocarps are the dominant woody tree family in the Indo-Malayan tropics, and they have the dubious distinction of being the world's main source of hardwood timber. Detailed distribution data is in greater abundance for the dipterocarps, a commercially valuable plant family, than for other less-prominent families. The dipterocarp data came from accounts of species distributions detailed in floras and monographs of the region and from herbarium specimens. Highlights of the study show that fine-scale data of one plant family can offer detailed information about which ecoregions support the greatest within-family biodiversity; however, the importance of selecting multiple indicator taxa with complementary distributions was shown to be necessary for conservation prioritization exercises. Recently, this study has been expanded to include 4,230 species from six additional plant families: the Bignoniaceae, Ericaceae, Euphorbiaceae, Fagaceae, Leguminosae, and Rosaceae (Krupnick and Kress 2003). Plate 10.5 shows plant species richness and endemism of these species in 84 ecoregions of the Indo-Pacific.

Fine-scale data collected for these purposes can then be compared to the estimates provided by taxonomic specialists. Figure 10.1 shows that the tabulated number of plant species culled from the floras strongly correlates with the estimated number for both the hot-spot approach (Myers et al. 2000) and the ecoregion approach (Wikramanayake et al. 2002). Individual families, however, do not always show similar patterns. The highest numbers of species per family in the Dipterocarpaceae, Euphorbiaceae, and Leguminosae are in the Sundaland hot spot and the lowland rain forests of Borneo and peninsular Malaysia ecoregions, whereas in the Bignoniaceae, Fagaceae, and Rosaceae the highest numbers of species are in the Indo-Burma hot spot and the subtropical forests of the northern Indochina ecoregion and the montane rain forests of the southern Annamites ecoregion. Diversity in the Ericaceae is greatest in the New Guinea major wilderness area and in the ecoregions of the Central Range montane rain forests and subalpine grasslands.

These results show that data based on verifiable reports of plant diversity extracted from published floras on a species-by-species basis are congruent with the widespread estimates based on the experience of scientific experts. The ranking of both the conservation hot spots and biodiversity ecoregions is thus supported and provides a further step in prioritizing the most outstanding regions of the earth for immediate conservation attention.

CONCLUSIONS

All regions are worthy of conservation efforts, yet by prioritizing hot spots, ecoregions, and centers of diversity, the most distinct species assemblages will garner

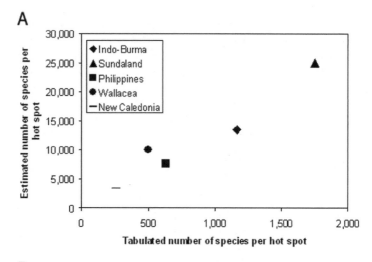

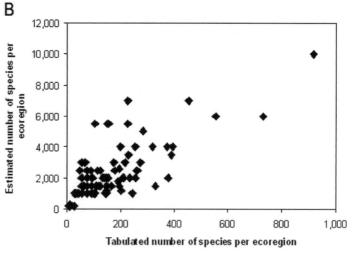

Figure 10.1 Correlations between tabulated and estimated number of species (A) per hot spot ($n = 5$, $r = 0.966$, $P < 0.009$) and (B) per ecoregion ($n = 84$, $r = 0.748$, $P < 0.001$) (after Krupnick and Kress 2003). The tabulated number of species is based on seven plant families (Bignoniaceae, Dipterocarpaceae, Ericaceae, Euphorbiaceae, Fagaceae, Leguminosae, and Rosaceae). The estimated number of plant species is a total for that region based on the original publication: A, Myers et al. 2000; B, Wikramanayake et al. 2002.

the immediate attention needed for preservation. Detailed and accurate biodiversity maps serve as an initial step in prioritization exercises. Conservation priorities can be based upon existing knowledge that may represent only a small part of total biodiversity. But care must be used in the interpretation of the data. Biodiversity maps can then be used for conservation work at the local and regional levels within priority hot spots and ecoregions.

Even though practical conservation priorities can be based on partial measures of biodiversity, it is imperative that we push forward to assemble the enormous amount of data found in the world's biological collections and ultimately complete an inventory of all organisms on earth. Accurate measures of species diversity and distribution should be the first goal when setting conservation priorities.

10.3 PHYLOGENETIC CONSIDERATIONS

M. Alejandra Jaramillo and Vicki A. Funk

PRESERVING BIODIVERSITY is the most important goal of conservation biology. Biodiversity defined as "the total complexity of life" is not measurable. As a result, species numbers are used as the default surrogate or proxy for biodiversity. In an effort to preserve the maximum diversity, some conservation biologists have focused on saving the largest number of species by setting goals based on species richness or by concentrating on areas with high levels of endemism (e.g., hot spots; Myers et al. 2000). Relying on species counts and numbers of endemics to establish a value for diversity depends on a large number of assumptions, two of which relate particularly to systematics: the assumption that taxonomic knowledge (including distributions) is adequate for most regions of the world and for most groups of organisms, and the assumption that all species are equivalent. However, many systematists would argue that neither of these two assumptions is correct. There is much we do not know about species distributions—many species await description, and many more species are unknown to science. The second assumption, that all species are equivalent, will be discussed in this section. It is clear that species are indeed nonequivalent; some have argued that more divergent species (genetically, taxonomically, or ecologically) make a larger contribution to the overall biodiversity than closely related species (Vane-Wright et al. 1991; Faith 1992). Therefore, whenever possible, it is essential to consider the evolutionary history of the taxa (species, genera, etc.) in order to rank geographic regions, taxa, or populations according to their conservation priority. Phylogenies can contribute in four ways: diversity indexes, vicariant histories, evolutionary process, and predicting hybridization risk after translocation.

DIVERSITY INDEXES

An appropriate measure is needed in order to identify areas of highest importance—a measure that is different from species totals, a way to measure diversity that is not based on simple numbers. Such diversity measures should quantify not only the numbers of species (or other taxa), but also their evenness—how the individuals are distributed among species, and their disparity—how varied are the species in the sample. Some argue that measures based on biological attributes, such as ecology and genes, are more desirable because they are better indicators of biological diversity (Noss 1990). Because the goal is to quantify evolutionary history rather than the amount of differentiation, however, this option does not fulfill the need. We seek not only to conserve current biodiversity but also future diversity, and so it is important to maximize the amount of information preserved and to select carefully which taxa might be the most important for the future evolution within a group. Modern systematics offers a number of analytical tools that permit measures of phylogenetic distinctiveness among the species and that allow us to identify a set of taxa or geographic regions that include the greatest biodiversity (Vane-Wright et al. 1991). Several phylogenetic diversity indexes have been proposed, falling into three categories: node-based or cladistic estimates (CD: Vane-Wright et al. 1991; Nixon and Wheeler 1992), genetic-distance estimates (GD: Crozier 1992), and feature-based or phylogenetic estimates (PD: Faith 1992). All of these indexes include information about the relatedness of the species in the sample (CD, PD) or level of genetic divergence among them (GD, PD), aiming to maximize the total diversity preserved. These indexes prioritize the uniqueness of the taxa being considered. Feature-based estimates have the advantage of including both aspects of a phylogenetic tree—topology and divergence among taxa—giving priority to highly divergent taxa that are either on basal or long branches (fig. 10.2). Given the relative scarcity of phylogenetic data available for specific regions and taxa, only a few studies have used these measurements (Crandall 1998; Virolainen et al. 1999; Whiting et al. 2000; Polasky et al. 2001; Posadas et al. 2001; Virolainen et al. 2001; Pérez-Losada et al. 2002; Sechrest et al. 2002; Jaramillo 2001). As the number of phylogenies available increases and as analytical and technical tools improve, the pace and accuracy of these studies will increase.

Phylogenetic diversity measures were illustrated from both the theoretical and practical viewpoint by the early proponents of phylogenetic diversity indexes (Vane-Wright et al. 1991; Faith 1992). The best-known example is the bumblebees in the Bombus sibiricus group, where the number of species and phylogenetic diversity (Vane-Wright 1991; Faith 1992) were estimated and three different reserve areas were compared using PD and the total number of species. The analyses showed that the area with the least number of species had the maximum PD because the species in that area were well-distributed across the phylogeny (Faith 1992). Another well-known example is that of the tuatara (Sphenodon). If all species of reptiles are equally weighted, the two species of Sphenodon represent only 0.03% of all extant reptiles, but they represent 50% of their evolutionary history (May 1990), given that the tuataras are the sister clade to all other reptiles. These two cases exemplify how species counts, when used alone, are a poor representation of phylogenetic diversity.

Some studies have used morphological phylogenies and taxonomic classifications to obtain an estimate of cladistic diversity (CD) and compare the efficiency of the CD index to "species richness" and "endemism" in selecting conservation re-

CD

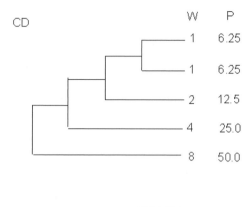

W	P
1	6.25
1	6.25
2	12.5
4	25.0
8	50.0

PD

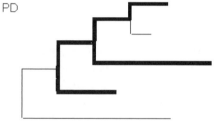

Figure 10.2 Cladistic diversity (CD, upper panel) applies equal weighting (W) to sister clades. Based on the W values, the percentage contribution of each taxon (P) can be calculated (Vane-Wright et al. 1991). Phylogenetic diversity (PD, lower panel) is calculated as the sum of branch lengths across the minimum spanning path (in bold; Faith 1992).

serves. An analysis of 177 species of plants in the herb-rich boreal-lake forests in Finland compared the efficiency of different measures (species richness, phylogenetic diversity, and restricted-range diversity) in selecting a reserve network representing total diversity. The authors emphasized that phylogenetic diversity was a poor predictor of total species richness because it does not take into account rare and sparsely distributed species if they are not phylogenetically distinct (Virolainen et al. 1999; Virolainen et al. 2001). A morphological phylogenetic analysis was used to compare the importance of endemisms and phylogeny in the conservation of South African orchids (Linder 1995). This study suggests that conserving "metaspecies," taxa that form unresolved assemblages with monophyletic clades imbedded in them (Donoghue 1985), may be better than choosing recent endemics growing on ephemeral habitats. Metaspecies have been able to speciate into other habitats in the past and have great potential of doing so in the future. In another attempt to take into account rare species, Posadas et al. (2001) proposed a measure that includes taxon distinctness and endemicity. In their analysis of southern South America, the authors considered 115 species of arthropods and angiosperms, distributed in 12 biogeographic regions. They found that the most important region when taxa distinctness and endemicity were considered, Santiago (in the central portion of Chile), was also the region of highest species richness. These studies both illustrate that CD can be important and complementary to species-richness data. However, they tried to maximize the number of species preserved by using CD as an indicator of species richness, which minimizes the use of CD and misses the point that CD is an important index for determining areas for conservation.

Molecular phylogenetics has burst upon the systematic scene in the last ten years with the use of the polymerase chain reaction (PCR) techniques that immensely facilitate the acquisition of molecular sequence data. Since the speed of production has greatly increased, the number of molecular phylogenies has surpassed the number previously prepared by using morphological characters, and molecular techniques are the leading provider of cladograms. In the past five years some studies have used genetic information in the estimation of phylogenetic diversity indexes. For instance, Crandall and his collaborators have used molecular phylogenetics of the crayfish in the Ozark plateaus (Crandall 1998) and Australia (Whiting et al. 2000) to establish conservation priorities. In both studies the crayfish data suggested that the conservation priorities selected using maximum PD were very similar to those chosen using species-richness indexes. Similarly, a molecular phylogeny of the plant genus *Piper* (Piperaceae, black pepper and its allies) was used to determine areas of high diversity in the Chocó region of Colombia (Jaramillo 2001). This study showed that PD was positively correlated with species richness but negatively correlated with endemism. The *Piper* analysis concluded that species richness can be a surrogate for PD as was shown by the crayfish studies. It also suggested that focusing on endemic species as conservation targets cannot be the main goal in biodiversity conservation, because endemics do not always make a large contribution to the phylogenetic diversity of an area. The broadest application of these methods is a study using carnivore and primate molecular phylogenies and their worldwide distribution to test the importance of biodiversity hot spots for conservation (Sechrest et al. 2002). The authors found that the taxa from these regions represent more than 70% of the evolutionary history of the groups evaluated.

From this small sample of cases that used phylogenetic measures, we can see how data from evolutionary histories is very important in determining conservation areas, because they are estimates of feature or information diversity. These studies also indicate that in some cases species richness can be a good surrogate for CD and PD, which is encouraging, given the scarcity of phylogenetic data, the high costs of generating it, and the little time we have for making conservation decisions. Some authors still give great importance to endemic and rare species (Virolainen et al. 1999; Posadas et al. 2001; Virolainen et al. 2001). However, two studies (Linder 1995; Jaramillo 2001) seem to suggest that endemicity is not invariably the most important factor, because endemic species do not always make a large contribution to feature diversity. It was hypothesized that the relatively low importance of endemics on the coast of South Africa and the Chocó region is a product of the recent speciation in the region, making most endemics young and closely related to each other. For an area that has relictual endemics, however, we predict that endemic species would make a larger contribution to the total diversity of the locality. The difference in the relative importance of endemics in the feature-based indexes depends on the evolutionary history of the biota of each region. Therefore one cannot a priori dismiss CD or PD and use only species richness as a surrogate. For the record, a recent study has shown that PD and the alternative ED (environmental distance) do not function well in some situations (Araújo et al. 2001).

IDENTIFICATION OF AREAS OF SHARED VICARIANT HISTORY

The intraspecific phylogeographies of codistributed species can be used to identify areas that share common vicariant history. One identifies evolutionarily distinct communities or areas and searches for repeating patterns among these areas in different groups of taxa (Nelson and Platnick 1981). Rosen (1978) illustrated how this could be accomplished using poeciliid fish genera, but he did not suggest that such areas should be the basis for conservation planning. Phylogenies were generated for many lineages of organisms endemic to the Hawaiian islands. Each island was treated as a separate area(s). The repeated pattern observed among the islands revealed the importance of the different areas for conservation (Wagner and Funk 1995).

The search for groups of species that share the same geographic distribution has been extended by the concept of evolutionarily significant units, within single species (Moritz 1994). Using congruence among haplotype phylogenies (mostly cpDNA or mtDNA, for plants and animals respectively), it is possible to identify recent vicariance events that have segregated the communities into distinct geographic units. A good example of this method comes from the tropical forest of northeast Queensland, Australia. Moritz and his collaborators have studied a large number of animal taxa from the region, showing that several species have similar geographic structures, with distinct northern and southern populations (Moritz and Faith 1998; Joseph et al. 1995). In order to prioritize among those areas, Moritz and Faith used a PD approach and proposed that if only two areas can be conserved, a subregion from the north and another from the south should be selected in order to maximize preserved genetic diversity. There are no similar examples in the plant literature, although some phylogeographic studies in northern Europe and northwestern North America have shown that most diversity is derived from Pleistocene refugia (Soltis et al. 1992; Manos et al. 1999). However, they have not discussed conservation issues explicitly. This approach can be extended to different areas to organisms of restricted and extended distribution that can be undergoing geographic differentiation that we need to take into account when making conservation decisions.

PRESERVING THE COMPONENTS OF EVOLUTIONARY PROCESSES

Phylogenies can also be enlightening about evolutionary processes and can provide information about how to preserve the potential for evolutionary change at either the population level or the species level. At the population level, phylogeographic studies can be useful in understanding the bridge between intra- and interspecific evolution and in inferring speciation events. Phylogenies can distinguish among three major factors that can spatially differentiate the lineages: restricted gene flow, range expansion (i.e., colonization), and past fragmentation (Templeton 1998). At the species level, tree-shape information can help distinguish different diversification processes and their correlation to extinction rates (Heard and Mooers 2000).

Intraspecific phylogenies reflect population histories, helping to reconstruct the processes that have shaped the current distribution of genetic diversity. Information about evolutionary processes is essential in making judicious conservation decisions. In a study of African bovids, it was found that while the populations of impala and wildebeest were fragmented by the Rift Valley in Tanzania, this barrier was not important for the buffalo populations that presented recurrent gene flow (Templeton and Beorgiadis 1996). According to these results there is no genetic rationale to stop the translocation of buffalos for repopulation. Translocation of impala and wildebeest individuals should be avoided, however, because it can disrupt local adaptation. These conclusions were not possible without the genetic analysis.

Phylogenies at the species or supraspecific level can provide information about diversification and extinction processes. Heard and Mooers (2000) demonstrated that the loss of diversity is associated with two factors: the diversification processes that produced the clades in consideration and the patterns of extinction risk across species. Given that phylogenetic trees are often unbalanced because of the differences in speciation rates across lineages (Guyer and Slowinski 1991), conservationists are better off selecting priorities guided by a minimum-loss algorithm (Heard and Mooers 2000). Additionally, given that extinction is not random, organisms belonging to certain clades or with particular biological conditions are more vulnerable. In the case of bird species, it has been shown that extinction risk depends on family affiliation, body size. and fecundity (Bennett and Owens 1997). In many plant groups there is no comparative study that evaluates extinction risk.

RISK OF HYBRIDIZATION WITH INVASIVE SPECIES

Species introductions into nonnative habitat impose a big danger to biodiversity conservation. Most studies have emphasized the ecological effects of invasions; however, the diversity lost by hybridization poses a great risk that is largely underestimated. Species introduced from different continents, in general distantly related to local taxa, produce ecological effects like competition, predation, and parasitism. Closely related taxa are prone to hybridize with local species and erode the genetic diversity of a population. Given these conditions, phylogenies are the best tool to predict the probability of hybridization after species translocation.

A recent review of invasions in freshwater habitats in North America showed that phylogenetic relationships are perhaps the best way to predict hybridization (Perry et al. 2002). Recent genetic studies using molecular markers confirmed that hybridization has occurred for more than a decade between the rusty crayfish (*Orconectes rusticus*) and the resident species *O. propinquus* in northern Wisconsin and Michigan lakes (Perry et al. 2001). Although this hybridization had been suggested by earlier morphological studies (Smith 1981), only molecular markers can test the extent of hybridization and its consequences. Although only one case has been documented genetically, the observation of intermediate phenotypes suggest that hybridization may be occurring in at least another five pairs of crayfish species, all of it among closely related species (Crandall and Fitzpatrick 1996).

A recent example in the plant kingdom shows the potential of invasive species. Consider the uproar over the possibility of transgenic DNA from genetically modified maize in local varieties of the crop in Oaxaca, Mexico (Quist and Chapela 2001). Corn is wind-pollinated, and so cross-pollination among fields is common. The appearance of genes from the genetically engineered maize in native varieties has many worried. Mexico is the center for native varieties of corn and seeds; the contamination of the genome of the native species has dire implications for the future of native maize. There is some controversy over the report (Butler 2002), but should this introgression prove to be true it would bear out the worst fears of the opponents of genetically altered crops and illustrate the danger of such crops as they are introduced into cultivation.

Four different scenarios have been presented where phylogenetic information is critical to conservation efforts. One of the largest barriers to making phylogenetic (especially molecular phylogenetics) methods more widely used is time and cost of obtaining the data. The speed of producing molecular phylogenies is increasing, however, and once the phylogenies are available, they are invaluable as conservation tools. For instance, in diversity indexes where the number of species can, in some cases, be a good surrogate of PD, a phylogeny is important in choosing priorities. Phylogenies illustrate the patterns of diversification of organisms at different taxonomic levels; thus they are helpful in understanding the ecological and geologic processes that have been responsible for the generation of diversity and as such should be preserved. Most examples presented here are from animals; it is clear that there is a real need to conduct similar studies in plants and to use these to compare evolutionary histories of the flora and fauna of different regions of interest.

LITERATURE CITED

Ab'Sáber, A. N. 1977. Os domínios morfoclimáticos na América do sul: primeira aproximação. Geomorfologia 53:1–23.

All Species Foundation. 2004. All Species Inventory. http://www.all-species.org/.

Araújo, M. B., Humphries, C. J., Densham, P. J., Lampinen, R., Hagemeijer, W. J. M., Mitchell-Jones, A. J., and Gasc, J. P. 2001. Would environmental diversity be a good surrogate for species diversity? Ecography 14:103–110.

Belbin, L. 1993. Environmental representativeness: regional partitioning and reserve selection. Biological Conservation 66:223–230.

———. 1995. A multivariate approach to the selection of biological reserves. Biodiversity and Conservation 4:951–963.

Bennett, P. M., and Owens, I. P. F. 1997. Variation in extinction risk among birds: chance or evolutionary predisposition? Proceedings of the Royal Society B 264:401–408.

Brugman, M. L., and Species 2000 Secretariat. 2004. Catalogue of Life Programme. Species 2000. http://www.sp2000.org/.

Busby, J. R. 1986. A biogeoclimatic analysis of *Nothofagus cunninghamii* (Hook.) Oerst. in southeastern Australia. Australian Journal of Ecology 11:1–7.

Butler, D. 2002. Alleged flaws in gene-transfer paper spark row over the genetically modified maize. Nature 415:948–949.

Carpenter, G., Gillison, A. N., and Winter, J. 1993. DOMAIN: a flexible modelling procedure for mapping potential distributions of plants and animals. Biodiversity and Conservation 2:667–680.

Conabio. 2004. The Constitutional President of Mexico. http://www.conabio.gob.mx/.

Crandall, K. A. 1998. Conservation phylogenetics of Ozark crayfishes: assigning priorities for aquatic habitat protection. Biological Conservation 84:107–117.

Crandall, K. A., and Fitzpatrick, J. F. J. 1996. Crayfish molecular systematics: using a combination of procedures to estimate phylogeny. Systematic Biology 45:1–26.

Crozier, R. H. 1992. Genetic diversity and the agony of choice. Biological Conservation 61: 11–15.

Davis, F. W., Stoms, D. M., and Andelman, S. 1999. Systematic reserve selection in the USA: an example from the Columbia Plateau ecoregion. Parks 9:31–41.

Donoghue, M. J. 1985. A critique of the biological species concept and recommendations for a phylogenetic alternative. Bryologist 88:172–181.

Edwards, J. L., Lane, M. A., and Nielsen, E. S. 2000. Interoperability of biodiversity databases: biodiversity information on every desktop. Science 289:2312–2314.

Faith, D. P. 1992. Conservation evaluation and phylogenetic diversity. Biological Conservation 61:1–10.

———. 2003. Environmental diversity (ED) as surrogate information for species-level biodiversity. Ecography 26:374–379.

Faith, D. P., Margules, C. R., Walker, P. A., Stein, J., and Natera, G. 2001. Practical application of biodiversity surrogates and percentage targets for conservation in Papua New Guinea. Pacific Conservation Biology 6:289–303.

Faith, D. P., and Walker, P. A. 1996. Environmental diversity: on the best possible use of surrogate data for assessing the relative biodiversity of sets of areas. Biodiversity and Conservation 5:399–415.

Ferrier, S. 1997. Biodiversity data for reserve selection: making best use of incomplete information. Pp. 315–329 in Pigram, J. J., and Sundell, R. C., eds., National Parks and Protected Areas: Selection, Delimitation, and Management. Centre for Water Policy Research, Armidale, Australia.

Ferrier, S., Pressey, R. L., and Barrett, T. W. 2000. A new predictor of the irreplaceability of areas for achieving a conservation goal, its application to real-world planning, and a research agenda for further refinement. Biological Conservation 93:303–325.

Fjeldså, J. 2000. The relevance of systematics in choosing priority areas for global conservation. Environmental Conservation 27:67–75.

Flather, C. H., Wilson, K. R., Dean, D. J., and McComb, W. 1997. Identifying gaps in conservation networks: of indicators and uncertainty in geographic-based analyses. Ecological Applications 7:531–542.

Funk, V. A., and Richardson, K. 2002. Biological specimen data: use it or lose it. Systematic Biology 51:303–316.

Funk, V. A., Zermoglio, M. F., and Nassir, N. 1999. Testing the use of specimen based collecting data and GIS in biodiversity exploration and conservation decision-making in Guyana. Biodiversity and Conservation 8:727–751.

Gaston, K. J., and Williams, P. H. 1993. Mapping the world's species: the higher taxon approach. Biodiversity Letters 1:2–8.

GBIF Secretariat. 2004. Global Biodiversity Information Facility. http://www.gbif.org.

Govaerts, R. 2001. How many species of seed plants are there? Taxon 50:1085–1090.

Guyer, C., and Slowinski, J. B. 1991. Comparison of observed phylogenetic topologies with null expectations among three monophyletic lineages. Evolution 45:340–350.

Heard, S. B., and Mooers, A. Ø. 2000. Phylogenetically patterned speciation rates and extinction risks change the loss of evolutionary history during extinctions. Proceedings of the Royal Society B 267:613–620.

Heyer, W. R., Coddington, J., Kress, W. J., Acevedo, P., Cole, D., Erwin, T. L., Meggers, B. J., Pogue, M., Thorington, R. W., Vari, R. P., Weitzman, M. J., and Weitzman, S. H. 1999. Amazonian biotic data and conservation decisions. Ciencia e Cultura 51:372–385.

Hollowell, T., Funk, V. A., and Kelloff, C. L. 2004. Biological Diversity of the Guianas. Smithsonian National Museum of Natural History. http://www.mnh.si.edu/biodiversity/bdg/.

Jaramillo, M. A. 2001. Divergence and diversification in the genus *Piper* and the order Piperales. Ph.D. diss., Duke University.

Joseph, L., Moritz, C., and Hugall, A. 1995. Molecular data support vicariance as a source of diversity in rainforests. Proceedings of the Royal Society B 260:177–182.

Kress, W. J., Heyer, W. R., Acevedo, P., Coddington, J., Cole, D., Erwin, T. L., Meggers, B. J., Pogue, M., Thorington, R. W., Vari, R. P., Weitzman, M. J., and Weitzman, S. H. 1998. Amazonian biodiversity: assessing conservation priorities with taxonomic data. Biodiversity and Conservation 7:1577–1587.

Krupnick, G. A., and Kress, W. J. 2003. Hotspots and ecoregions: a test of conservation priorities using taxonomic data. Biodiversity and Conservation 12:2237–2253.

Krupnick, G. A., and Rubis, J. 2002. Plant richness and endemism in the Indo-Pacific: Dipterocarpaceae. Pp. 92–98 in Wikramanayake, E. D., Dinerstein, E., Loucks, C. J., Olson, D. M., Morrison, J., Lamoreux, J., McKnight, M., and Hedao, P., Terrestrial Ecoregions of the Indo-Pacific: A Conservation Assessment. Island Press, Washington, DC.

Linder, H. P. 1995. Setting conservation priorities: the importance of endemism and phylogeny in the southern African orchid genus *Herschelia*. Conservation Biology 9:585–595.

Lombard, A. T., Cowling, R. M., Pressey, R. L., and Mustart, P. J. 1997. Reserve selection in a species-rich and fragmented landscape on the Agulhas Plain, South Africa. Conservation Biology 11:1101–1116.

Lombard, A. T., Hilton-Taylor, C., Rebelo, A. G., Pressey, R. L., and Cowling, R. M. 1999. Reserve selection in the Succulent Karoo, South Africa: coping with high composition turnover. Plant Ecology 142:35–55.

Mackey, B. G., Nix, H. A., Hutchinson, M. F., MacMahon, J. P., and Fleming, P. M. 1988. Assessing the representativeness of places for conservation reservation and heritage listing. Environmental Management 12:501–514.

Mackey, B. G., Nix, H. A., Stein, J. A., and Cork, S. E. 1989. Assessing the representativeness of the wet tropics of Queensland world heritage property. Biological Conservation 50:279–303.

Manos, P. S., Doyle, J. J., and Nixon, K. C. 1999. Phylogeny, biogeography, and processes of molecular differentiation in *Quercus* subgenus *Quercus* (Fagaceae). Molecular Phylogenetics and Evolution 12:333–349.

Margules, C. R., and Austin, M. P. 1994. Biological models for monitoring species decline: the construction and use of data bases. Philosophical Transactions of the Royal Society B 344:69–75.

Margules, C. R., Nicholls, A. O., and Pressey, R. L. 1988. Selecting networks of reserves to maximize biological diversity. Biological Conservation 43:63–76.

May, R. M. 1990. Taxonomy as destiny. Nature 347:129–130.

Mittermeier, R. A., and Werner, T. B. 1990. Wealth of plants and animals unites "megadiversity" countries. Tropicus 4:4–5.

Moritz, C. 1994. Defining "evolutionary significant units" for conservation. Trends in Ecology and Evolution 9:373–375.

Moritz, C., and Faith, D. P. 1998. Comparative phylogeography and the identification of genetically divergent areas for conservation. Molecular Ecology 7:419–429.

Moritz, C., Richardson, K. S., Ferrier, S., Monteith, G. B., Stanisic, J., Williams, S. E., and Whiffen, T. 2001. Biogeographic concordance and efficiency of taxon indicators for establishing conservation priority for a tropical rainforest biota. Proceedings of the Royal Society B 268:1875–1881.

Myers, N. 1988. Threatened biotas: "hotspots" in tropical forests. Environmentalist 8:1–20.

———. 1990. The biodiversity challenge: expanded hot-spot analysis. Environmentalist 10:243–256.

Myers, N., Mittermeier, R. A., Mittermeier, C. G., da Fonseca, G. A. B., and Kent, J. 2000. Biodiversity hotspots for conservation priorities. Nature 403:853–858.

Nelson, B. W., Ferreira, C. A. C., da Silva, M. F., and Kawasaki, M. L. 1990. Endemism centres, refugia, and botanical collection density in Brazilian Amazonia. Nature 345:714–716.

Nelson, G., and Platnick, N. 1981. Systematics and Biogeography: Cladistics and Vicariance. Columbia University Press, New York.

New South Wales National Park Service. 1999. C-Plan Module, Version 3.03. C-Plan Support, New South Wales National Parks and Wildlife Service. http://www.ozemail.com/~cplan.

Nix, H. A. 1982. Environmental determinants of biogeography and evolution in Terra Australis. Pp. 47–66 in Baker, W. R., and Greenslade, P. J. M., eds., Evolution of the Flora and Fauna of Australia. Peacock Publishers, Adelaide.

Nixon, K. C., and Wheeler, Q. D. 1992. Measures of phylogenetic diversity. Pp. 216–234 in Novacek, M. J., and Wheeler, Q. D., eds., Extinction and Phylogeny. Columbia University Press, New York.

Noss, R. F. 1990. Indicators for monitoring biodiversity. Conservation Biology 4:355–364.

Olson, D. M., Dinerstein, E., Wikramanayake, E. D., Burgess, N. D., Powell, G. V. N., Underwood, E. C., D'Amico, J. A., Itoua, I., Strand, H. E., Morrison, J. C., Loucks, C. J., Allnutt, T. F., Ricketts, T. H., Kura, Y., Lamoreux, J. F., Wettengel, W. W., Hedao, P., and

Kassem, K. R. 2001. Terrestrial ecoregions of the world: a new map of life on earth. Bio-Science 51:933–938.

Panzer, R., and Schwartz, M. 1998. Effectiveness of a vegetation-based approach to insect conservation. Conservation Biology 12:693–702.

Pearson, D. L., and Cassola, F. 1992. World-wide species richness patterns of tiger beetles (Coleoptera: Cicindelidae): indicator taxon for biodiversity and conservation studies. Conservation Biology 6:376–391.

Pérez-Losada, M., Jara, C. G., Bond-Backup, G., and Crandall, K. A. 2002. Conservation phylogenetics of Chilean freshwater crabs *Aegla* (Anomura, Aeglidae): assigning priorities for aquatic habitat protection. Biological Conservation 105:345–353.

Perry, W. L., Feder, J. L., and Lodge, D. M. 2001. Hybrid zone dynamics and species replacement between *Orconectes* crayfishes in northern Wisconsin lake. Evolution 55:1153–1166.

Perry, W. L., Lodge, D. M., and Feder, J. L. 2002. Importance of hybridization between indigenous and non-indigenous freshwater species: an over-looked threat to North American biodiversity. Systematic Biology 51:255–275.

Pharo, E. J., Beattie, A. J., and Pressey, R. L. 2000. Effectiveness of using vascular plants to select reserves for bryophytes and lichens. Biological Conservation 96:371–378.

Pitman, N. C. A., and Jørgensen, P. M. 2002. Estimating the size of the world's threatened flora. Science 298:989.

Polasky, S., Csuti, B., Vossler, C. A., and Meyers, S. M. 2001. A comparison of taxonomic distinctness versus richness as criteria for setting conservation priorities for North American birds. Biological Conservation 97:99–105.

Posadas, P., Miranda, D. R., and Crisci, J. V. 2001. Using phylogenetic diversity measures to set priorities in conservation: an example from southern South America. Conservation Biology 15:1325–1334.

Prance, G. T., Beentje, H., Dransfield, J., and Johns, R. 2000. The tropical flora remains undercollected. Annals of the Missouri Botanical Garden 87:67–71.

Prendergast, J. R., Quinn, R. M., Lawton, B. C., Eversham, B. C., and Gibbons, D. W. 1993. Rare species: the coincidence of diversity hotspots and conservation strategies. Nature 365:335–337.

Pressey, R. L. 1994. Ad hoc reservations: forward or backward steps in developing representative reserve systems? Conservation Biology 8:662–668.

Pressey, R. L., Humphries, C. J., Margules, C. R., Vane-Wright, R. I., and Williams, P. H. 1993. Beyond opportunism: key principles for systematic reserve selection. Trends in Ecology and Evolution 8:124–128.

Pressey, R. L., Johnson, I. R., and Wilson, P. D. 1994. Shades of irreplaceability: towards a measure of the contribution of sites to a reservation goal. Biodiversity and Conservation 3:242–262.

Pressey, R. L., and Logan, V. S. 1995. Reserve coverage and requirements in relation to partitioning and generalization of land classes: analyses for western New South Wales. Conservation Biology 9:1506–1517.

Quist, D., and Chapela, I. H. 2001. Transgenic DNA introgressed into traditional maize landraces in Oaxaca, Mexico. Nature 414:541–543.

Rebelo, A. G. 1994. Using the Proteaceae to design a nature reserve network and determine priorities for the Cape Floristic Region. Pp. 375–396 in Forey, P. L., Humphries, C. J., and Vane-Wright, R. I., eds., Systematics and Conservation Evaluation. Clarendon Press, Oxford.

Reid, W. V. 1998. Biodiversity hotspots. Trends in Ecology and Evolution 13:275–280.

Richardson, K. S., and Funk, V. A. 1999. An approach to designing a protected area system in Guyana. Parks 9:7–16.

Ricketts, T. H., Dinerstein, E., Olson, D. M., Loucks, C. J., Eichbaum. W., DellaSala, D., Kavanagh, D., Hedao, P., Hurley, P. T., Carney, K. M., Abell, R., and Walters, S. 1999. Terrestrial Ecoregions of North America: A Conservation Assessment. Island Press, Washington, DC.

Rosen, D. E. 1978. Vicariant patterns and historical explanation in biogeography. Systematic Zoology 2:159–188.

Sechrest, W., Brooks, T. M., da Fonseca, G. A. B., Konstant, W. R., Mittermeier, R. A., Purvis, A., Rylands, A. B., and Gittleman, J. L. 2002. Hotspots and the conservation of evolutionary history. Proceedings of the National Academy of Sciences USA 99:2067–2071.

Smith, D. G. 1981. Evidence for hybridization between two crayfish species (Decapoda, Cambaridae, *Orconectes*) with a comment on the phenomenon in cambarid crayfish. American Midland Naturalist 105:405–407.

Soltis, D. E., Soltis, P. S., Kuzoff, R. K. and Tucker, T. L. 1992. Geographic structuring of chloroplast DNA genotypes in *Tiarella trifoliata* (Saxifragaceae). Plant Systematics and Evolution 181:203–216.

Templeton, A. R. 1998. Nested clade analyses of phylogeographic data: testing hypotheses about gene flow and population history. Molecular Ecology 7:381–397.

Templeton, A. R., and Beorgiadis, N. J. 1996. A landscape approach to conservation genetics: conserving evolutionary processes in the African Bovidae. Pp. 398–430 in Avise, J. C., and Hamrick, J. L., eds., Conservation Genetics: Case Histories from Nature. Chapman and Hall, New York.

Ter Steege, H. 1998. The use of forest inventory data for a national protected area strategy in Guyana. Biodiversity and Conservation 7:1457–1483.

Ter Steege, H., Jansen-Jacobs, M. J., and Datadin, V. K. 2000. Can botanical collections assist in a national protected area strategy in Guyana? Biodiversity and Conservation 9:215–240.

Vane-Wright, R. I., Humphries, C. J., and Williams, P. H. 1991. What to protect? Systematics and the agony of choice. Biological Conservation 55:235–254.

Virolainen, K. M., Nattinen, K., Suhonen, J., and Kuitunen, M. 2001. Selecting herb-rich forest networks to protect different measures of biodiversity. Ecological Applications 11:411–420.

Virolainen, K. M., Virola, T., Suhonen, J., Kuitunen, M., Lammi, A., and Siikamäki, P. 1999. Selecting networks of nature reserves: methods do affect the long-term outcome. Proceedings of the Royal Society B 266:1141–1146.

Voss, R. S., and Emmons, L. H. 1996. Mammalian diversity in Neotropical lowland rainforest: a preliminary assessment. Bulletin of the American Museum of Natural History 230:1–115.

Wagner, W. L., and Funk, V. A. 1995. Hawaiian Biogeography: Evolution on a Hotspot Archipelago. Smithsonian Institution Press, Washington, DC.

Wessels, K. J., Freitag, S., and Van Jaarsveld, A. S. 1999. The use of land facets as biodiversity surrogates during reserve selection at a local scale. Biological Conservation 89:21–38.

Whiting, A. S., Lawler, S. H., Horwitz, P., and Crandall, K. A. 2000. Biogeographic regionalization of Australia: assigning conservation priorities based on endemic freshwater crayfish phylogenetics. Animal Conservation 3:155–163.

Wikramanayake, E. D., Dinerstein, E., Loucks, C. J., Olson, D. M., Morrison, J., Lamoreux, J., McKnight, M., and Hedao, P. 2002. Terrestrial Ecoregions of the Indo-Pacific: A Conservation Assessment. Island Press, Washington, DC.

Williams, P. H., and Gaston, K. J. 1994. Measuring more of biodiversity: can higher-taxon richness predict wholesale species richness? Biological Conservation 67:211–217.

Williams, P. H., Prance, G. T., Humphries, C. J., and Edwards, K. S. 1996. Promise and problems in applying quantitative complementary areas for representing the diversity of some Neotropical plants (families Dichapetalaceae, Lecythidaceae, Caryocaraceae, Chrysobalanaceae, and Proteaceae). Biological Journal of the Linnean Society 58:125–157.

Workshop 90: Biological Priorities for Conservation in Amazonia. 1991. Conservation International, Washington, DC. Map and explanatory text.

WWF and IUCN (World Wildlife Fund and World Conservation Union). 1994–1997. Centres of Plant Diversity: A Guide and Strategy for Their Conservation. 3 vols. IUCN Publication Unit, Cambridge.

CHAPTER 11

ASSESSING CONSERVATION STATUS

How do we evaluate which plant species, populations, and communities are in need of conservation or protection? Conservation assessment is based upon a variety of measurements ranging from quantifying genetic variability within a species to appraising species diversity within a habitat. At the species level, the field of conservation genetics concentrates on determining the levels of genetic diversity within a species, which is then used to assess the evolutionary fitness of threatened species and their adaptability to changing environments. The first section of this chapter focuses on the modern technology and tools of molecular genetics that are essential to understanding the patterns and processes of plant diversification and conservation.

The second section explores how the status of species conservation is assessed at the global level. A species is considered threatened when the size and distribution of its global population falls to nonsustainable numbers. To evaluate and quantify the conservation status of the earth's plant and animal species, the World Conservation Union (IUCN) developed the Red List Programme. The lists resulting from this effort are comprehensive inventories of species threatened with extinction or already extinct. As a dynamic catalogue, the Red List adds, removes, and changes the threat category of monitored species as their conservation status changes over time.

To assess conservation status at the community and ecosystem level, biological diversity and species uniqueness is evaluated through the activities of rapid biological inventory teams. Plant communities that are targeted for evaluation are usually located in threatened regions of the world with estimated high levels of diversity. These surveys evaluate the diversity, uniqueness, and biological importance of large wilderness areas, based on particular groups of organisms that are indicative of habitat types and conditions. The final section of this chapter describes the history and future of these rapid assessment programs.

11.1 GENETIC ASSESSMENT METHODS FOR PLANT CONSERVATION BIOLOGY

Elizabeth A. Zimmer

GENETIC MEASURES of biodiversity have become readily available to the field biologist over the past decade. Genetic analyses have been applied to reintroductions as well as to ex situ and in situ preservation strategies. Optimally, a conservation genetics program is concerned with both the retention of organismal fitness and adaptability to environmental change (Schemske et al. 1994; Crandall et al. 2000), as well as with the preservation of major historical lineages within taxonomic units (Moritz 2002; see also section 10.3). For the former, process-oriented studies, quantitative genetic methods and reciprocal crossing experiments have been used to measure phenotypic trait heritability and fitness differences. Molecular phylogenetic and phylogeographic comparisons employ markers to detect patterns of genetic variation, which result in interspecific and intraspecific genealogies, respectively. For example, detection and quantification of molecular variation across populations of endangered plant species would allow conservationists to choose for preservation those individuals that have the greatest degree of genetic potential as expressed by genetic distance or phylogenetic diversity scores (Gemmill et al. 1998; Jaramillo 2001).

This section presents practical considerations in the application of molecular genetic techniques to elucidate the patterns and processes defining biodiversity. These include protocols for collecting plant tissues for genetic screening, types of molecular markers used in various species and population comparisons, and some examples of conservation biology studies that have used genetic markers. For a discussion of the appropriateness of using genetic diversity measures in conservation planning, see the references cited above.

COLLECTION AND MOLECULAR ISOLATION METHODS

For most plant genetic assessment methods, sampling of leaf tissue is the most suitable and least destructive choice. Seeds may also be a good source of macromolecules, especially when leaf extracts have been shown to contain large quantities of mucilaginous or polyphenolic compounds. Whenever possible for interspecific studies, a voucher specimen containing vegetative and reproductive material should also be made and deposited in a local herbarium. For intraspecific studies, a voucher should be prepared for each geographically or phenotypically distinct population.

The optimal starting material for both plant isozyme and DNA marker surveys is fresh (or quick-frozen in liquid nitrogen) tissue. Isozyme studies, in fact, require fresh or quick-frozen material. In cases where the field site is quite remote, the logistics of carrying a liquid nitrogen tank or a field isolation kit may be prohibitive, and express mailing of fresh tissue back to the laboratory impractical. Therefore, the most commonly used method for collecting plant tissue in the field for later DNA isolation is to dehydrate one to several grams of fresh tissue in a plastic zip bag with a 10- to 15-fold quantity of silica gel containing a small amount of indicator silica. Dry herbarium specimens may also be used for DNA isolation, although the quantity and quality of the DNA obtained may be inferior to that from frozen or silica-dried material. On the other hand, use of herbarium specimens obtained over a number of years may provide evidence for changes in diversity for species undergoing genetic bottlenecks or range contraction. Additional information on plant collection techniques suitable for molecular studies is available online (Nickrent 2002), with an excellent introduction to the subject.

For both protein and nucleic-acid comparisons, the first step is preparation of a tissue extract. In order to get good breakage of plant cell walls, it is useful to grind the tissue to a fine powder after freezing one to several grams wet weight of tissue in liquid nitrogen. The grinding may be done manually in a mortar and pestle or with a commercially available tissue homogenizer (e.g., the Bio 101 Fast Prep Cell Disrupter system). In the case of protein isozyme surveys, the powder is rehydrated in buffer solutions that will retain the enzymatic activity necessary for the staining step after starch gel electrophoresis. Chemicals that retard "browning" of the extracts are used for species with significant amounts of polyphenolics, such as ferns. A detailed set of protocols for isozyme work can be found in Soltis et al. 1983. The remainder of this section will focus on DNA methods of genetic assessment; isozyme data is analyzed in a manner similar to DNA fragment data, either as frequency information or as discrete characters (Hamrick and Godt 1990, 1996; Rieseberg and Swensen 1996).

For DNA comparisons, frozen plant powders are extracted with solutions containing the detergent CTAB and chelating agents that inhibit nuclease action (Doyle and Doyle 1987). The extraction step is followed either by an organic extraction or by adsorption to an insoluble matrix available from Qiagen. In the former case, the aqueous fraction is subjected to ethanol precipitation, and then the DNA is brought up in a small volume of sterile water. With the Qiagen kits, the DNA is eluted from the matrix into sterile water after a series of ethanol wash steps. Samples are stored in a $-20°C$ or $-80°C$ freezer. DNA is quantified using a spectrophotometer or fluorimeter, and the quality of the DNA is assessed relative to DNA fragment standards on agarose gels stained with fluorescent dyes.

GENETIC MARKERS USEFUL FOR CONSERVATION WORK

Subsequent to extraction, DNA variation is measured either by comparison of known gene regions or of anonymous markers. Currently, the most commonly used techniques all involve applications of the polymerase chain reaction (PCR), the same method used in forensic studies of human samples. PCR can reproduce segments of the genetic material of any organism or mixtures of organisms in quantities necessary to detect DNA differences. A layperson's introduction to PCR was published in 1996 as part of the Breakthroughs in Bioscience series developed by the Federation of American Societies for Experimental Biology (see Breakthroughs in Bioscience 2003). For a more extensive, technical introduction to the subject, a variety of books are available (see PCR 1998).

Where phylogenetic assessment of samples is called for in conservation biology (e.g., in detecting interspecific hybridization or in developing strategies for maximizing adaptive genetic diversity among taxa or geographic areas), DNA sequencing of rapidly evolving genes or gene regions is the preferred comparative method. Among the most commonly used sequences are the internal transcribed spacers (ITS) and external transcribed spacers (ETS) of nuclear ribosomal DNA (Baldwin and Markos 1998) and the chloroplast *trnL* intron and spacer regions (Zimmer et al. 2002); these can be sequenced directly after PCR amplification using plant-specific DNA primers. Some nuclear gene introns (e.g., for the starch synthetase/*waxy* locus [Mason-Gamer et al. 1998] and the housekeeping gene *G3pdh* [Olsen and Schaal 1999; Schaal and Olsen 2000]) also have been used effectively in interspecific and intraspecific comparisons, but nucleotide polymorphism may require that these PCR products be further "purified" in agarose gels or by plasmid cloning of individual allelic copies. Sequencing of purified PCR products is performed on instruments that employ laser detection of fluorescent DNA fragments; these instruments are directly linked to computers that automatically provide chromatograms and resultant base calls. The DNA sequences from each species or individual are aligned relative to each other and to short conserved primer stretches. The alignment sequence matrix is then exported to one or several phylogenetic program packages that can produce an evolutionary tree.

One example of the application of a phylogenetic approach to conservation biology is provided by Jaramillo's study of the genus *Piper* in the Chocó region of Colombia (Jaramillo 2001; see also section 10.3). Tissue samples for 53 species of *Piper* in the Chocó were collected and sequenced for the ITS of nuclear ribosomal genes. A single evolutionary tree was obtained from the data matrix including the Chocó species and outgroup taxa from other geographic regions of the world. From this tree, phylogenetic diversity indexes were estimated, and four regions were proposed as having a high priority for conservation.

Another application of DNA phylogenies relevant to plant conservation biology is their use in identifying interspecific hybrids. Such hybrids occur frequently in plants and can confound species delimitation. Olsen and Schaal (1999), using *G3pdh* intron sequences, explored the diversity of cultivated cassava (*Manihot esculenta*) and its wild relatives. They were able to identify the likely origin of cultivated cassava in the southern Amazon basin and provided evidence that other *Manihot* species, possibly capable of introgression into the crop, were not contributing to the genetic diversity seen in cultivated and wild populations of *M. esculenta*.

With recently differentiated species or populations, the genes and spacers readily available for comparative sequencing may be unable to detect any genetic variation. In those cases, methods that use anonymous markers or "DNA fingerprints" often provide the necessary resolving power. These include RAPD (randomly amplified polymorphic DNA), ISSR (inter-simple sequence repeat), AFLP (amplified fragment length polymorphism), and SSR (simple sequence repeat, also called microsatellite marker). In the first two methods, single DNA primers are used in the PCR reaction; in the latter two, pairs of primers are employed. For the first three methods, high-quality DNA samples (i.e., having high molecular weight with no inhibitory molecules) are essential in order to ensure reproducibility of the fragment profiles. RAPD and ISSR profiling methods are the most economical in terms of both the time needed for technique development and the cost of screening large numbers of samples. AFLP profiling is more labor-intensive, since prior to PCR amplification steps, the DNA must be treated with restriction endonucleases and ligated to special DNA primers. AFLP profiles tend to be the most reproducible of the three techniques, however, and have been increasingly popular in population genetic surveys. For all three of the above methods, fragment patterns are scored for presence or absence of particular bands, and genealogies are produced using genetic-distance algorithms.

Development of microsatellite markers is the most time-consuming and costly of the methods involving DNA fragment comparisons. These simple sequence repeats are perhaps the most polymorphic regions of an organism's genome and are scattered throughout the DNA. Before one can survey a population for a particular microsatellite marker, however, randomly sheared genomic DNA must be cloned into plasmid vectors and individual clones screened for the presence of di- or trinucleotide repeats. For clones that test positive, the sequences flanking the SSRs must be determined, and primers must be designed to amplify regions of DNA flanking the repeats. Finally, the primer pairs must be shown to be polymorphic for the samples being compared. Typically, five to ten microsatellite primer pairs are used in a survey, so 10 to 30 clones must be sequenced and tested prior to the final PCR screening. It may take up to six months to develop a suite of microsatellite markers for a project. This extra effort may be warranted, however, as SSR, unlike RAPD, ISSR, and AFLP, provides more easily analyzed, codominant markers and recovers the highest numbers of polymorphisms (Milbourne et al. 1998). In certain cases, where the species of interest is closely related to a model system species (e.g., *Helianthus annuus* or *Arabidopsis thaliana* in molecular genetics studies or *Mimulus* or *Aquilegia* in ecological ones), microsatellites developed for a study in the model organism may be useful in surveying populations of its close relatives. The journal *Molecular Ecology* maintains a database of published microsatellite primer pairs on its Web site (see Molecular Ecology Notes).

Fragment markers have been used to detect genetic variation among plant populations and species, patterns of plant reproduction, gene flow between populations, and interspecific hybridization. Specifically, with respect to conservation applications, fragment screening may provide measures of overall diversity in small populations or germ-plasm collections and may help in the evaluation of procedures used to collect, preserve, or reintroduce plants (Schemske et al. 1994; van de Wiel and Vosman 1998).

In the case of the endangered Mauna Kea silversword on the island of Hawaii, microsatellite markers were used to demonstrate that a population crash in the species (the remnant population was reduced to 50 individuals) due to an increase in the number of alien ungulates did not significantly reduce heterozygosity or number of alleles present relative to the more abundant Haleakala silversword with populations exceeding 60,000 individuals (Friar et al. 2000; see also chapter 9). However, the reintroduction regime initially used, which had not screened the remnant population for genetic variation, did indicate that the Mauna Kea outplanted populations were significantly less genetically diverse than the remnant natural population. Thus, although the reintroduction regime employed for the Mauna Kea silversword increased the number of individuals present, the loss in genetic diversity among the outplants may result in population extinction due to inbreeding depression unless new reintroductions better represent the remnant population's diversity.

These authors also used the same microsatellite primer pairs to generate markers in a survey of the remnant populations (totaling <1,000 individuals) of the closely related Mauna Loa silversword on Kauai (Friar et al. 2001). They found that the three remnant populations were not genetically depauperate relative to each other or to other silversword populations. They also found that there was significant genetic differentiation among the remnants, which was reflected also in ecological and morphological observations. Their results provided a strategy for reintroduction that includes sampling the diversity across a particular remnant population but avoids interpopulation mixing in the outplants.

In conclusion, genetic methods have significant applications to a diversity of problems faced by the conservation biologist. Choice of a particular method will depend on the likely relatedness of the individuals being compared and the resources available to implement a particular technique. As genetic tools become increasingly automated and cost-effective, their use will become a routine component of germ-plasm sampling and reintroduction strategies.

11.2 SPECIES ASSESSMENT: THE IUCN RED LIST

Gary A. Krupnick

CATALOGUES OF THREATENED, endangered, and extinct species are useful tools for assessing conservation status at the species level. Because species are the principal unit of conservation concern, priorities are determined based upon their level of threat. Producing catalogues that list all known plant species from a given reserve, county, state, or country, or a subset of these data from a few plant families, is the first step in summarizing the conservation status of a habitat or ecoregion. Detailed information typically includes the taxonomic name (families, genera, and species), synonyms, author, distribution, and endangerment status of each species.

Assessing the threatened status of individual plant species and compiling a list of all known threatened plant species has served as an instrument for setting conservation priorities. Other approaches have included identifying threatened habitats that have high levels of biodiversity and composing lists of rare and endemic species at the national scale.

One successful approach in cataloguing threatened species has been the Red List Programme, developed by the World Conservation Union (IUCN). This program was developed as a first step in summarizing the conservation status on a global level of the earth's species. Their publications, including the *2000 IUCN Red List of Threatened Species* (Hilton-Taylor 2000), contains the names, conservation status, distributions, and sources of all recognized threatened plants worldwide. The Red List has become an essential tool in conservation planning and management, used by many scientists, conservation organizations, and governments.

EARLY HISTORY

The first published list of threatened plants appeared in 1970 as the *Red Data Book 5: Angiospermae* (Melville 1970–1971). This book, fifth in a series of Red Data books, followed the published accounts of threatened mammals, birds, amphibians and reptiles, and fish. This initial exploratory study of threatened plants resulted in two sets of loose-leaf sheets covering 118 plant species, and the startling prediction that a minimum of 20,000 plant species were in need of some form of protection to ensure their safety and continued existence. As the book title implies, ferns (Pteridophyta), conifers and cycads (Gymnospermae), nonvascular plants (such as mosses and liverworts), and lichens were not included in this initial list.

The next episode in the accumulation of names of threatened plants appeared in the *Report on Endangered and Threatened Plant Species of the United States* (Smithsonian Institution 1975). In 1973, the U.S. Congress passed the Endangered Species Act. This act directed the Smithsonian Institution to prepare a list of all endangered and threatened plant species in the nation, to review methods of adequately conserving these species, and to report the institution's recommendations to Congress. The report provided the government with a list of endangered, threatened, recently extinct, and exploited native plants species of the United States. The list comprised approximately 10% of the known flora, with 100 recently extinct species, 761 endangered species, and 1,238 threatened species.

The example set by the United States was followed in Europe with the publication of the *List of Rare, Threatened, and Endemic Plants for the Countries of Europe* (Lucas and Walters 1976), presented to the Council of Europe. The list included approximately 2,000 plant species, and was the first list to cover an entire continent.

THE IUCN RED LIST

The first comprehensive listing of threatened plants at a global scale appeared in the *1997 IUCN Red List of Threatened Plants* (Walter and Gillett 1998). The book was the result of a 20-year effort by a unique coalition of scientists, conservation organizations, botanical gardens, and museums, with major input from the Smithsonian Institution, the Nature Conservancy, Environment Australia, CSIRO, the National Botanical Institute (South Africa), the Royal Botanic Gardens at Kew and Edinburgh, and the New York Botanical Garden. The list includes records for 33,798 threatened or extinct species—roughly 12.5% of the world's known vascular plants. One interesting finding is that an astoundingly high 91% of the listed species are limited in their geographic distribution to single countries. Like the first Red Data books, only vascular plants were included in the list. Furthermore, for many parts of Africa, Asia, the Caribbean, and South America, the data are either patchy or lacking. These gaps in information reflect the deficiency of taxonomic and conservation knowledge for these geographic areas.

In 2000, the Red List began a new era in cataloguing threatened species with the electronic publication of the *2000 IUCN Red List of Threatened Species* (Hilton-Taylor 2000). The new list not only incorporates plants and animals into a single Red List, but was the first to be available in CD-ROM format and as an electronic publication on the World Wide Web (IUCN 2003), allowing for efficient updates on an annual basis. Bryophytes were included in the list for the first time and thus broadened the conservation scope of the plant kingdom by including nonvascular plant species.

While the *1997 IUCN Red List of Threatened Plants* is based on the pre-1994 IUCN Red List categories, drawn up by the IUCN Species Survival Commission (SSC), the *2000 IUCN Red List of Threatened Species* is based on the revised 1994 IUCN Red List categories, a scientifically rigorous and objective approach that determines risks of extinction. Since the revised categories are applicable to all species, the *2000 IUCN Red List of Threatened Species* saw the merging of the botanical and zoological worlds. The 2000 Red List was created by combining two publications: *The World List of Threatened Trees* (Oldfield et al. 1998) and the *1996 IUCN Red List of Threatened Animals* (Baillie and Groombridge 1996). Both publications used the 1994 IUCN Red List categories. One limitation of the 2000 Red List is that only woody plant species were considered, and thus the 2000 version lists fewer plant species than the 1997 version. As the listings in the 1997 version are subjected to the more rigorous categories of threat, all plant species identified as threatened will eventually be added to the new Red List.

RED LIST CATEGORIES

The IUCN Red List categories are intended to be an easily and widely understood system for classifying species at high risk of global extinction. The general aim of the system is to provide an explicit, objective framework for the classification of the broadest range of species (plants, animals, and fungi) according to their extinction risk. Before 1994, a more subjective threatened species system was used in the IUCN Red Data books and Red List. These categories had been in place, with some modification, for almost 30 years. Beginning in 1989, a more objective approach was requested from the IUCN SSC Steering Committee. In 1994, the IUCN council adopted the new Red List system, which has the following aims: (1) to provide a system that can be applied consistently by different people, (2) to improve objectivity by providing users with clear guidance, (3) to provide a system that will facilitate comparisons across different taxa, and (4) to give people using threatened-species lists a better understanding of how species were classified.

Since 1994, the IUCN Red List categories have become widely recognized internationally, and are now used in publications produced by IUCN as well as other governmental and nongovernmental organizations. This version was used for the 1996 *IUCN Red List of Threatened Animals, The World List of Threatened Trees,* and the 2000 *IUCN Red List of Threatened Species.*

The nine Red List categories can be applied to any taxonomic unit at or below the species level. There are seven categories of threat: "extinct," "extinct in the wild," "critically endangered," "endangered," "vulnerable," "near threatened," and "least concern." Threatened species fall in the critically endangered, endangered, or vulnerable categories. A taxon is extinct (EX) when there is no reasonable doubt that the last individual has died. A taxon is extinct in the wild (EW) when it is known only to survive in cultivation or in captivity, or has a naturalized population well outside the past range. A taxon is critically endangered (CR), endangered (EN), or vulnerable (VU) when the best available evidence indicates that it meets any one of the five criteria established for CR, EN, or VU, respectively. These five criteria are based on quantitative values: declining population, small distribution and decline or fluctuation, small population size and decline, very small or restricted population, and quantitative analysis indicating the probability of extinction. For example, when a population is in decline and is estimated at fewer than 250 mature individuals, the taxon is placed in the CR category. A taxon with a declining population of fewer than 2,500 mature individuals is placed in the EN category, and one with a declining population of fewer than 10,000 mature individuals is placed in the VU category. A listing in CR, EN, or VU identifies the taxon as threatened.

Species that are neither extinct nor threatened are placed in one of four other categories. A taxon is near threatened (NT) when it has been evaluated against the criteria but does not qualify for CR, EN, or VU now, but is close to qualifying for or is likely to qualify for a threatened category in the near future. A taxon is least concern (LC) when it does not qualify for CR, EN, VU, or NT. Widespread and abundant taxa are included in this category. A taxon is data deficient (DD) when there is inadequate information to make a direct, or indirect, assessment of its risk of extinction based on its distribution or population status. Finally, a taxon is not evaluated (NE) when it has not yet been evaluated against the criteria.

APPROACHES

Red List authorities take the responsibility for approving the category assigned to each species. Each authority acts as a peer-review group by accepting, accepting with modification, or rejecting an assessment. Authorities work together in geographic, thematic, or taxonomic plant specialist groups. For example, geographic regional groups include the Eastern Africa Plant Specialist Group, the Japanese Plant Specialist Group, and the South Atlantic Island Plant Specialist Group. In comparison, thematic groups include those that specialize in ecological characteristics (e.g., the Cactus and Succulent Specialist Group, the Carnivorous Plant Specialist Group); habitats (e.g., the Temperate Broadleaved Tree Specialist Group); and other characteristics important to conservation (e.g., the Medicinal Plant Specialist Group). Examples of groups that specialize in taxonomic categories include the Bryophyte Specialist Group, the Cycad Specialist Group, and the Orchid Specialist Group. Additionally, botanical gardens, herbaria, and other academic institutions offer their support, answering questions on taxonomy, distribution, and other conservation concerns. Moreover, they maintain collections of threatened species in cultivation and house seed banks for future reintroduction efforts.

CURRENT FINDINGS

In the *2003 IUCN Red List of Threatened Species,* 9,706 plant species are listed, with 106 species listed as extinct or extinct in the wild, and 6,774 threatened. Since the list primarily comes from *The World List of Threatened Trees,* which emphasizes woody plants, this number is far less than the 33,798 threatened plant species from the 1997 Red List. In fact, of all known plant species, less than 1% of mosses, 4% of dicotyledons, and 1% of monocotyledons were assessed for threatened status. At 93%, the gymnosperms are the only major plant group to be almost completely evaluated. Thus, the most current Red List underestimates the problem. The countries in the 2003 Red List with the highest number of threatened plant species are Ecuador (975 species), Malaysia (683 species), Indonesia (383 species), Brazil (381 species), and Sri Lanka (280 species). The plant families with the most threatened species are the Leguminosae (548 species), Dipterocarpaceae (369 species), and Compositae (305 species). In contrast, the 1997 Red List has 2,553 threatened species of composites and 1,779 threatened species of orchids.

LIMITATIONS

As with many taxonomic catalogues, limitations exist in interpreting summary statistics from the Red List, and these should be made evident before interpreting the data. One obvious limitation is that not all plant species are known. It has been estimated that there are between 220,000 and 420,000 extant plant species (Prance et al. 2000; Govaerts 2001; Scotland and Wortley 2003), but only 250,000 have been discovered and described. Most of the undiscovered species exist in areas that have never been surveyed by plant taxonomists. Areas that have inconsistent data or that lack data completely include many parts of tropical Africa, Asia, and South America. In addition, many microhabitats are not easily accessible (e.g., tree canopies), and thus additional plant species await discovery. By emphasizing what is lacking, concerted efforts in the future may result in a better understanding of these biologically rich and essential areas.

USEFULNESS FOR CONSERVATION PURPOSES

Catalogues and checklists of endangered species can be used in a variety of ways for conservation purposes. First and foremost is the preservation of species in their native habitats. By identifying which species are threatened and documenting population sizes, distributions, and the probability of extinction, efforts can be made to save highly endangered species. These efforts include identifying critical habitats where the species exist and putting in place a means of protection. Lists of threatened species can also serve as an effective communication tool when explaining to people the importance of conservation and the consequences of extinction.

The main purpose of threatened species lists is "to catalogue and highlight those taxa that are facing a higher risk of global extinction" (IUCN 2003). Lately, threatened species lists have been used to fulfill important political, social, and scientific needs—purposes beyond their original intent (Possingham et al. 2002). These purposes include setting priorities for resource allocation for species recovery, setting priorities for reserve selection, constraining development or exploitation, and indicating changes in the status of biodiversity. Possingham et al. (2002, 503) critiques these extraneous uses, because "the lists were not designed for any one of these purposes, and consequently perform some of them poorly." They argue that threatened-species lists can be a part of the contributing information, but that other tools are necessary in dealing effectively with these tasks.

Red List assessments have been used in a variety of biological diversity studies, including taxonomic treatments (e.g., Torres et al. 2000; Carine and Scotland 2000; Tye and Jager 2000), biogeographic surveys (e.g., Dulloo et al. 1999), ecological field studies (e.g., Pfab and Witkowski 1999; Simon et al. 2001), and management plans (e.g., Soehartono and Newton 2000; Figueiredo and Gascoigne 2001; Vanderpoorten et al. 2001). In addition, many of today's checklists include Red List assessments. For instance, Wagner et al. (1999) compiled a side-by-side comparison of the conservation status rankings of rare and endangered Hawaiian plant species as assessed by the Smithsonian Institution, the U.S. Fish and Wildlife Service, the Hawaii Natural Heritage Program of the Nature Conservancy of Hawaii, and the 1997 *IUCN Red List of Threatened Plants*. The comprehensive list of vascular plants included 904 of 1,342 taxa native to Hawaii that are currently recognized with an "at risk" or "of concern" rating in at least one of the four systems.

Commercially exploited threatened species can gain additional protection through urgent listings in international agreements, such as the Convention on

International Trade in Endangered Species of Wild Fauna and Flora (CITES). The aim of CITES is to ensure that the international trade of wild animals and plants does not threaten their survival. Roughly 25,000 species are protected by CITES. In some cases, only a subspecies or geographically separate population of a species is listed. In other cases, whole groups, such as cacti and orchids, are included.

11.3 COMMUNITY ASSESSMENT: RAPID ASSESSMENT TEAMS

William S. Alverson

THE RAPID BIOLOGICAL INVENTORY (RBI) program began in 1999 at the Field Museum in Chicago. Its goal is to catalyze effective action for conservation in threatened regions of high biological diversity and uniqueness. The RBI field teams focus primarily on groups of organisms that indicate habitat type and condition and can be surveyed quickly and accurately. These inventories do not attempt to produce an exhaustive list of organisms. Rather, they use a time-effective, integrated approach to identify the important biological communities in the site or region of interest and to determine whether these communities are of outstanding quality and significance in a regional or global context. In-country scientists are central to the field team; their experience is critical for understanding areas with little or no history of scientific exploration. After each inventory, protection of the natural communities and further research rely on initiatives from local scientists and conservationists. Once an inventory has been completed—typically within a month—members of the RBI team work closely with local and international decision makers who can set priorities and guide conservation action in the host country.

The Rapid Assessment Programs (RAP) of Conservation International, in Washington, DC, were initiated in 1989. In addition to terrestrial inventories (in which Field Museum scientists were core participants for nearly ten years), Conservation International carries out marine and freshwater surveys, such as AquaRAP (also in collaboration with the Field Museum). Currently, RAP inventory sites are chosen from priority lists of Conservation International "hot spots" and "corridors" (Myers et al. 2000). Likewise, RBI sites are selected in poorly known regions with high species richness and endemism, where new biological information will contribute directly and immediately to conservation action. RBI and RAP staff are committed to producing and distributing the inventory reports quickly and to putting the information to work without delay (e.g., see Rapid Biological Inventories 2004; Rapid Assessment Program 2004).

THE METHODS

This section summarizes field methods used by botanists working with RBI and RAP teams, based on a review of published inventory reports (Parker and Bailey 1991; Parker and Carr 1993; Parker et al. 1993a; Parker et al. 1993b; Parker et al. 1993c; Foster et al. 1994; Foster et al. 2001; Schulenberg and Awbrey 1997a, 1997b; Killeen and Schulenberg 1998; Mack 1998; Chernoff and Willink 1999; Schulenberg et al. 1999; Alverson et al. 2000; Alverson et al. 2001; Alverson et al. 2003; Bestelmeyer and Alonso 2000; Mack and Alonso 2000; Alonso et al. 2001; Chernoff et al. 2001; Montambault 2002; Pitman et al. 2002; Stotz et al. 2003). The examples derive from the RBI program because of the direct participation and greater familiarity of the author.

Satellite Imagery
The areas to be inventoried are very large, from several thousand to over a million hectares in size, so most inventories start with good satellite imagery of the region. With these images, one can distinguish major habitat types—upland forests, floodplain forests, forests dominated by bamboo, savannas—as well as plan the logistics for accessing study sites on the ground. Many human-induced changes in the landscape are clearly visible from these images, including most roads and clearings for agriculture or cattle. For example, the RBI program currently uses Landsat-7 Thematic Mapper (EMT+) wavelength bands 4, 5, and 3, each with a 30-m spatial resolution, to produce printed and electronic images for a large-scale examination of each region, including the specific areas to be inventoried on the ground (see plate 11.1).

Overflights
Whenever possible, rapid inventory teams spend two to ten hours or more flying over each inventory site in small planes or helicopters, both before and after work on the ground. Overflights are often undertaken by a subset of the field team, who can detect more subtle elements in the landscape (such as distributions of individual tree species, areas subject to landslide and slippage, smaller forest roads and trails, and sometimes even the presence of certain bird species) and refine the habitat classification created from the satellite images. Flights prior to the inventory allow the team to plan access routes to field sites—by foot, truck, boat, or air—and maximize the number of habitats sampled. Digital video is a very welcome recent addition to overflights. The many hours of footage captured and archived for these poorly known areas documents their current condition and serves as another source of information for habitat classification.

On the Ground: Qualitative Assessment
The satellite images and overflights provide a basis for stratifying on-the-ground sampling efforts. Simply put, botanical survey teams want to visit as many different habitat types as possible, within the time allocated for the survey. Team members often spend the first day at each new site walking trails, taking notes on plant com-

munity composition, vegetative structure, and substrate types, as well as starting a list of species observed within the site. This initial time at sites also allows the field team to determine which type of vegetation transects, if any, are appropriate for use during the inventory. In some cases, the diversity of habitats within the inventory area is so great that the botanical survey teams conclude that quantitative sampling is both impractical and highly inadequate. Instead, the teams simply explore all accessible habitats as thoroughly as possible, recording and collecting all species encountered in the process. Qualitative observations also provide a crucial context for understanding the physical nature of the landscape, the age and ecological dynamics of component habitats, and their relationship to human activities.

On the Ground: Transects

In many cases transects are feasible and useful. Various types have been used during the 12-year combined history of RAP and RBI surveys. Until 1994, "Gentry" transects were often used: these are 2 by 50 m (0.01 ha) in size. Uncommonly, 0.1- or 1.0-ha plots have been used, but these tend to require additional work not concurrent with the field survey itself. In the more recent RAP and RBI surveys, the botanists have used "variable transects." With the variable-transect method, one samples a standard number of plants at every location, rather than a standard area. Foster et al. (1998) discusses the rationale and details of the method. By eliminating the need to measure precisely a standard area, and by other streamlining techniques, botanists using the variable-transect method can significantly increase the number of individual plants sampled and identified each day.

During RBIs in terrestrial Neotropical habitats, these variable transects typically are used as follows: For emergent trees (>60 cm DBH, i.e., diameter at breast height) and canopy trees (>30 cm DBH), all individuals encountered along a 20-m-wide strip are recorded to species (or morphospecies) until a certain number—say 50— is reached. For medium, subcanopy trees (10–30 cm DBH), 5-m-wide strips are used; and for shrubs (1–10 cm DBH), 1-m-wide strips. For herbs, segments that are 1 by 5 m segments along the shrub transect are used, with each species represented no more than once per transect. For example, during a three-week trip to the Cordillera Azul in central Peru, the RBI team conducted 16 transects in 6 areas, for a total of 1,660 individuals sampled (Foster et al. 2001). These variable transects sacrifice some potential to compare sites on an area basis (though the total area of each transect is usually recorded). But having many variable transects spread across a greater variety of habitats and sampling a greater number of individual plants provides a better assessment of the species richness and character of a large area than does a smaller number of precise, area-based transects confined to fewer habitats.

On the Ground: Documentation

When legal and practical, voucher specimens are made for species encountered in the transects. Additional, fertile material and specimens representing species not encountered in the transects are also collected whenever possible. Typically, one or two duplicates of each collection are deposited in herbaria within the host country, and one or two duplicates are exported to the U.S. host institution and the specialist for the group. Field teams also rely heavily on photo documentation using 35-mm film or digital cameras. This allows the confirmation of well-known species as well as subsequent identification of species unfamiliar to the team while in the field. Rarely, permission to collect plants is denied by the in-country natural resource agency, even if collections would be deposited only in the host country's herbarium. In this case, field notes and photographs serve as the only source of documentation. Collections of research material for anatomical, morphological, or molecular analyses are not routine because of time constraints.

After Fieldwork
Transect data, species lists, and summary and specific data on species richness and vegetative composition and structure are made public in print and digital formats by both RAP and RBI. Robin Foster, a member of the RBI team, uses the thousands of plant photographs taken during each field inventory to build rapid color guides, that is, printed, laminated field guides to each area for use by park guards, students, local residents, and visitors. Foster and others also are developing a searchable, online database using these images (see Tropical Plant Guides 2004).

The field reports are published as quickly as possible, so that if the inventory area is of sufficiently high quality, local conservationists can immediately seek additional protection, typically by formal designation of the inventory area as a park or conservation refuge. RBI and RAP personnel select inventory sites that have a high conservation potential. That is, they appear to be rich in species and endemics, seem to have unique biological properties, and can directly and tangibly benefit from the additional information provided by an inventory. Thus, the primary objective of the field biologists is to verify (and document) whether the inventory site is indeed home to significant biological diversity that warrants long-term protection. On occasion, despite seeming potential, an area fails to meet these criteria and is not recommended for further conservation action. For example, a forested area in the central Chuquisaca region of Bolivia was not considered to be a conservation priority because it was severely overgrazed and invaded by an exotic species of *Citrus* (Schulenberg and Awbrey 1997b).

THE FUTURE

A central challenge is to improve the success rate at which RAP and RBI field inventories directly catalyze successful conservation action, whether by designation of park or wildlife refuge status, or by some other significant improvement in conservation management of the areas inventoried. In the short term, the rapid inventory programs are striving to improve the georeferencing of plant collections, transect data, photos, and video, and the delivery of these and other inventory information in print and searchable, digital form.

A longer-term challenge is to explore how current inventory methods might be modified, without compromising their efficiency and efficacy, to allow better comparisons of species richness. For example, Condit et al. (1996, 1998) recommend the use of species-individual curves for comparison of the species richness of trees and shrubs in 50-ha plots. They also consider the extension of their results to large areas, up to 13.2 million hectares, or roughly ten times the size of the larger areas assessed by the RBI team. Using reasonable assumptions about the slope of species-individual curves, it would be possible to rank the species richness of rapid biological survey areas if (1) a large number of woody stems—40,000 or more—are scored to species, and (2) the areas being compared have fairly uniform climates and habitats. Current RBI efforts fall far short of these requirements, but the gap may narrow if further evaluation of empirical data from field studies serves to relax the assumptions required for such comparisons. Readers should keep in mind, however, that the conservation decisions relevant to these inventory areas are not being made on the basis of species richness (or endemism, or political opportunity) alone, so the main contribution of such changes in methodology and theory would be to increase the scientific and research value of the inventories—a laudable goal.

LITERATURE CITED

Alonso, L. E., Alonso, A., Schulenberg, T. S., and Dallmeier, F., eds. 2001. Biological and Social Assessments of the Cordillera de Vilcabamba, Peru. RAP Working Papers 12 and SI/MAB Series 6. Conservation International, Washington, DC.

Alverson, W. S., Moskovits, D. K., and Halm, I. C., eds. 2003. Bolivia: Pando, Federico Román. Rapid Biological Inventories Report 06. Field Museum, Chicago.

Alverson, W. S., Moskovits, D. K., and Shopland, J. M., eds. 2000. Bolivia: Pando, Río Tahuamanu. Rapid Biological Inventories Report 01. Field Museum, Chicago.

Alverson, W. S., Rodríguez, L. O., and Moskovits, D. K., eds. 2001. Perú: Biabo Cordillera Azul. Rapid Biological Inventories Report 02. Field Museum, Chicago.

Baillie, J., and Groombridge, B., comps. and eds. 1996. 1996 IUCN Red List of Threatened Animals. IUCN, Gland, Switzerland, and Cambridge.

Baldwin, B., and Markos, S. 1998. Phylogenetic utility of the external transcribed spacer (ETS) of 18S-26S rDNA: congruence of ETS and ITS trees of *Calycadenia* (Compositae). Molecular Phylogenetics and Evolution 10:449–463.

Bestelmeyer, B. T., and Alonso, L. E., eds. 2000. A Biological Assessment of Laguna del Tigre National Park, Petén, Guatemala. RAP Bulletin of Biological Assessment 16. Conservation International, Washington, DC.

Breakthroughs in Bioscience. 2003. FASEB Office of Public Affairs. http://www.faseb.org/opar/break/.

Carine, M. A., and Scotland, R. W. 2000. The taxonomy and biology of *Stenosiphonium nees* (Acanthaceae). Botanical Journal of the Linnean Society 133:101–128.

Chernoff, B., and Willink, P. W., eds. 1999. A Biological Assessment of the Aquatic Ecosystems of the Upper Río Orthon Basin, Pando, Bolivia. RAP Bulletin of Biological Assessment 15. Conservation International, Washington, DC.

Chernoff, B., Willink, P. W., and Montambault, J. R., eds. 2001. A Biological Assessment of the Aquatic Ecosystems of the Río Paraguay Basin, Alto Paraguay, Paraguay. RAP Bulletin of Biological Assessment 19. Conservation International, Washington, DC.

Condit, R., Foster, R. B., Hubbell, S. P., Sukumar, R., Leigh, E. G., Manokaran, N., Loo de Lao, S., Lafrankie, J. V., and Ashton, P. S. 1998. Assessing forest diversity on small plots: calibration using species-individual curves from 50-hectare plots. Pp. 247–268 in Dallmeier, F., and Comiskey, J. A., eds., Forest Biodiversity Research, Monitoring, and Modelling: Conceptual Background and Old World Case Studies. MAB Series vol. 20. UNESCO, Paris.

Condit, R., Hubbell, S. P., LaFrankie, J. V., Sukumar, R., Manokaran, N., Foster, R. B., and Ashton, P. S. 1996. Species-area and species-individual relationships for tropical trees: a comparison of three 50-ha plots. Journal of Ecology 84:549–562.

Crandall, K. A., Bininda-Emonds, O. R. P., Mace, G. M., and Wayne, R. K. 2000. Considering evolutionary processes in conservation biology. Trends in Ecology and Evolution 15: 290–295.

Database of Primers. 2004. Molecular Ecology Notes. Blackwell Science. http://tomato.bio .trinity.edu/home.html.

Doyle, J. J., and Doyle, J. L. 1987. A rapid DNA isolation procedure for small quantities of fresh leaf tissue. Phytochemical Bulletin 19:11–15.

Dulloo, M. E., Maxted, N., Guarino, L., Florens, D., Newbury, H. J., and Lloyd, B. V. F. 1999. Ecogeographic survey of the genus *Coffea* in the Mascarene Islands. Botanical Journal of the Linnean Society 131:263–284.

EarthExplorer. 2002. U.S. Geological Survey. http://earthexplorer.usgs.gov.

Figueiredo, E., and Gascoigne, A. 2001. Conservation of pteridophytes in São Tomé e Príncipe (Gulf of Guinea). Biodiversity and Conservation 10:45–68.

Foster, R. B., Beltrán, H., and Alverson, W. S. 2001. Flora and vegetation. Pp. 124–141 in Alverson, W. S., Rodríguez, L. O., and Moskovits, D. K., eds., Perú: Biabo Cordillera Azul. Rapid Biological Inventories Report 02. Field Museum, Chicago.

Foster, R. B., Hernández, N. C., Kakudidi, E. K., and Burnham, R. J. 1998. Rapid assessment of tropical plant communities using variable transects: an informal and practical guide. Field Museum. http://www.fmnh.org/rbi/pdfs/VarTrans.pdf.

Foster, R. B., Parker, T. A. III, Gentry, A. H., Emmons, L. H., Chicchón, A., Schulenberg, T., Rodríquez, L., Lamas, G., Ortega, H., Icochea, J., Wust, W., Romo, M., Alban Castillo, J., Phillips, O., Reynel, C., Kratter, A., Donahue, P. K., and Barkley, L. J. 1994. The Tambopata-Candamo Reserved Zone of Southeastern Perú: A Biological Assessment. RAP Working Papers 6. Conservation International, Washington, DC.

Friar, E. A., Boose, D. L., LaDoux, T., Roalson, E. H., and Robichaux, R. H. 2001. Population structure in the endangered Mauna Loa silversword, *Argyroxiphium kauense* (Asteraceae), and its bearing on reintroduction. Molecular Ecology 10:1657–1663.

Friar, E. A., LaDoux, T., Roalson, E. H., and Robichaux, R. H. 2000. Microsatellite analysis of a population crash and bottleneck in the Mauna Kea silversword, *Argyroxiphium sandwicense* ssp. *sandwicense* (Asteraceae), and its implications for reintroduction. Molecular Ecology 9:2027–2034.

Gemmill, C. E. C., Ranker, T. A., Ragone, D., Perlman, S. P., and Wood, K. R. 1998. Conservation genetics of the endangered endemic Hawaiian genus *Brighamia* (Campanulaceae). American Journal of Botany 85:528–540.

Govaerts, R. 2001. How many species of seed plants are there? Taxon 50:1085–1090.

Hamrick, J. L., and Godt, M. J. 1990. Allozyme diversity in plant species. Pp. 43–63 in Brown, A. H. D., Clegg, M. T., Kahler, A. L., and Weir, B. S., eds., Plant Population Genetics, Breeding, and Genetic Resources. Sinauer Associates, Sunderland, Massachusetts.

————. 1996. Conservation genetics of endemic plant species. Pp. 281–384 in Avise, J. C., and Hamrick, J. L., eds., Conservation Genetics: Case Histories from Nature. Chapman and Hall, New York.

Hilton-Taylor, C., comp. 2000. 2000 IUCN Red List of Threatened Species. IUCN, Gland, Switzerland, and Cambridge.

IUCN (World Conservation Union). 2003. 2003 IUCN Red List of Threatened Species. IUCN Species Survival Commission. http://www.redlist.org/.

Jaramillo, M. A. 2001. Divergence and diversification in the genus *Piper* and the order Piperales. Ph.D. diss., Duke University.

Killeen, T. J., and Schulenberg, T. S., eds. 1998. A Biological Assessment of Parque Nacional Noel Kempff Mercado, Bolivia. RAP Working Papers 10. Conservation International, Washington, DC.

Lucas, G. L., and Walters, S. M. 1976. List of Rare, Threatened, and Endemic Plants for the Countries of Europe. IUCN Threatened Plants Committee Secretariat, Royal Botanic Gardens, Kew, Surrey, United Kingdom.

Mack, A. L., ed. 1998. A Biological Assessment of the Lakekamu Basin, Papua New Guinea. RAP Working Papers 9. Conservation International, Washington, DC.

Mack, A. L., and Alonso, L. E., eds. 2000. A Biological Assessment of the Wapoga River Area of Northwestern Irian Jaya, Indonesia. RAP Bulletin of Biological Assessment 14. Conservation International, Washington, DC.

Mason-Gamer, R. J., Weil, C., and Kellogg, E. A. 1998. Granule-bound starch synthase: structure, function, and phylogenetic utility. Molecular Biology and Evolution 15:1658–1673.

Melville, R. 1970–1971. Red Data Book 5: Angiospermae. IUCN, Morges, Switzerland.

Milbourne, D., Russell, J., and Waugh, R. 1998. Comparison of molecular marker assays in inbreeding (barley) and outbreeding (potato) species. Pp. 371–381 in Karp, A., Isaac, P. G., and Ingram, D. S., eds., Molecular Tools for Screening Biodiversity. Chapman and Hall, London.

Montambault, J. R., ed. 2002. Informes de las Evaluaciones Biológicas Pampas del Health, Perú, Alto Madidi, Bolivia, y Pando, Bolivia. RAP Bulletin of Biological Assessment 24. Conservation International, Washington, DC.

Moritz, C. 2002. Strategies to protect biological diversity and the evolutionary processes that sustain it. Systematic Biology 51:238–254.

Myers, N., Mittermeier, R. A., Mittermeier, C. G., da Fonseca, G. A. B., and Kent, J. 2000. Biodiversity hotspots for conservation priorities. Nature 403:853–858.

Nickrent, D. 2002. Instructions for Shipping Plant Material for DNA Analysis. Southern Illinois University Carbondale. http://www.science.siu.edu/plant-biology/faculty/nickrent/Plant.DNA.html.

Oldfield, S., Lusty, C., and MacKinven, A. 1998. The World List of Threatened Trees. World Conservation Press, Cambridge.

Olsen, K., and Schaal, B. A. 1999. Evidence on the origin of cassava: phylogeography of *Manihot esculenta*. Proceedings of the National Academy of Sciences USA 96:5586–5591.

Parker, T. A. III, and Bailey, B., eds. 1991. A Biological Assessment of the Alto Madidi Region and Adjacent Areas of Northwest Bolivia, May 18–June 15, 1990. RAP Working Papers 1. Conservation International, Washington, DC.

Parker, T. A. III, and Carr, J. L., eds. 1993. Status of Forest Remnants in the Cordillera de la Costa and Adjacent Areas of Southwestern Ecuador. RAP Working Papers 2. Conservation International, Washington, DC.

Parker, T. A. III, Foster, R. B., Emmons, L. H., Freed, P., Forsyth, A. B., Hoffman, B., and Gill, B. D., eds. 1993a. A Biological Assessment of the Kanuku Mountain Region of Southwestern Guyana. RAP Working Papers 5. Conservation International, Washington, DC.

Parker, T. A. III, Gentry, A. H., Foster, R. B., Emmons, L. H., and Remsen, J. V. Jr., eds. 1993b. The Lowland Dry Forests of Santa Cruz, Bolivia: A Global Conservation Priority. RAP Working Papers 4. Conservation International, Washington, DC.

Parker, T. A. III, Holst, B. K., Emmons, L. H., and Meyer, J. R., eds. 1993c. A Biological Assessment of the Columbia River Forest Reserve, Toledo District, Belize. RAP Working Papers 3. Conservation International, Washington, DC.

PCR: Highly Recommended Books. 1998. Atlantis Enterprises. http://www.a-ten.com/alz/pcr.htm.

Pfab, M. F., and Witkowski, E. T. F. 1999. Contrasting effects of herbivory on plant size and reproductive performance in two populations of the critically endangered species *Euphorbia clivicola* R. A. Dyer. Plant Ecology 145:317–325.

Pitman, N., Moskovits, D. K., Alverson, W. S., and Borman, R., eds. 2002. Ecuador: Serranías Cofán–Bermejo, Sinangoe. Rapid Biological Inventories Report 03. Field Museum, Chicago.

Possingham, H. P., Andelman, S. J., Burgman, M. A., Medellín, R. A., Master, L. L., and Keith, D. A. 2002. Limits to the use of threatened species lists. Trends in Ecology and Evolution 17:503–507.

Prance, G. T., Beentje, H., Dransfield, J., and Johns, R. 2000. The tropical flora remains undercollected. Annals of the Missouri Botanical Garden 87:67–71.

Rapid Assessment Program. 2004. Conservation International. http://www.biodiversityscience.org/xp/CABS/research/rap/aboutrap.xml.

Rapid Biological Inventories. 2004. Field Museum. http://www.fmnh.org/rbi.

Rieseberg, L. H., and Swensen, S. M. 1996. Conservation genetics of endangered island plants. Pp. 305–331 in Avise, J. C., and Hamrick, J. L., eds., Conservation Genetics: Case Histories from Nature. Chapman and Hall, New York.

Schaal, B., and Olsen, K. 2000. Gene genealogies and population variation in plants. Proceedings of the National Academy of Sciences USA 97:7024–7029.

Schemske, D. W., Husband, B. C., Ruckelshaus, M. H., Goodwillie, C., Parker, I. M., and Bishop, J. G. 1994. Evaluating approaches to the conservation of rare and endangered plants. Ecology 75:584–606.

Schulenberg, T. S., and Awbrey, K., eds. 1997a. The Cordillera del Cóndor Region of Ecuador and Peru: A Biological Assessment. RAP Working Papers 7. Conservation International, Washington, DC.

———. 1997b. A Rapid Assessment of the Humid Forests of South Central Chuquisaca, Bolivia. RAP Working Papers 8. Conservation International, Washington, DC.

Schulenberg, T. S., Short, C. A., and Stephenson, P. J., eds. 1999. A biological assessment of Parc National de la Marahoué, Côte d'Ivoire. RAP Bulletin of Biological Assessment Working Papers 13. Conservation International, Washington, DC.

Scotland, R. W., and Wortley, A. H. 2003. How many species of seed plants are there? Taxon 52:101–104.

Simon, J., Bosch, M., Molero, J., and Blanché, C. 2001. Conservation biology of the Pyrenean larkspur (*Delphinium montanum*): a case of conflict of plant versus animal conservation? Biological Conservation 98:305–314.

Smithsonian Institution. 1975. Report on Endangered and Threatened Plant Species of the United States. Government Printing Office, Washington, DC.

Soehartono, T., and Newton, A. C. 2000. Conservation and sustainable use of tropical trees in the genus *Aquilaria*. Part 1, Status and distribution in Indonesia. Biological Conservation 96:83–94.

Soltis, D. E., Haufler, C. H., Darrow, D. C., and Gastony, G. J. 1983. Starch gel electrophoresis of ferns: a compilation of grinding buffers, gel and electrode buffers, and staining schedules. American Fern Journal 73:9–27.

Stotz, D. F., Harris, E. J., Moskovits, D. K., Hao, K., Yi, S., and Adelmann G. W., eds. 2003. China: Yunnan, Southern Gaoligongshan. Rapid Biological Inventories Report 04. Field Museum, Chicago.

Torres, N., Saez, L., Rossello, J. A., and Blanche, C. 2000. A new *Delphinium* subsp. from Formentera (Balearic Islands). Botanical Journal of the Linnean Society 133:371–377.

Tropical Plant Guides. 2004. Field Museum. http://www.fieldmuseum.org/plantguides.

Tye, A., and Jager, H. 2000. *Galvezia leucantha* subsp. *porphyrantha* (Scrophulariaceae), a new shrub snapdragon endemic to Santiago Island, Galapagos, Ecuador. Novon 10:164–168.

Vanderpoorten, A., Sotiaux, A., and Sotiaux, O. 2001. Integrating bryophytes into a forest management plan: lessons from grid-mapping in the forest of Soignes (Belgium). Cryptogamie Bryologie 22:217–230.

van de Wiel, C., and Vosman, B. 1998. Molecular analysis of variation in *Lactuca*. Pp. 388–393 in Karp, A., Isaac, P. G., and Ingram, D. S., eds., Molecular Tools for Screening Biodiversity. Chapman and Hall, London.

Wagner, W. L., Bruegmann, M. M., Herbst, D. M., and Lau, J. Q. C. 1999. Hawaiian vascular plants at risk: 1999. Bishop Museum Occasional Papers 60:1–58.

Walter, K. S., and Gillett, H. J., eds. 1998. 1997 IUCN Red List of Threatened Plants. IUCN, Gland, Switzerland, and Cambridge.

Zimmer, E. A., Roalson, E. H., Skog, L. E., Boggan, J. K., and Idnurm, A. 2002. Phylogenetic relationships in the Gesnerioideae (Gesneriaceae) based on nrDNA ITS and cpDNA *trn*L-F and *trn*E-T spacer region sequences. American Journal of Botany 89:296–311.

CHAPTER 12

MANAGEMENT STRATEGIES

A multipronged strategy for achieving successful conservation must maintain and manage plant diversity in both terrestrial and aquatic environments at the genetic, population, and species levels. It is generally agreed that the prevention of species extinction resulting from destruction of natural habitats can best be accomplished through the long-term protection of in situ habitats in a minimally disturbed state. However, the maintenance of plant genetic diversity in botanical gardens and other "gene banks" has traditionally played and will continue to play an important role in conservation. As the rate of extinction of plant species in the wild continues to increase, rare and endangered taxa and their genes can be successfully maintained in botanical gardens, germ-plasm banks, and DNA repositories. The adoption of the Global Strategy for Plant Conservation by the Convention on Biological Diversity was a critical step in furthering these efforts. International agreements to conserve plant genetic diversity and the role of ex situ conservation are discussed in the first section of this chapter.

Successful in situ conservation protects plants from threats to their native habitats. However, the level of management required to maintain these habitats is directly proportional to the amount of human disturbances that threatens the plant communities. The second section of this chapter presents one model of sustainable management in marine environments, using tropical reef ecosystems as a case study. This case study illustrates the complexity of developing these management practices. The major tenets of the proposed model are that (1) competition for space and light is important in determining the relative abundances of photosynthetic organisms and (2) competition for these resources is most often controlled by top-down factors, such as grazing, and bottom-up factors, such as nutrient levels.

The third section of the chapter applies the proposed management model to a complex community of tropical seagrasses by exploring the roles of grazers and nutrients for native marine flora. Such management models and their application to in situ habitats may require many years before their effectiveness can be adequately evaluated.

12.1 EX SITU CONSERVATION OF PLANTS

Stephen Blackmore

PUBLIC PERCEPTION of the importance of plant conservation, or even of the basic need for it, has lagged behind awareness of the necessity of conserving animals, especially the conspicuous megafauna. It is widely appreciated that zoos play an active role in the ex situ conservation of animals, especially through captive breeding programs. The success of such programs is sometimes a contentious issue, however, particularly if they are judged in terms of their effectiveness in restoring viable populations in the wild. In contrast, public perception that plants may be threatened in the wild is usually confined to a general awareness of the threats to important ecosystems such as tropical rain forests or to plants such as orchids that are perceived as exotic and unusual.

Consequently, it is not surprising that plant-oriented conservation initiatives and organizations, such as the Plant Committee of the World Conservation Union (IUCN), tend to be more recently established than those concerned with animals. Although the historical reasons for this animal bias are readily appreciated, the sequence of events that have led belatedly to the emergence of plant conservation is scarcely logical. It is universally agreed that conserving species in situ is the ideal objective, mainly because only effectively functioning ecosystems can preserve the food chains and myriad of other complex interactions that make up the web of life. This is one of the reasons why the Convention on Biological Diversity (CBD) has adopted the "ecosystem approach." If the plant biodiversity that forms the base of all terrestrial and most marine ecosystems is lost or becomes too restricted, it is difficult to ensure the survival of the animal species higher up the food web that were the original focus of conservation efforts. Logically, in situ conservation efforts have to start with plants.

What then of ex situ plant conservation? Perhaps when one considers the low visibility of plant conservation in general it is not surprising that ex situ plant conservation has had an even lower profile. Indeed, there are those who in their enthusiasm to argue for the preservation of wilderness areas or other priorities have actually argued against ex situ plant conservation. It has sometimes been proposed that if we begin to accept the continued survival of some plant species only in ex situ collections, then political leaders in countries around the world may decide to dispense with in situ conservation efforts. The *International Agenda for Botanic Gardens in Conservation* (Wyse Jackson and Sutherland 2000, 27) steers a careful course in relation to this sensitivity, stating that "the purpose of ex situ conservation is to provide protective custody. It is justifiable only as part of an overall conservation strategy to ensure that species ultimately survive in the wild." The motivation behind this concern is laudable but perhaps misguided, given the actual levels of threat facing many plant species in the wild. Recent estimates have suggested that at least a third of all flowering-plant species are threatened with extinction in the wild in the present century (and some estimates put the figure as high as two-thirds of plants threatened with extinction). The wholesale destruction of plant habitats and the erosion of diversity within distinctive kinds of vegetation around the world make it churlish to see ex situ conservation of plants as a last resort. Furthermore, the basic biology of plants makes them far better targets for ex situ conservation than animals and increases the likelihood of success over corresponding programs for animals. Plants are far more easily produced in large numbers, whether by vegetative or sexual reproduction, and have dormant stages of their life cycles (seeds or spores) that can often be stored relatively easily for long periods. These properties of plants, including the property of totipotency of plant cells, makes it possible to generate whole plants in ways that are currently impossible in animals. Consider the relative ease with which giant pandas and bamboos, their principal source of food, can reproduce whether in the wild or in captivity.

Fortunately, botanical gardens have a long tradition of sharing and distributing their collections, and because many of these are plants threatened in the wild, it can be argued that they always have made a significant contribution to ex situ conservation. The circulation of lists of material available for exchange between botanical gardens is a well-established tradition. Recent years, especially since the introduction of the CBD, have created a more tightly regulated environment for such exchanges but have also seen the emergence of more strategic approaches to ex situ conservation. Strategic developments have been greatly facilitated by the establishment of regional and international networks such as Botanic Gardens Conservation International (BGCI), and the International Association of Botanic Gardens (IABG). Some particularly important recent outcomes of the trend toward concerted action have been the development of the *International Agenda for Botanic Gardens in Conservation* (Wyse Jackson and Sutherland 2000) and the recent Global Strategy for Plant Conservation of the CBD.

The recent *International Review of the Ex Situ Collections of the Botanic Gardens of the World* (Wyse Jackson 2002) documents the impressive extent of the living collections held in the world's botanical gardens. It highlights the collective ability of these organizations to contribute to the specific goals of the CBD and, more generally, to a sustainable future both for plants and people. The *International Review* documents 2,178 botanical gardens in 153 countries, of which more than 500 are in western Europe, more than 350 in North America, and more than 200 in East and Southeast Asia. Interestingly, more than half of the world's botanical gardens have been established within the last 50 years, and new gardens continue to be established at a rate of about ten a year. Most of the more recently established botanical gardens have conservation and public education as their primary aims, whereas the world's oldest were typically established as physic gardens allied to the use of plants in medicine. In addition to some 42 million preserved plant specimens held in their respective herbaria, the world's botanical gardens are thought to cultivate more than 6 million accessions of living plants. However, the *International Review* considered that the actual number of accessions might be as much as 50% higher. These accessions represent more than 80,000 species of vascular plants, which is somewhere between a quarter and a third of all known plants. Not all of these plants can be regarded as ex situ collections, because many botanical gardens include areas of native vegetation that are often important sites for in situ plant conservation. Nevertheless, a significant proportion of the world's vascular plants are, in effect, already represented within botanical gardens.

How can such a resource best be further developed and put to use? At the 16th International Botanical Congress in St. Louis in 1999, a resolution was passed that called for a global strategy to address the challenges of plant conservation. Botanical gardens and other interested parties responded to the call, and early in 2000 Botanic Gardens Conservation International convened a meeting of experts at the Jardin Viera y Claviyo in Gran Canaria, itself an important center for the in situ and ex situ conservation of endemic Canarian plants and threatened species of palms from around the world. The outcome was the *Gran Canaria Declaration* (Convention on Biological Diversity Web site) a document that was presented to the fifth Conference of the Parties of the CBD at their meeting in Nairobi a few months later. The Conference of the Parties welcomed the *Gran Canaria Declaration* and called for further work to develop a Global Strategy for Plant Conservation (GSPC). With input from conservation nongovernmental organizations, international agencies, and other stakeholders, a strategy was drawn up and subsequently adopted by the sixth Conference of the Parties, meeting in The Hague in 2002 (the full text and background for the strategy can be found at the CBD Web site). The GSPC marks an important new development for the CBD because it contains, for the first time in any of the CBD work programs, specific targets for achievement within a set time period of ten years. The targets are grouped under five major thematic headings (see table 12.1).

Although only target 8 refers explicitly to ex situ conservation, several other targets have direct relevance to the effective implementation of such efforts. For example, the targets under "Understanding and documenting plant diversity" relate primarily to the research efforts needed to underpin plant conservation. In the absence of a complete inventory of the world's plants and a preliminary assessment of their conservation status, we have no baseline for their conservation. It is estimated that about 30% of threatened plants (i.e., plants that have been assessed and placed in formal IUCN categories according to their status in the wild), some 10,000 species, are currently held in various kinds of ex situ collection from seed banks and tissue-culture collections to gardens, but that as few as 2% are the subject of recovery or restoration efforts. It follows that, to achieve the ultimate goal of successful in situ conservation to be achieved, far greater priority needs to be given to species and habitat restoration programs for threatened species.

The proportion of species to be brought into ex situ conservation under target 8 is clearly challenging, but not impossible, provided sufficient resources are made available and efforts are well coordinated. To bring a further 10,000 threatened species into ex situ collections ought to be readily achievable over the next ten years.

Table 12.1 Global Strategy for Plant Conservation

Understanding and documenting plant diversity
  1. A widely accessible working list of known plant species, as a step toward a complete world flora
  2. A preliminary assessment of the conservation status of all known plant species, at national, regional, and international levels
  3. Development of models with protocols for plant conservation and sustainable use, based on research and practical experience

Conserving plant diversity
  4. At least 10% of each of the world's ecological regions effectively conserved
  5. Protection of 50% of the most important areas for plant diversity assured
  6. At least 30% of production lands managed consistently with the conservation of plant diversity
  7. 60% of the world's threatened species conserved in situ
  8. 60% of threatened plant species in accessible ex situ collections, preferably in the country of origin, and 10% of them included in recovery and restoration programs
  9. 70% of the genetic diversity of crops and other major socioeconomically valuable plant species conserved, and associated indigenous and local knowledge maintained
  10. Management plans in place for at least 100 major alien species that threaten plants, plant communities, and associated habitats and ecosystems

Using plant diversity sustainably
  11. No species of wild flora endangered by international trade
  12. 30% of plant-based products derived from sources that are sustainably managed
  13. Halting the decline of plant resources and of associated indigenous and local knowledge, innovations, and practices that support sustainable livelihoods, local food security, and health care

Promoting education and awareness about plant diversity
  14. The importance of plant diversity and the need for its conservation incorporated into communication, educational, and public-awareness programs

Building capacity for the conservation of plant diversity
  15. The number of trained people working with appropriate facilities in plant conservation increased, according to national needs, to achieve the targets of this strategy
  16. Networks for plant conservation activities established or strengthened at national, regional, and international levels

Notes: The 16 global targets for the year 2010. For terms and technical rationale, see appendix to Decision VI/9 at Convention on Biological Diversity Web site.

However, it is widely appreciated that the 30,000 threatened plants so far assessed under the IUCN categories for endangered plants are only a proportion of the plants actually threatened in the wild. Estimates suggest that between one-third and two-thirds of plants are threatened with extinction in the wild in the present century, primarily as a result of the loss of their natural habitats. The urgency of target 2 aims to address this lack of information on the threats and status of plants by providing a preliminary assessment of their status. It might be argued that the ex situ collections of the world must at least double the total number of species held, from around 80,000 to 160,000 species, a little over half of the currently described species (although the precise number of known plant species is itself a matter of uncertainty in the absence of a world inventory). It is interesting that the Chinese Academy of Sciences has recently defined a target of doubling the number of species cultivated in the Beijing Institute of Botany's living collections (Huang et al. 2002). The need for additional accessions is emphasized in the plans of many individual institutions and in the *International Agenda for Botanic Gardens in Conservation*. The accompanying text for target 8 of the global strategy suggests that critically endangered species should be given highest priority and proposes a target of 90% of these species being brought into ex situ conservation.

Target 9, which focuses on the plants most used by humanity as crops, medicines, or sources of other economically important benefits, effectively prioritizes the conservation of such plants, including their ex situ conservation. The final three targets also bear directly on the success of ex situ efforts. As emphasized in this chapter, wider awareness of the need for plant conservation lags behind appreciation of the need to conserve animals. Critically the accompanying text for this target (see the CBD Web site) emphasizes the need to increase the awareness of policy-makers to the issues, and not just the public at large.

Target 15 recognizes the need to increase the human resources involved in plant conservation. At present these are very limited in the developing countries, and this imbalance of expertise in plant conservation mirrors the more general imbalance of expertise that is a concern to the Global Taxonomy Initiative (see the CBD Web site) and the CBD itself. In part the solution to the lack of human capacity in plant conservation depends upon the success of target 16, which concerns itself with national, regional, and international networks to enhance communication, training, and the adoption of best practice. It is encouraging to note that the drafting of the GSPC has already drawn together several key communities with expertise relevant to plant conservation, including agencies such as the IUCN, concerned with the assessment of threats and with monitoring the status of plant populations in the wild; nongovernmental organizations concerned with gaining public support for conservation efforts; and botanical gardens.

The philosophy of the GSPC is that plants are best conserved in situ in naturally functioning ecosystems. It recognizes that this ideal cannot always be achieved, however, and that, even in protected areas, invasive species, fire, and other hazards can continue to threaten the survival of plants. A pragmatic perspective would be to recognize that until programs of habitat restoration and eradication of invasive species have succeeded, for the majority of threatened plants, ex situ conservation (ideally close to their natural habitat, in the country of origin) provides necessary insurance. However, it is important that the genetic diversity found in wild populations is adequately represented in these collections, especially if the ex situ material forms the basis of subsequent reintroductions to the wild. Increasingly, practical conservation biology concerns itself with a need to understand biodiversity at the genetic level, not just the species level, and the emerging field of conservation genetics has an important part to play in the future.

12.2 A PROPOSED SUSTAINABLE CORAL-REEF MANAGEMENT MODEL

Mark M. Littler and Diane S. Littler

DUE TO THE GROWING problems associated with coastal eutrophication and destructive fishing along tropical and subtropical shores, the responses of coral reefs and macroalgae to both nutrient enrichment and release from predation have been repeatedly cited as priority areas in need of intense management. The relative dominance model (RDM) proposed here suggests a useful perspective to resource managers attempting to protect coral reefs and similar coastal systems from eutrophication, destructive fishing, and initiation of harmful algal blooms. Thus, this approach is timely in proposing a framework and guidelines for improved understanding and sustainable management of critical reef ecosystems. Unfortunately, the recurrent role of modern humankind on coral reefs has been to decrease herbivorous fishes (Littler et al. 1991; Littler et al. 1993; Hughes 1994) through trapping, netting, poisoning, and dynamiting, while simultaneously adding nutrients via sewage and agricultural eutrophication (Littler et al. 1991; Littler et al. 1993; Goreau et al. 1997; Lapointe et al. 1997). Unless these anthropogenic effects are curbed, induced shifts from coral to fleshy-algal domination are anticipated to expand geographically at an accelerated pace (Nixon 1995).

COMPLEXITY

Because of the long history of environmental stability within tropical zones, coral reefs have evolved astounding levels of complexity and biological diversity. This complex nature of tropical reef ecosystems makes identifying causative factors difficult and has contributed to the considerable controversy concerning the roles of herbivory (top-down factors) versus nutrients (bottom-up factors). Turbulent water motion and the many uniquely specialized benthic algae and photosynthetic symbionts dominating tropical reefs are responsible for some of the most productive natural ecosystems known. Four groups of benthic primary producers are responsible for the bulk of coral-reef productivity: cnidarian corals (containing symbiotic algae), crustose coralline algae, algal turfs, and frondose macroalgae. Of these, photosynthetic corals create much of the structural complexity and, with coralline algae, are primarily responsible for accretion of calcium carbonate ($CaCO_3$) into the reef matrix, making them the most desirable functional groups from a management perspective.

TOP-DOWN AND BOTTOM-UP CONTROLS

A basic objective in plant ecology is to understand the mechanisms by which natural and anthropogenic factors may maintain or alter structure and interactions in biotic communities. Anthropogenic eutrophication and destructive fishing are the most frequently cited factors correlated with the marked global decline in tropical-reef communities over the past two decades (see reviews in Ginsberg 1993; Birkeland 1997; papers in Szmant 2001). The concepts "top-down" and "bottom-up" controls have been used (e.g., Atkinson and Grigg 1984; Carpenter et al. 1985) to describe mechanisms where either the actions of predators or resource availability regulate the structure of aquatic communities. These factors provide a useful perspective to assess and manage the interactive mechanisms controlling stable states and phase shifts among the dominant functional groups of primary producers on tropical reefs.

In healthy tropical reefs, nutrient concentrations are extremely low, and attachment space is preempted by a broad diversity of epilithic organisms. Given these conditions, the major tenets of the RDM are (1) that competition for space and light is important in determining the relative abundances of major benthic photosynthetic organisms, and (2) that the outcome of competition for these resources is most often controlled by the complex interactions of biological factors and environmental factors. As proposed by Grime (1979) for terrestrial plants and expanded for marine macroalgae (Littler and Littler 1984a; Steneck and Dethier 1994), primary-producer abundances and evolutionary strategies are controlled by physical disturbances (i.e., factors that remove biomass) coupled with physiological stresses (i.e., factors that limit metabolic production). In the conceptual RDM (fig. 12.1), grazing (top-down) physically reduces biomass of fleshy algae, and nutrients (bottom-up) control production. The complex interactions between herbivory and nutrients are most dramatically impacted by large-scale disturbances such as tropical storms (e.g., Done 1992a), warming events (e.g., Macintyre and Glynn 1990; Lough 1994), diseases (e.g., Santavy and Peters 1997; Bruno et al. 2003), and predator outbreaks (e.g., Cameron 1977); however, these accelerate the ultimate long-term phase shifts postulated in the RDM. Such stochastic events selectively eliminate the longer-lived organisms in favor of fast-growing early-successional macroalgae, which are competitively superior following disturbances.

On undisturbed oligotrophic coral-reef habitats, the effects of top-down physi-

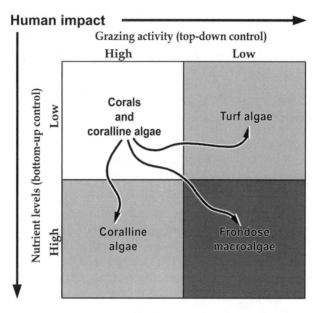

Figure 12.1 The competition-based relative dominance model (RDM) modified from Littler and Littler 1984a. All of the functional indicator groups are present all of the time on coral reefs, but dominate most often under the interacting conditions indicated by the four compartments. The complex interacting vectors of long-term eutrophication and declining herbivory (either naturally or anthropogenically derived) are postulated to produce competitive shifts (arrows) away from coral and coralline domination on pristine reefs toward various phases of algal dominance. Hypothetically, one vector can partially offset the other (e.g., high herbivory may delay the impact of elevated nutrients, or low nutrients may offset the impact of reduced herbivory). Latent trajectories are most often catalyzed or accelerated by large-scale stochastic disturbances such as tropical storms, warming events, diseases, and predator outbreaks. Degree of desirability, from a management perspective, is shown by light to dark shading.

cal controls via intense herbivory prevail, resulting in overcompensation by grazers; whereas bottom-up stimulation of productivity is minimal, due to lack of nutrient availability. Under persistent elevated nutrients, however, consistent coral declines can occur, concomitant with algal increases that may lead to profound long-term effects throughout all combinations of herbivory. Changes in bottom-up controls and their interactions not only alter the dominance patterns of primary-producer groups, but also can have even longer-term consequences mediated through structural transformations and chemical modifications to reef systems and their resident fish populations. In other words, excessive nutrient enrichment not only increases the productivity and biomass of weedy macroalgae, but, over the long term, may lead to coral habitat degradation through reduced spatial heterogeneity by overgrowth and nighttime anoxic conditions.

The proposed management model addresses the considerable complexity of coral-reef systems. Much of the overall diversity at the primary-producer level is afforded by the interaction of opposing herbivory and nutrient controls. Because of the sensitive nature of direct and indirect interacting factors, coral reefs are particularly vulnerable to anthropogenic reversal effects that decrease top-down controls and increase bottom-up controls, dramatically altering community dynamics. For example, insufficient nutrients may act directly to limit fleshy-algal domination; conversely, abundant nutrients enhance fleshy-algal growth, with the opposite effect on reef-building corals. Furthermore, the effects of controls can be indirect by influencing competition. Competition between algae and corals can be direct (e.g., overgrowth) or indirect (e.g., preemption of substrate). Low nutrients and high herbivory also act indirectly on fleshy algae through reduced competitive abilities, whereas lowered herbivory and elevated nutrients also indirectly affect corals and coralline algae by favoring fleshy-algal competition. Other ecologically important bottom-up factors, such as light regime, abrasion, allelopathy, and sediment smothering, also can be indirect side effects of algal competition.

STATUS OF KNOWLEDGE

The relevant data on top-down versus bottom-up controls consist of short-term caging or feeding experiments, as well as circumstantial evidence (e.g., Hallock et al. 1993), correlative biogeography surveys (e.g., Littler et al. 1991; Verheij 1993), physiological assays (e.g., Littler and Littler 1990; Lapointe et al. 1997), and long-term manipulative studies. Top-down control by abundant populations of large mobile herbivores is particularly well studied for coral reefs, beginning over four decades ago with the caging study of Stephenson and Searles (1960). As examples, Sammarco et al. (1974), Ogden and Lobel (1978), Sammarco (1980), Carpenter (1986), Lewis (1986), Morrisson (1988), and numerous other workers (see review by McCook 2001) have all demonstrated that lowering herbivory without changing nutrient inputs most often results in rapid increases in algal turfs. Such low mats are unique in containing an abundance of nitrogen-fixing blue-green algae (Cyanobacteria) that can enrich other low-growing members of the turf community (Adey and Goertemiller 1987; Adey 1998). A sequence of phase shifts in algal form groups (from crustose corallines to algal turfs and, finally, to frondose macroalgae) as a function of declining herbivory was noted by Steneck (1989), who also pointed out that the biomass of an alga or functional group is ultimately the result of its rate of productivity relative to the rate at which it is removed by herbivores (see also Hatcher and Larkum 1983; Carpenter 1986; Russ 1987).

Although nutrient data are usually lacking in coral-reef field herbivory studies, natural background levels in conjunction with ample water motion are often assumed to exceed levels limiting to macroalgal growth. As pointed out by Lewis (1986), large frondose macroalgae do occur in oligotrophic reef areas of low herbivory (see also Littler et al. 1986; McCook 2001); however, many of these perennating forms occupy microhabitats that generate increased current acceleration, such as the reef crest and tops of patch-reef rocks, implicating higher nutrient fluxes (Atkinson et al. 2001). Also, some of the large perennial macroalgae are relatively slow growing and, therefore, can maintain large biomass under low nutrient concentrations. A further consideration is the presence of substantial anthropogenic nitrogen sources in rainfall worldwide (Vitousek et al. 1997a). The decrease in coral cover (Pollock 1928), and associated rise in frondose algae (Doty 1971) and coralline algae (Littler 1971), on the reef flat at Waikiki, Hawaii, was the first phase shift from coral to macroalgal domination that was postulated to result from increases in eutrophication (Littler 1973). Spatial and temporal patterns of nutrients also have been shown to covary with algal biomass (Adey et al. 1977; Hatcher and Hatcher 1981; Hatcher and Larkum 1983). The primary production and growth of algal turfs (Hatcher and Larkum 1983; Williams and Carpenter 1988), frondose macroalgae (Lapointe 1987; Littler et al. 1991), and coralline algae (Littler 1973) generally increase with higher nutrient availability on undisturbed reef systems, suggesting limitation by nutrients.

Shifts from coral dominance to algal dominance that indicate linkages with chronic nutrient loading are exemplified by case studies worldwide (e.g., Littler 1973; Banner 1974; Weiss and Goddard 1977; Mergener 1981; Tomascik and Sander 1985, 1987; Cuet et al. 1988; Lapointe and O'Connell 1989; Bell 1992; Littler et al. 1993; Lapointe et al. 1994, 1997; Goreau et al. 1997; Bruno et al. 2003). Other coral to algal phase shifts on various types of degraded reefs have also been reviewed (Marszalek 1981; McClanahan and Shafir 1990; Done 1992b; Knowlton 1992; Hallock et al. 1993; Gardner et al. 2003).

Herbivory patterns, like nutrient levels, alone do not always explain the distributions and abundances of benthic algae on coral reefs (Adey et al. 1977; Hay 1981; Hatcher and Larkum 1983; Hatcher 1983; Carpenter 1986). For example, several studies found no significant correlation between grazing intensity and algal biomass (e.g., Hatcher 1981; Schmitt 1997; Lirman and Biber 2000). A dramatic increase in algal biomass due to eutrophication was reported (Fishelson 1973) without any concomitant reduction in herbivore populations. The importance of the very low nutrient levels involved in eutrophication (either natural or anthropogenic) has only recently come to light (Bell 1992; Lapointe et al. 1997; Small and Adey 2001). Such low levels can sometimes correlate with the phase shifts from corals toward macroalgal dominance without changing herbivory. These kinds of biotic phase shifts also have been attributed to overfishing (e.g., see Hughes 1994), in concert with cultural eutrophication (Goreau et al. 1997; Lapointe et al. 1997).

Smith et al. (2001) rigorously conducted the first appropriate experimental test of the RDM in a natural oligotrophic coral-reef setting, in conjunction with natural successional and competitive bouts, to determine dominance among the major benthic producer groups over an adequate time scale. The results precisely fitted all tenets of the RDM, confirming its efficacy as an important management tool. In contrast, a sophisticated nutrient-enrichment experiment (Larkum and Koop 1997; Encore Group 2001) did not produce results relevant to the RDM because ambient nutrient levels were well above threshold concentrations and the experimental organisms were isolated on raised grids, precluding natural encroachment, overgrowth, or other key competitive interactions critical to testing the RDM. Another short-term study produced equivocal findings that showed herbivory effects, but not nutrient effects (Miller et al. 1999); however, the choice of enrichment (chlorinated tree stakes) was unfortunate. Highly diverse living model systems of coral-reef communities (i.e., microcosms), operated for decades (Small and Adey 2001), have profoundly demonstrated that minute increases in nitrogen and phosphorus reduce coral growth. In addition, such systems require an abundance of grazers to maintain a high coral and algal diversity (see section 12.3). Therefore, both the processes of productivity (bottom-up) and those of disturbance (top-down) must be appreciated to manage the mechanisms that mediate the competitive interactions that determine reef health.

INTERPRETATION OF FUNCTIONAL INDICATOR GROUPS

The fast growth and turnover rates of algae compared to other reef organisms suggest their use as early-warning indicators of reef degradation. The three algal representatives of ubiquitous form and function groups are increasingly encountered as dominants on coral reefs, particularly those subjected to human activities (e.g., Lapointe 1989). From a management perspective (fig. 12.1), (1) a predominance of corals and calcareous coralline algae relative to frondose macroalgae and algal turfs would indicate a desirable healthy state reflecting low nutrients and high herbivory; (2) an abundance of frondose macroalgae indicates the least desirable condition of elevated nutrient levels and reduced herbivory, possibly reflecting pollution in concert with destructive fishing practices; (3) high coverage of coralline algae could indicate problems with elevated nutrients, but healthy high-herbivory levels; and (4) domination by turf algae suggests desirably low nutrient levels, but an inadequate herbivory component. As with any environmental indicator group, however, knowledge of distribution, variability and natural history is needed to justify its application. To provide such information, the four functional indicator groups have been characterized in detail.

Macroalgae

With an increase in nutrients, the growth of undesirable fleshy algae is favored over the slower-growing but highly desirable corals (Genin et al. 1995; Miller and Hay 1996; Lapointe et al. 1997), and the latter become inhibited by competition for space and light. On some healthy oligotrophic coral reefs, even very low nutrient increases may exceed critical levels that can shift relative dominances by stimulating macroalgal production. Birkeland (1977) noted that filamentous and frondose algae can outcompete corals (but see also McCook et al. 2001), some of which are inhibited under elevated nutrient levels (reviewed in Marubini and Davies 1996). Fast-growing algae are not just opportunists that depend on disturbances to release space resources from established longer-lived populations, but become the superior competitors when provided with abundant nutrients (Birkeland 1977). As a result, frondose macroalgae as a group are now generally recognized as harmful to the longevity of coral reefs because of the linkage between excessive blooms and coastal eutrophication (ECOHAB 1995). Potential competitive dominance of macroalgae is inferred from their overshadowing canopy heights, as well as from inverse correlations in abundances between algae and the other producer groups (Lewis 1986), particularly at higher nutrient concentrations (e.g., Littler et al. 1993; Lapointe et al. 1997). Turbulent water motion driven by wind and wave action can be sufficient to reduce boundary-layer diffusion gradients and can increase delivery rates to support considerable macroalgal growth (e.g., Atkinson and Bilger 1992), but the abundant herbivores may mask these effects. The fleshy-macroalgal form group has proven to be particularly vulnerable to herbivory (see Hay 1981; Littler et al. 1983a; Littler et al. 1983b) and becomes abundant only where grazing is low or herbivores become swamped by excessive algal growth. Such overcompensation by herbivory may explain some of the reported cases (e.g., Crossland et al. 1984; Szmant 1997) of specific corals surviving high-nutrient reef environments.

Crustose Coralline Algae

Members of the crustose coralline algae tend to be slow-growing, competitively inferior (relative to corals, turfs, and frondose macroalgae) understory taxa abundant in most reef systems (Littler 1972). Crustose corallines generally are conspicuous, but not dominant, under low concentrations of nutrients and high levels of herbivory (Littler et al. 1991). Accordingly, they do well under both low and elevated nutrients. Therefore, crustose coralline algae do not require elevated nutrients, as might be interpreted from the RDM (fig. 12.1); instead, their rise to dominance is largely controlled indirectly by the factors influencing the abundances of the other groups, primarily corals and fleshy macroalgae. The key point is that crustose corallines predominate mainly by default (i.e., under conditions of minimal competition), where either elevated nutrients inhibit corals or intense herbivory removes fleshy algae.

Turf Algae

Turf algae tend to become dominant under minimal inhibitory top-down and minimal bottom-up controls. Their relatively small size and rapid perennation result in moderate losses to herbivory at low grazing pressures. They have opportunistic life-history characteristics, including the ability to maintain substantial nutrient uptake and growth rates under low-nutrient conditions (Rosenberg and Ramus 1984), and contain an abundance of nitrogen-fixing cyanobacteria (Adey and Goertemiller 1987; Adey 1998) that can enrich other low-growing members of the turf community. Microcosm studies have consistently shown that nutrient increases can thicken algal turfs and increase the cyanobacteria component, while lessening the overall productivity (Adey and Goertemiller 1987; Adey 1998). Algal turfs have been shown to be favored under reduced nutrient-loading rates (Fong et al. 1987) or episodic nutrient pulses (Fujita et al. 1988) and can form massive horizontal mats. Numerous studies have shown the expansion of algal turfs, not macroalgae, resulting from the removal of grazers in a wide variety of sites worldwide (e.g., Vine 1974; Hatcher and Larkum 1983; Sammarco 1983; Lewis 1986; Klumpp et al. 1987; Carpenter 1988; Littler and Littler 1997).

Reef-Building Corals (Cnidaria)
Because of their three-dimensional heterogeneity, which provides habitat for other reef organisms, their roles in producing the carbonate structure of reefs, and their aesthetic qualities, corals are the most desirable components of biotic reefs. The vertical structure and horizontal canopies of branching forms allow abundant populations of shade-dwelling crustose coralline algae to co-occur. Reef-building corals, while preyed upon by a few omnivorous fishes and specialist invertebrates (e.g., the crown of thorns sea star), generally achieve dominance under the top-down control of intense herbivory (Lewis 1986; Lirman 2001) and extremely low nutrient concentrations (Bell 1992; Lapointe et al. 1993). Massive corals are resistant to grazing at the highest levels of herbivory. Hard mound-shaped forms show little colony mortality under high grazing pressure, even though occasionally rasped by parrot fish. In contrast, some delicately branched corals such as *Porites porites* are quite palatable and readily eaten by parrot fish (Littler et al. 1989; Miller and Hay 1998). However, some corals are inhibited by increases in nitrate, ammonium or orthophosphate (see Townsley cited in Doty 1969; Stambler et al. 1991; Muller-Parker et al. 1994; Marubini and Davies 1996; Hoegh-Guldberg et al. 1997). Nutrient inhibition of coral larval settlement has been shown for *Acropora longicyathis* (Ward and Harrison 1997).

CONCLUSIONS

The recent increased awareness of coral-reef degradation worldwide (see Ginsberg 1993; chapters in Birkeland 1997), particularly from coastal pollution (e.g., Windom 1992; Bell 1992) and destructive fishing (e.g., Hughes 1994), makes the RDM timely and important. To effectively manage the mechanisms that mediate the competitive interactions within ecosystems, the processes of productivity (bottom-up) and disturbance (top-down) must be considered. The RDM addresses the roles of top-down and bottom-up controls in the benthic community structure of coral reefs. The model provides a management perspective for the mechanisms that initiate and sustain harmful blooms of algae that degrade tropical coral-reef communities. For example, if managers see a transition from coral to coralline algae, then they should attempt to limit nutrients; if a transition to turf algae occurs, grazer populations should be augmented; and a shift to macroalgal domination indicates that both excessive nutrification and destructive fishing should be curtailed. Because of global-scale degradation of coral-reef ecosystems (e.g., Ginsberg 1994; Wilkinson 1999), we emphasize the need to obtain relevant information on nutrient and herbivory thresholds for bottom-up and top-down controls, respectively. This section has evaluated the essential literature and addressed this need by providing new management insights for ascertaining and monitoring the nutrient and herbivore status of coral reefs.

12.3 APPLICATION OF A SEAGRASS
MANAGEMENT MODEL

Mark M. Littler and Diane S. Littler

MANY LARGE SEAGRASS systems, such as Florida Bay in south Florida, are undergoing severe degradation, with controversy raging as to the possible causes (e.g., water-quality issues of hypersalinity versus eutrophication). To date, no one has considered the possible implication of destructive fishing (e.g., overharvesting of large herbivorous conch [*Strombus* spp.]) as well as the trapping of herbivorous pinfish (Atlantic, Sparidae) and the netting of rabbitfish (Pacific, Siganidae) and mullet (pantropical, Mugilidae) for seagrass fitness. The protection of beneficial herbivores could present a serious conservation and management issue that is presently receiving little attention.

Aside from catastrophic events such as hurricanes, the major direct source of seagrass mortality and degradation is excessive overgrowth by filamentous and frondose (fleshy) forms of epiphytic algae. Epiphyte loading has been documented to decrease the productivity of seagrasses (e.g., Gacia et al. 1999), as well as to inhibit both subterranean (e.g., Tomasko and Lapointe 1991) and emergent (e.g., Howard and Short 1986) growth. Since epiphytes diminish the light energy and nutrients reaching the host plant, they may indirectly influence seagrass abundance, distribution, and productivity, as well as both sexual and vegetative reproduction (Orth and Montfrans 1984). This phenomenon is especially pronounced in south Florida and the Florida Keys on grass flats impacted by humans where filamentous and frondose algae overgrow and smother the rooted seagrasses (Tomasko and Lapointe 1991). The result is reduced seagrass cover, biomass, habitat diversity, and biological diversity. Although the problem is controversial, most scientists agree that it is largely related to anthropogenic effects on an interactive complex of factors that threatens the pristine conditions under which seagrass systems flourish.

A corollary of the relative dominance model (RDM) for biotic reefs (see section 12.2; Littler and Littler 1984a) is proposed here as a testable paradigm for the major structuring components of tropical seagrass ecosystems (fig. 12.2). These vast grass beds represent the shallow, sedimentary-bottom biological equivalent of coral reefs. Healthy seagrass ecosystems, where marine vascular plants provide more than 50% of the cover, occur where nutrient pollution and destructive fishing (by hand-collecting conch and poisoning, netting, trapping, or dynamiting fish) are either low or absent.

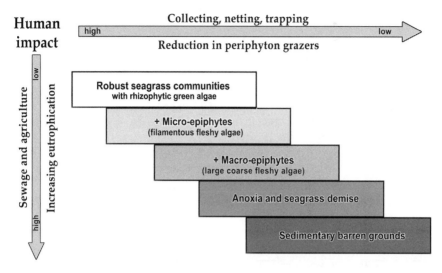

Figure 12.2 Predicted changes in the relative dominance of seagrasses, rhizophytic algae, and epiphytes as a result of the predominant forcing functions of declining water-column quality (eutrophication) and herbivory. All four functional indicator groups of primary producers are present all the time in seagrass beds, but dominate most often under the interacting conditions indicated. The complex interacting vectors of long-term water-column eutrophication and declining herbivory (either naturally or anthropogenically derived) are postulated to produce competitive shifts away from seagrass/rhizophytic algae domination on healthy grass beds toward various phases of epiphytic algal dominance. Hypothetically, one vector can partially offset the other (e.g., high herbivory may delay the impact of elevated water-column nutrients, or low water-column nutrients may offset the impact of reduced herbivory). Latent trajectories are most often catalyzed or accelerated by large-scale stochastic disturbances such as tropical storms, salinity fluctuations, diseases, and global-warming phenomena. Degree of desirability, from a management perspective, is shown by light to dark shading.

NATURAL HISTORY OF THE SEAGRASS ECOSYSTEM

Seagrasses (Magnoliophyta) are the only submerged marine plants having true roots, stems, and leaves, and contain an abundance of vascular tissues as well as inconspicuous flowers. Seagrasses are derived from two monocot plant families, the Hydrocharitaceae and the Potamogetonaceae, and are the only flowering plants to have colonized (presumably from terrestrial estuaries) the depths of the oceans, occurring down to 40 m deep. Seagrasses have a coarse, fibrous, grasslike texture and are apple green or grass green. Like all higher-plant groups, they contain many of the same pigments as their relatives, the Chlorophyta (green algae). Species of the strap-shaped forms are often distinguished by the number of veins running lengthwise in the blades. In oval-bladed species, the numbers and angles of lateral veinlets branching from the main vein or midrib are diagnostic. Most species have a well-developed runner (rhizome) system that binds and consolidates sedimentary bottoms, thereby adding protection and a sheltered environment for the many organisms that live in the stabilized sand, thick foliage, or extensive root systems. There are only about 48 species in 12 genera worldwide; nevertheless, seagrasses play major roles in relatively calm, tropical marine environments. Areas of high seagrass endemism are not known for the tropics, and *Halophila decipiens* is the only pantropical seagrass.

Like mangrove forest trees, seagrass species themselves are not diverse; however, these relatively large marine plants cover vast areas, forming grass flats with tremendous biomass and primary productivity. Most of the considerable seagrass community diversity resides in the epiphytic biota, particularly algal epiphytes but including invertebrates. The blades of the strap-shaped seagrasses are literally conveyor belts of various stages in early to late succession, owing to their basal growth by intercalary meristems (like terrestrial grasses). In the healthiest of systems, sediment-dwelling, siphonaceous (lacking cellular cross walls), rhizomatous forms (= rhizophytes) of Bryopsidales (green seaweeds), such as *Halimeda*, *Avrainvillea*, *Udotea*, *Penicillus*, *Rhipocephalus*, and *Caulerpa*, are ubiquitously present, scattered among the grass blades. Conspicuously abundant and often dominant, seagrasses form vast meadows in sandy or silty shallows, and certain species, such as *Thalassodendron ciliatum*, can even overgrow hard carbonate substrates. The most luxurious and spatially complex seagrass beds occur in clear shallow waters and serve as habitats and nursery grounds for juvenile and adult stages of a myriad of epiphytes, fishes, and invertebrates. They are the feeding grounds of some of the most sought-after sport fishes (e.g., tarpon, bonefish, and permit), and this high-dollar catch-and-release fishery provides an attractive tourism incentive to managers of tropical marine resources worldwide. Tropical seagrasses also serve as important food sources for "charismatic" large animals, such as sea turtles, parrot fish, manatees, and dugongs (sea cows). The seeds of *Enhalus* species are gathered and eaten by South Pacific islanders.

ECOLOGY OF THE SEAGRASS ECOSYSTEM

Plant communities are regulated by a combination of top-down controls, involving the activities of predators (herbivores and carnivores), and bottom-up factors, related to resource availability (McQueen et al. 1989). Changes in herbivore populations can cascade through the entire food web (Carpenter et al. 1985), triggering complex interactions depending on the strength (Levitan 1987) and frequency (Threlkeld 1988) of the disturbances (physical removal). In the case of terrestrial plants, Grime (1979) proposed that communities are regulated by the interactions of (1) physical forces that remove biomass and (2) limiting resources that control productivity, in conjunction with (3) competitive interactions. The Grime model was adapted to marine plant communities (Littler and Littler 1984b; Steneck and Dethier 1994) by adding a component for (4) physiological stress. In the management model proposed in figure 12.2, bottom-up nutrient levels enhance plant productivity, whereas the grazing activities of predatory herbivores physically remove plant biomass. These two primary factors hypothetically interact over long time spans to maintain stable states or to cause phase shifts in tropical seagrass ecosystems. It is postulated that low water-column nutrient levels coupled with high abundances of epiphyte grazers, such as mullet, pinfish, rabbitfish (Pacific only), sea urchins, and gastropods, maintain low standing stocks of competitively superior fleshy epiphytic algae and lead to the relative dominance of seagrasses, such as the turtle grass *Thalassia testudinum*.

Small-scale human perturbations such as anchor and propeller damage or larger-scale activities such as salinity changes and sedimentation also tend to eliminate the slower-growing deeply rooted seagrasses in favor of more ephemeral microscopic or opportunistic macrophytic forms of algae. Aside from uncontrollable catastrophic events such as tropical storms, salinity fluctuations due to drought and flooding, diseases, or global-warming trends, the major reason for seagrass degradation and mortality, from a management perspective, would appear to be excessive overgrowth by fleshy epiphytic algae. Given an overabundance of water-column nutrients (bottom-up), a reduction in grazing (top-down) resulting from destructive fishing or natural causes (e.g., diseases) could potentially shift the relative dominance within seagrass beds to a condition dominated by large masses of microscopic filamentous algae or ultimately to complete inundation by larger frondose macroalgae (fig. 12.2).

MANAGEMENT MODEL

The proposed management model (fig. 12.2) uses the four groups of indicator plants reviewed above to predict the health of a given seagrass ecosystem. The most desirable condition is indicated by clean seagrass stands such as *Thalassia* with diverse green algal rhizophytes (rooted forms) contributing dense cover and biomass in clear waters and anchored in aerobic sediments. Less healthy would be thinner seagrass beds with increased epiphytic loads of small filamentous algal forms. Massive inundation by long filamentous algal forms along with large coarse fleshy epiphytes (or even free-lying unattached fleshy forms) and sparseness of both seagrass and rhizophytic green algal populations would characterize seagrass systems possibly on the verge of collapse, leading ultimately to anoxia and to sedimentary barren grounds.

Large mobile herbivorous fishes such as parrot fish (Scaridae), surgeonfish (Acanthuridae), and rudderfish (Kyphosidae), while beneficial to coral-reef systems, are deadly to palatable seagrasses, as evidenced by ubiquitous "halo effects" (Ogden et al. 1973; see also Littler et al. 1983a for fish-preference ranking). In one study, experimentally elevated nutrient levels increased the palatability and attractiveness of enriched seagrass plots to large herbivorous parrot fish (McGlathery 1995). Fortunately, in healthy systems these fishes are prevented by predatory birds (e.g., osprey, pelicans, herons, cormorants) and carnivorous fish (e.g., barracuda, snapper, grouper, jacks) from straying far from protective reef cover into the relatively open waters over shallow seagrass beds. Thus, seagrass flats with natural carnivore populations are not subject to devastation by powerful grazers. Pristine seagrass beds are, as mentioned earlier, home to large schools of herbivorous rabbitfish, mullet, and pinfish, which feed extensively on epiphytic and filamentous periphyton algae (Odum 1970; Darcy 1985; Gilmore 1988). Numerous gastropods, including the large conchs in the genus *Strombus*, also share this periphyton resource (Stoner and Waite 1991). Herbivorous sea urchins and smaller mesofauna also have the potential to ameliorate the detrimental impact of elevated water-column nutrients associated with epiphyte overgrowth on seagrass communities (Orth and Montfrans 1984; Brawley 1992). The importance of such mesograzer/periphyton interactions seems clear; however, further tests of their impact are needed because of the complexity of food-web interactions in general. Therefore, certain herbivorous fishes and invertebrates lacking powerful biting apparatuses (Orth and Montfrans 1984; Klumpp et al. 1992) have the potential to regulate the more delicate, but harmful, epiphytic algal overgrowth, which should increase seagrass productivity, growth, and reproduction.

The important role of herbivory in eliminating harmful blooms of epiphytes from plant hosts has been demonstrated for frondose algae (e.g., Brawley 1992) and reef-building calcareous algae (Littler et al. 1995), as well as seagrasses (Howard and Short 1986; Sand-Jensen and Borum 1991; Gacia et al. 1999). Under increasing eutrophication (fig. 12.2), a reduction in grazing from overfishing or natural causes (e.g., diseases) could potentially shift the relative dominance within healthy seagrass ecosystems to a condition dominated by microscopic filamentous algae or ultimately to complete inundation by larger frondose macroalgae. Filamentous species are always present naturally but, hypothetically, are cropped to low levels by herbivorous fishes, sea urchins, and gastropods. Severe reductions in grazing could allow the algal biomass to accumulate to an upper limit determined by the second major factor, water-column water quality.

The remarkable feature of seagrasses that allows them to thrive under such nutrient-impoverished water-column conditions is their rooted subterranean system, which gives them access to the relatively nutrient-rich sediment pore waters and which confers a competitive advantage over epiphytic fleshy algae. In seagrass systems, an increase in nutrient supply to the water column leads to increased epiphyte loads on the blades (Sand-Jensen 1977; Cambridge and McComb 1984; Twilley et al. 1985; Silberstein et al. 1986; Tomasko and Lapointe 1991), causing a shift in patterns of primary productivity. Overgrowth by epiphytes reduces light and increases boundary-layer diffusion gradients, inhibiting nutrient and gaseous exchange as well as limiting light energy available for photosynthesis. This effect dramatically reduces both seagrass growth (Kiorboe 1980; Kemp et al. 1983; Short and Short 1984; Gacia et al. 1999) and reproduction (Orth and Montfrans 1984). Thus, low water-column nutrient levels coupled with high levels of periphyton feeders, mostly conch, sea urchins, rabbitfish, mullet, and pinfish, maintain low standing stocks of competitively superior fleshy epiphytic algae and lead to the relative dominance of robust seagrasses such as the turtle grass *Thalassia testudinum*.

Herbivory could also represent an important natural route for the export of nutrients from seagrass beds. Herbivore excretions are exported to surrounding waters by currents. Because the intrinsic production rate of filamentous and frondose algae is much greater than the lower-producing seagrasses (Littler 1980), the algae can theoretically overgrow and outcompete seagrasses as the dominant space-occupying organisms. Since algae can withstand anoxic conditions that prevail at night during bloom conditions, seagrasses and sessile animals often undergo higher mortality than the algae (Tomascik and Sander 1987; Bell 1991).

CASE STUDY

The interactions between groups of producers (seagrasses and epiphytes) in a nutrient-rich estuarine system were investigated to assess the potential alleviating effects of macroherbivores (this case study is from Gacia et al. 1999). The concept that herbivorous fish can regulate epiphytic algal overgrowth, thereby enhancing primary production and growth of the dominant seagrass (*Thalassia*), was tested. An exclosure experiment was carried out within a monospecific *T. testudinum* meadow in the Indian River Lagoon (IRL), Florida. The Gacia et al. study involved multifaceted approaches including characterization of environmental parameters, particularly those that might be conducive to enhanced epiphytic algal growth; documentation of the abundances of the two predominant herbivorous fish species; assessment of the indirect role of grazers on seagrass primary production and biomass accumulation; and experimental tests of the direct effects of fish grazing on epiphyte biomass, species composition, and relative abundance. The last involved four 2.0-m^2 exclusion cages (fences) of 2-cm^2 plastic-coated wire mesh, four open 2.0-m^2 plots (controls), and four 2.0-m^2 two-sided cages (cage controls) to control for cage artifacts such as current and light (see fig. 12.3).

It was estimated by Gacia et al. (1999) that within the IRL study area there is an annual load of epiphytic algae of 0.022 g organic dry mass (ODM) per *Thalassia* shoot (blade). Given a mean annual shoot density of 616 shoots per square meter and a shoot turnover rate of 4.1 per year, the minimum annual production of macrophytic epiphytes was estimated at 55 g ODM per square meter per year. These values represent a significant macroepiphyte biomass that falls within the range of total epiphyte production for seagrass beds in Florida Bay (Frankovich and Zieman 1994), while being 40% lower than comparable data provided for *Thalassia hemprichii* beds in Papua New Guinea (Heijs 1987).

The pinfish *Logodon rhomboides* is a ubiquitous omnivorous species in Florida seagrass beds and has a relatively homogeneous distribution in both space and time (Gilmore 1988). Darcy (1985) estimated that *L. rhomboides* has a subsistence feeding rate of 5.75% of body mass per day, and at least 65% of this consumption is algae for the fish sizes excluded from the cages (body mass >90 g per individual). From these data, it was estimated that the potential algal demand needed to sustain the population of pinfish at the IRL study site would be 0.1 g ODM per square meter per day, or about 36 g ODM per square meter per year, which is about 60% of the conservative estimate of epiphyte turnover (55 g ODM per square meter per year). Large schools of herbivorous/detritivorous mullet commonly reside and graze

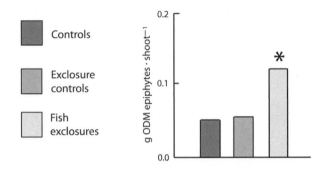

Figure 12.3 Epiphyte growth (mean grams of organic dry mass [weight] per *Thalassia* shoot) from 4 January to 13 March 1995 in the different treatment plots (modified from Gacia et al. 1999). Asterisk indicates significantly higher values for the epiphytic community inside the cage treatments compared to controls (ANOVA, $P < 0.005$).

within seagrass beds throughout the world. The estimated daily algal consumption by striped mullet (*Mugil cephalus*) individuals longer than 20 cm is 9.8% of the body mass (Odum 1970), with a gut-content turnover rate for this size class of five times per day. The estimated algal demand by striped mullet in the immediate IRL study area would be about 6.4 g ODM per square meter per day. Therefore, the total algal biomass required to support the combined demands of the two predominant fish grazers in the seagrass bed studied by Gacia et al. would be about 6.5 g ODM of algal epiphytes per square meter per day.

Sudden blooms in biomass of epiphytic algae correlated (Gacia et al. 1999) with seasonal spikes in dissolved inorganic nitrogen and soluble reactive phosphorus. The nutrient concentrations consistently recorded in the IRL far exceeded the threshold levels conducive to macroalgal proliferation in other tropical seagrass and coral-reef ecosystems (i.e., 0.2 μM soluble reactive phosphorus and 1.0 μM dissolved inorganic nitrogen; Bell 1992; Lapointe et al. 1997). Crossland et al. (1984) also correlated dissolved inorganic nitrogen above the almost undetectable threshold levels of 1.2 μM and soluble reactive phosphorus of 0.22 μM for macroalgal-dominated high-latitude communities of Western Australia. During the early spring, blooms of ephemeral green algal species, mostly of the genera *Cladophora* and *Enteromorpha,* reached peak abundances, with their maximum epiphytic biomass occurring inside the fish-exclusion cages (fig. 12.3). Herbivore pressure on these delicate chlorophytes was critical during the early spring bloom when the epiphyte assemblage was dominated by fleshy forms. These ephemeral forms are delicate, filamentous and thin-tubular species that bloom under eutrophic conditions (Littler and Arnold 1982) and are easily grazed (Littler et al. 1983a). The fish-exclusion experiment (fig. 12.3) strongly supported the hypothesis that epiphytic biomass on seagrass blades would be significantly reduced in the presence of grazers. As predicted, the increased epiphyte loads in the grazer-exclusion cages had an inhibitory effect on leaf growth of *Thalassia* (fig. 12.4A); leaf initiation also was sig-

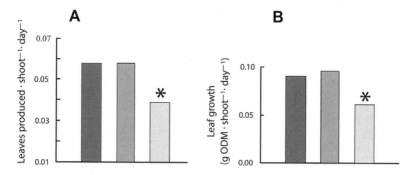

Figure 12.4 Seagrass growth (new *Thalassia* leaves and biomass) from 4 January to 13 March 1995 in the different treatment plots (key same as for fig. 12.3; modified from Gacia et al. 1999). Asterisks indicate significantly lower production of new leaves and new biomass for the plants growing inside the fish exclosure cages (ANOVA, $P < 0.005$) than for accessible plants in the controls.

nificantly reduced (fig. 12.4B) for the nongrazed plants compared to that in the treatments exposed to grazers.

Eutrophication of coastal waters is now seen as one of the most pervasive, worldwide anthropogenic impacts (National Research Council 1994; Vitousek et al. 1997a, 1997b; Jackson et al. 2000; Tilman et al. 2001). Unless major social, economic, and political measures are taken, the escalation of the problem is forecast to worsen in the next decades (Nixon 1995). In light of the growing recognition of the consequences of destructive fishing and increasing pollution on seagrass ecosystems globally (National Research Council 1994), management approaches should include (1) monitoring of herbivore, algal, and seagrass stocks, (2) inventories and assays of the health of herbivore stocks (see section 12.2), (3) characterization of water-column nutrient concentrations and epiphyte tissue analyses for C:N:P ratios, and (4) bioassays using epiphyte physiological responses (i.e., productivity, growth rate, biomass) to experimental nutrient pulses (methods referenced in Lapointe et al. 2004).

LITERATURE CITED

Adey, W. H. 1998. Coral reefs: algal structured and mediated ecosystems in shallow, turbulent, alkaline waters. Journal of Phycology 34:393–406.

Adey, W. H., Adey, P., Burke, R., and Kaufman, L. 1977. The Holocene reef systems of eastern Martinique. Atoll Research Bulletin 281:1–40.

Adey, W. H., and Goertemiller, T. 1987. Coral reef algal turfs: master producers in nutrient poor seas. Phycologia 26:374–386.

Atkinson, M. J., and Bilger, R. W. 1992. Effects of water velocity on phosphate uptake in coral reef-flat communities. Limnology and Oceanography 37:273–279.

Atkinson, M. J., Falter, J. L., and Hearn, C. J. 2001. Nutrient dynamics in the Biosphere 2 coral reef mesocosm: water velocity controls NH_4 and PO_4 uptake. Coral Reefs 20:341–346.

Atkinson, M. J., and Grigg, R. W. 1984. Model of a coral reef ecosystem. Part 2, Gross and net benthic primary production at French Frigate Shoals, Hawaii. Coral Reefs 3:13–22.

Banner, A. H. 1974. Kaneohe Bay, Hawaii: urban pollution and a coral reef ecosystem. Proceedings of the Second International Coral Reef Symposium 2:685–702.

Bell, P. R. F. 1991. Status of eutrophication in the Great Barrier Reef Lagoon. Marine Pollution Bulletin 23:89–93.

———. 1992. Eutrophication and coral reefs: some examples in the Great Barrier Reef Lagoon. Water Resources 26:555–568.

Birkeland, C. 1977. The importance of rate of biomass accumulation in early successional stages of benthic communities to the survival of coral recruits. Proceedings of the Third International Coral Reef Symposium 1:15–21.

———. 1997. Life and Death of Coral Reefs. Chapman and Hall, New York.

Brawley, S. H. 1992. Mesoherbivores. Pp. 235–264 in John, D. M., Hawkins, S. S., and Price, J. H., eds., Plant-Animal Interactions in the Marine Benthos. Systematic Association Special Volume. Clarendon Press, Oxford.

Bruno, J. F., Petes, L. E., Harvell, C. D., and Hettinger, A. 2003. Nutrient enrichment can increase the severity of coral diseases. Ecology Letters 6:1056–1061.

Cambridge, M. L., and McComb, A. J. 1984. The loss of seagrass in Cockburn Sound, Western Australia. Part 1, The time course and magnitude of seagrass decline in relation to industrial development. Aquatic Botany 20:229–243.

Cameron, A. M. 1977. *Acanthaster* and coral reefs: population outbreaks of a rare and specialized carnivore in a complex high-diversity system. Proceedings of the Third International Coral Reef Symposium 1:193–200.

Carpenter, R. C. 1986. Partitioning herbivory and its effects on coral reef algal communities. Ecological Monographs 56:345–363.

———. 1988. Mass mortality of a Caribbean sea urchin: immediate effects on community metabolism and other herbivores. Proceedings of the National Academy of Science USA 85:511–514.

Carpenter, S. R., Kitchell, J. F., and Hodgson, J. R. 1985. Cascading trophic interactions and lake productivity. BioScience 35:634–639.

Convention on Biological Diversity. Secretariat of the Convention on Biological Diversity. http://www.biodiv.org.

Crossland, C. J., Hatcher, B. G., Atkinson, M. J., and Smith, S. V. 1984. Dissolved nutrients of a high-latitude coral reef, Houtman Abrolhos Island, Western Australia. Marine Ecology Progress Series 14:159–163.

Cuet, P., Naim, O., Faure, G., and Conan, J. Y. 1988. Nutrient-rich groundwater impact on benthic communities of La Saline fringing reef (Reunion Island, Indian Ocean): preliminary results. Proceedings of the Sixth International Coral Reef Symposium 2:207–212.

Darcy, G. H. 1985. Synopsis of biological data on the pinfish, *Lagodon rhomboides* (Pisces: Sparidae). NOAA Technical Report NMFS23. FAO Fisheries, Synopsis. Seattle.

Done, T. J. 1992a. Effects of tropical cyclone waves on ecological and geomorphological structures on the Great Barrier Reef. Continental Shelf Research 12:859–872.

———. 1992b. Phase shifts in coral reef communities and their ecological significance. Hydrobiologia 247:121–132.

Doty, M. S. 1969. The ecology of Honaunau Bay, Hawaii. Hawaii Botanical Science Paper no. 14. University of Hawaii, Manoa.

———. 1971. Physical factors in the production of tropical benthic marine algae. Pp. 99–121 in Costlow, J. D. Jr., ed., Fertility of the Sea, vol. 1. Gordon and Breach, New York.

ECOHAB (Ecology and Oceanography of Harmful Algal Blooms). 1995. The ecology and oceanography of harmful algal blooms: a national research agenda. Proceedings of the National Workshop. Woods Hole Oceanographic Institution, Woods Hole, Massachusetts.

Encore Group. 2001. The effect of nutrient enrichment on coral reefs: synthesis of results and conclusions. Marine Pollution Bulletin 42:91–120.

Fishelson, L. 1973. Ecology of coral reefs in the Gulf of Aqaba (Red Sea) influenced by pollution. Oecologia 12:55–67.

Fong, P., Rudnicki, R., and Zedler, J. B. 1987. Algal community response to nitrogen and phosphorus loading in experimental mesocosms: management recommendations for southern California lagoons. Report of the California State Water Control Board, San Diego.

Frankovich, T. A., and Zieman, J. C. 1994. Total epiphyte and epiphytic carbonate production on *Thalassia testudinum* across Florida Bay. Bulletin of Marine Science 54:679–695.

Fujita, R. M., Wheeler, P. A., and Edwards, R. L. 1988. Metabolic regulation of ammonium uptake by *Ulva rigida* (Chlorophyta): a compartmental analysis of the rate-limiting step for uptake. Journal of Phycology 24:560–566.

Gacia, E., Littler, M. M., and Littler, D. S. 1999. An experimental test of the capacity of food web interactions (fish-epiphytes-seagrass) to alleviate the consequences of eutrophication on seagrass communities. Estuarine, Coastal, and Shelf Science 48:757–766.

Gardner, T. A., Côte, I. M., Gill, J. A., Grant, A., and Watkinson, A. R. 2003. Long-term region wide declines in Caribbean coral reefs. Science 301:958–960.

Genin, A., Lazar, G., and Brenner, S. 1995. Vertical mixing and coral death in the Red Sea following the eruption of Mt. Pinatubo. Nature 377:507–510.

Gilmore, R. G. 1988. Subtropical seagrass fish communities: population dynamics, species guilds, and microhabitat associations in the Indian River Lagoon, Florida. Ph.D. diss., Florida Institute of Technology, Melbourne.

Ginsberg, R. N., comp. 1993. Global Aspects of Coral Reefs: Health, Hazards, and History. University of Miami, Florida.

———. 1994. Proceedings of the Colloquium on Global Aspects of Coral Reefs: Health, Hazards, and History. University of Miami, Florida.

Goreau, T. J., Dalay, L., Ciappara, S., Brown, J., Dourke, S., and Thacker, K. 1997. Community-based whole-watershed and coastal zone management in Jamaica. Proceedings of the Eighth International Coral Reef Symposium 2:2093–2096.

Grime, J. P. 1979. Plant Strategies and Vegetation Processes. John Wiley and Sons, New York.

Hallock, P., Müller-Karger, F. E., and Hallas, J. C. 1993. Coral reef decline. National Geographic Research and Exploration 9:358–378.

Hatcher, A. I., and Hatcher, B. G. 1981. Seasonal and spatial variation in dissolved nitrogen in One Tree Reef Lagoon. Proceedings of the Fourth International Coral Reef Symposium 2:419–424.

Hatcher, B. G. 1981. The interaction between grazing organisms and the epilithic algal community of a coral reef: a quantitative assessment. Proceedings of the Fourth International Coral Reef Symposium 2:515–524.

———. 1983. Grazing in coral reef ecosystems. Pp. 164–179 in Barnes, D. J., ed., Perspectives on Coral Reefs. Australian Institution of Marine Science, Townsville, Australia.

Hatcher, B. G., and Larkum, A. W. D. 1983. An experimental analysis of factors controlling the standing crop of the epilithic algal community on a coral reef. Journal of Experimental Marine Biology and Ecology 69:61–84.

Hay, M. E. 1981. Spatial patterns of grazing intensity on a Caribbean barrier reef: herbivory and algal distribution. Aquatic Botany 11:97–109.

Heijs, F. M. L. 1987. Qualitative and quantitative aspects of the epiphytic component in a mixed seagrass meadow from Papua New Guinea. Aquatic Botany 27:363–383.

Hoegh-Guldberg, O., Takabayashi, M., and Moreno, G. 1997. The impact of long-term nutrient enrichment on coral calcification and growth. Proceedings of the Eighth International Coral Reef Symposium 1:861–866.

Howard, R. K., and Short, F. T. 1986. Seagrass growth and survivorship under the influence of epiphyte grazers. Aquatic Botany 24:287–302.

Huang, H., Han, X., Kang, L., Raven, P., Wyse Jackson, P., and Chen, Y. 2002. Conserving native plants in China. Science 297:935–936.

Hughes, T. P. 1994. Catastrophes, phase shifts, and large-scale degradation of a Caribbean coral reef. Science 265:1547–1551.

Jackson, L. E., Kurtz, J. C., and Fisher, W. S., eds. 2000. Evaluation Guidelines for Ecological Indicators. EPA/620/R-99/005. U.S. Environmental Protection Agency, Office of Research and Development, Research Triangle Park, North Carolina.

Kemp, W. M., Twilley, R. R., Stevenson, J. C., Boynton, W. R., and Means, J. C. 1983. The decline of submerged vascular plants in upper Chesapeake Bay: summary of results concerning possible causes. Marine Technology Society Journal 17:78–89.

Kiorboe, T. 1980. Production of *Ruppia cirrhosa* in mixed beds in Rinkobing Fjord (Denmark). Aquatic Botany 9:135–143.

Klumpp, D. W., McKinnon, D., and Daniel, P. 1987. Damselfish territories: zones of high productivity on coral reefs. Marine Ecology Progress Series 40:41–51.

Klumpp, D. W., Salita-Espinosa, J. T., and Fortes, M. D. 1992. The role of epiphytic periphyton and macroinvertebrate grazers in the trophic flux of a tropical seagrass community. Aquatic Botany 43:327–349.

Knowlton, N. 1992. Thresholds and multiple stable states in coral reef community dynamics. American Zoologist 32:674–679.

Lapointe, B. E. 1987. Phosphorus- and nitrogen-limited photosynthesis and growth of *Gracilaria tikvahiae* (Rhodophyceae) in the Florida Keys: an experimental field study. Marine Biology 93:561–568.

———. 1989. Caribbean coral reefs: are they becoming algal reefs? Sea Frontiers 35:82–91.

Lapointe, B. E., Barile, P. J., Yentsch, C. S., Littler, M. M., Littler, D. S., and Kakuk, B. 2004. The relative importance of nutrient enrichment and herbivory on macroalgal communities near Norman's Pond Cay, Exumas Cays, Bahamas: a "natural" enrichment experiment. Journal of Experimental Biology and Ecology 298:275–301.

Lapointe, B. E., Littler, M. M., and Littler, D. S. 1993. Modification of benthic community structure by natural eutrophication: the Belize barrier reef. Proceedings of the Seventh International Coral Reef Symposium 1:323–334.

———. 1997. Macroalgal overgrowth of fringing coral reefs at Discovery Bay, Jamaica: bottom-up versus top-down control. Proceedings of the Eighth International Coral Reef Symposium 1:927–932.

Lapointe, B. E., Matzie, W. R., and Clark, M. W. 1994. Phosphorus inputs and eutrophication on the Florida Reef Tract. Pp. 106–112 in Ginsberg, R. N., comp., Proceedings of the Colloquium on Global Aspects of Coral Reefs: Health, Hazards, and History. University of Miami, Miami, Florida.

Lapointe, B. E., and O'Connell, J. D. 1989. Nutrient-enhanced productivity of *Cladophora prolifera* in Harrington Sound, Bermuda: eutrophication of a confined phosphorus-limited marine ecosystem. Estuarine, Coastal, and Shelf Science 28:347–360.

Larkum, A. W. D., and Koop, K. 1997. ENCORE, algal productivity, and possible paradigm shifts. Proceedings of the Eighth International Coral Reef Symposium 1:881–884.

Levitan, C. 1987. Formal stability analysis of a plankton fresh-water community. Pp. 71–100 in Kerfoot, W. C., and Sih, A., eds., Predation: Direct and Indirect Impacts on Aquatic Communities. University Press of New England, Hanover, New Hampshire.

Lewis, S. M. 1986. The role of herbivorous fishes in the organization of a Caribbean reef community. Ecological Monographs 56:183–200.

Lirman, D. 2001. Competition between macroalgae and corals: effects of herbivore exclusion and increased algal biomass on coral survivorship and growth. Coral Reefs 19:392–399.

Lirman, D., and Biber, P. 2000. Seasonal dynamics of macroalgal communities of the northern Florida Reef Tract. Botanica Marina 43:305–314.

Littler, M. M. 1971. Standing stock measurements of crustose coralline algae (Rhodophyta) and other saxicolous organisms. Journal of Experimental Marine Biology and Ecology 6:91–99.

———. 1972. The crustose Corallinaceae. Oceanography and Marine Biology, Annual Review 10:311–347.

———. 1973. The population and community structure of Hawaiian fringing-reef crustose Corallinaceae (Rhodophyta, Cryptonemiales). Journal of Experimental Marine Biology and Ecology 11:103–120.

———. 1980. Morphological form and photosynthetic performances of marine macroalgae: tests of a functional/form hypothesis. Botanica Marina 22:161–165.

Littler, M. M., and Arnold, K. E. 1982. Primary productivity of marine macroalgal functional-form groups from southwestern North America. Journal of Phycology 18:307–311.

Littler, M. M., and Littler, D. S. 1984a. Models of tropical reef biogenesis: the contribution of algae. Pp. 323–364 in Round, F. E., and Chapman, D. J., eds., Progress in Phycological Research, vol. 3. Biopress, Bristol, United Kingdom.

———. 1984b. Relationships between macroalgal functional form groups and substrata stability in a subtropical rocky-intertidal system. Journal of Experimental Marine Biology and Ecology 74:13–34.

———. 1990. Productivity and nutrient relationships in psammophytic versus epilithic forms of Bryopsidales (Chlorophyta): comparisons based on a short-term physiological assay. Hydrobiologia 204/205:49–55.

———. 1997. Disease-induced mass mortality of crustose coralline algae on coral reefs provides rationale for the conservation of herbivorous fish stocks. Proceedings of the Eighth International Coral Reef Symposium 1:719–724.

Littler, M. M., Littler, D. S., and Lapointe, B. E. 1993. Modification of tropical reef community structure due to cultural eutrophication: the southwest coast of Martinique. Proceedings of the Seventh International Coral Reef Symposium 1:335–343.

Littler, M. M., Littler, D. S., and Taylor, P. R. 1983a. Evolutionary strategies in a tropical barrier reef system: functional-form groups of marine macroalgae. Journal of Phycology 19: 229–237.

———. 1995. Selective herbivore increases biomass of its prey: a chiton-coralline reef-building association. Ecology 76:1661–1681.

Littler, M. M., Littler, D. S., and Titlyanov, E. A. 1991. Comparisons of N- and P-limited productivity between high granitic islands vs. low carbonate atolls in the Seychelles Archipelago: a test of the relative-dominance paradigm. Coral Reefs 10:199–209.

Littler, M. M., Taylor, P. R., and Littler, D. S. 1983b. Algal resistance to herbivory on a Caribbean barrier reef. Coral Reefs 2:111–118.

———. 1986. Plant defense associations in the marine environment. Coral Reefs 5:63–71.

———. 1989. Complex interactions in the control of coral zonation on a Caribbean reef flat. Oecologia 80:331–340.

Lough, J. M. 1994. Climate variation and El Niño–Southern Oscillation events on the Great Barrier Reef, 1958–1987. Coral Reefs 13:181–195.

Macintyre, I. G., and Glynn, P. W. 1990. Upper limit of El Niño killoff. Coral Reefs 9:92.

Marszalek, D. S. 1981. Effects of sewage effluents on reef corals. Proceedings of the Fourth International Coral Reef Symposium 1:213. Abstract.

Marubini, F., and Davies, P. S. 1996. Nitrate increases zooxanthellae population density and reduces skeletogenesis in corals. Marine Biology 127:319–328.

McClanahan, T. R., and Shafir, S. H. 1990. Causes and consequences of sea-urchin abundance and diversity in Kenyan coral reef lagoons. Oecologia 83:362–370.

McCook, L. J. 2001. Competition between corals and algal turfs along a gradient of terrestrial influence in the nearshore central Great Barrier Reef. Coral Reefs 19:419–425.

McCook, L. J., Jompa, J., and Diaz-Pulido, G. 2001. Competition between corals and algae on coral reefs: a review of evidence and mechanisms. Coral Reefs 19:400–417.

McGlathery, K. J. 1995. Nutrient and grazing influences on a subtropical seagrass community. Marine Ecology Progress Series 122:239–252.

McQueen, D. J., Johannes, M. R. S., Post, J. R., Stewart, T. J., and Lean, D. R. S. 1989. Bottom-up and top-down impacts on freshwater pelagic community structure. Ecological Monographs 59:289–309.

Mergener, H. 1981. Man-made influences on and natural changes in the settlement of the Aqaba reefs (Red Sea). Proceedings of the Fourth International Coral Reef Symposium 1:193–207.

Miller, M. W., and Hay, M. E. 1996. Coral-seaweed-grazer-nutrient interactions on temperate reefs. Ecological Monographs 66:323–344.

———. 1998. Effects of fish predation and seaweed competition on the survival and growth of corals. Oecologia 113:231–238.

Miller, M. W., Hay, M. E., Miller, S. L., Malone, D., Sotka, E. E., and Szmant, A. 1999. Effects of nutrients versus herbivores on reef algae: a new method for manipulating nutrients on coral reefs. Limnology and Oceanography 44:1847–1861.

Morrisson, D. 1988. Comparing fish and urchin grazing in shallow and deeper coral reef algal communities. Ecology 69:1367–1382.

Muller-Parker, G., McCloskey, L. R., Høegh-Guldberg, O., and McAuley, P. J. 1994. Effects of ammonium enrichment on animal and algal biomass of the coral *Pocillopora damicornis*. Pacific Science 48:273–283.

National Research Council. 1994. Priorities for Coastal Ecosystem Science. National Academy Press, Washington, DC.

Nixon, S. W. 1995. Coastal marine eutrophication: a definition, social causes, and future concerns. Ophelia 41:199–219.

Odum, W. E. 1970. Utilization of the direct grazing and plant detritus food chains by the striped mullet *Mugil cephalus*. Pp. 222–240 in Steel, J. H., Marine Food Chains. University of California Press, Berkeley.

Ogden, J. C., Brown, R. A., and Salesky, N. 1973. Grazing by the echinoid *Diadema antillarum* Philippi: formation of halos around West Indian patch reefs. Science 182:715–717.

Ogden, J. C., and Lobel, P. S. 1978. The role of herbivorous fishes and urchins in coral reef communities. Environmental Biology of Fishes 3:49–63.

Orth, R. J., and Montfrans, J. van. 1984. Epiphyte-seagrass relationships with an emphasis on the role of micrograzing: a review. Aquatic Botany 18:43–69.

Pollock, J. B. 1928. Fringing and fossil coral reefs of Oahu. Bulletin of the Bernice P. Bishop Museum 55:1–56.

Rosenberg, G., and Ramus, J. 1984. Uptake of inorganic nitrogen and seaweed surface area:volume ratios. Aquatic Botany 19:65–72.

Russ, G. R. 1987. Is rate of removal of algae by grazers reduced inside territories of tropical damselfishes? Journal of Experimental Marine Biology and Ecology 110:1–17.

Sammarco, P. W. 1980. *Diadema* and its relationship to coral spat mortality: grazing, competition, and biological disturbance. Journal of Experimental Marine Biology and Ecology 45:245–272.

———. 1983. Effects of fish grazing and damselfish territoriality on coral reef algae. Part 1, Algal community structure. Marine Ecology Progress Series 13:1–14.

Sammarco, P. W., Levinton, J. S., and Ogden, J. C. 1974. Grazing and control of coral reef community structure by *Diadema antillarum* Philippi (Echinodermata: Echinoidea): a preliminary study. Journal of Marine Research 32:47–53.

Sand-Jensen, K. 1977. Effects of epiphytes on eelgrass photosynthesis. Aquatic Botany 3:55–63.

Sand-Jensen, K., and Borum, J. 1991. Interactions among phytoplankton, periphyton, and macrophytes in temperate freshwaters and estuaries. Aquatic Botany 41:137–175.

Santavy, D. L., and Peters, E. C. 1997. Microbial pests: coral disease in the western Atlantic. Proceedings of the Eighth International Coral Reef Symposium 1:607–612.

Schmitt, E. F. 1997. The influence of herbivorous fishes on coral reef communities with low sea urchin abundance: a study among reef community types and seasons in the Florida Keys. Ph.D. diss., University of Miami, Florida.

Short, F. T., and Short, C. A. 1984. The seagrass filter: purification of coastal water. Pp. 395–413 in Kennedy, V. S., ed., The Estuary as a Filter. Academic Press, New York.

Silberstein, K., Chiffings, A. W., and McComb, A. J. 1986 The loss of seagrass in Cockburn Sound, Western Australia. Part 3, The effect of epiphytes on productivity of *Posidonia australis* Hook. Aquatic Botany 24:355–371.

Small, A., and Adey, W. H. 2001. Reef corals, zooxanthellae, and free-living algae: a microcosm that demonstrates synergy between calcification and primary production. Ecological Engineering 16:443–457.

Smith, J. E., Smith, C. M., and Hunter, C. L. 2001. An experimental analysis of the effects of herbivory and nutrient enrichment on benthic community dynamics on a Hawaiian reef. Coral Reefs 19:332–342.

Stambler, N., Popper, N., Dubinsky, Z., and Stimson, J. 1991. Effects of nutrient enrichment and water motion on the coral *Pocillopora damicornis*. Pacific Science 45:299–307.

Steneck, R. S. 1989. Herbivory on coral reefs: a synthesis. Proceedings of the Sixth International Coral Reef Symposium 1:37–49.

Steneck, R. S., and Dethier, M. N. 1994. A functional group approach to the structure of algal-dominated communities. Oikos 69:476–498.

Stephenson, W., and Searles, R. B. 1960. Experimental studies on the ecology of intertidal environments at Heron Island. Part 1, Exclusion of fish from beach rock. Australian Journal of Marine and Freshwater Research 2:241–267.

Stoner, A. W., and Waite, J. M. 1991. Trophic biology of *Strombus gigas* in nursery habitats: diets and food sources in seagrass meadows. Journal of Molluscan Studies 57:451–460.

Szmant, A. M. 1997. Nutrient effects on coral reefs: a hypothesis on the importance of topographic and trophic complexity to reef nutrient dynamics. Proceedings of the Eighth International Coral Reef Symposium 2:1527–1532.

Szmant, A. S. 2001. Introduction to special issue, Coral Reef Algal Community Dynamics. Coral Reefs 19:299–302.

Threlkeld, S. T. 1988. Planktivory and planktivore biomass effects on zooplankton, phytoplankton, and the trophic cascade. Limnology and Oceanography 33:1362–1375.

Tilman, D., Fargione, J., Wolff, B., D'Antonio, C., Dobson, A., Howarth, R., Schindler, D., Schlesinger, W. H., Simberloff, D., and Swackhamer, D. 2001. Forecasting agriculturally driven global environmental change. Science 292:281–284.

Tomascik, T., and Sander, F. 1985. Effects of eutrophication on reef-building corals. Part 1, Growth rate of the reef-building coral *Montastrea annularis*. Marine Biology 87:143–155.

———. 1987. Effects of eutrophication on reef-building corals: reproduction of the reef-building coral *Porites porites*. Marine Biology 94:77–94.

Tomasko, D. A., and Lapointe, B. E. 1991. Productivity and biomass of *Thalassia testudinum* as related to water column nutrient availability and epiphyte levels: field observations and experimental studies. Marine Ecology Progress Series 75:9–17.

Twilley, R. R., Kemp, W. M., Staver, K. W., Stevenson, J. C., and Boynton, W. R. 1985. Nutrient enrichment of estuarine submerged vascular plant communities. Part l, Algal growth and effects on production of plants and associated communities. Marine Ecology Progress Series 23:179–141.

Verheij, E. 1993. Marine Plants of the Spermonde Archipelago, SW Sulawesi Indonesia: Aspects of Taxonomy, Floristics, and Ecology. Rijksherbarium, Leiden.

Vine, P. J. 1974. Effects of algal grazing and aggressive behavior of the fishes *Pomacentrus lividus* and *Acanthurus sohal* on coral reef ecology. Marine Biology 4:131–136.

Vitousek, P. M., Aber, J., Howarth, R. W., Likens, G. E., Matson, P. A., Schindler, D. W., Schlesinger, W. H., and Tilman, D. G. 1997a. Human alteration of the global nitrogen cycle: sources and consequences. Ecological Adaptation 7:737–750.

Vitousek, P. M., Mooney, H. A., Lubchenco, J., and Melillo, J. M. 1997b. Human domination of earth's ecosystems. Science 277:494–499.

Ward, S., and Harrison, P. L. 1997. The effects of elevated nutrient levels on settlement of coral larvae during the Encore experiment, Great Barrier Reef, Australia. Proceedings of the Eighth International Coral Reef Symposium 1:891–896.

Weiss, M. P., and Goddard, D. A. 1977. Man's impact on coastal reefs: an example from Venezuela. American Association of Petroleum Geologists, Studies in Geology 4:111–124.

Wilkinson, C. 1999. Global and local threats to coral reef functioning and existence: review and predictions. Marine and Freshwater Research 50:867–878.

Williams, S. L., and Carpenter, R. C. 1988. Nitrogen-limited primary productivity of coral reef algal turfs: potential contribution of ammonium excreted by *Diadema antillarum*. Marine Ecology Progress Series 47:145–152.

Windom, H. 1992. Contamination of the marine environment from land-based sources. Marine Pollution Bulletin 25:1–4.

Wittenberg, M., and Hunte, W. 1992. Effects of eutrophication and sedimentation on juvenile corals. Part 1, Abundance, mortality, and community structure. Marine Biology 112:131–138.

Wyse Jackson, P. S. 2002. International Review of the Ex Situ Collections of the Botanic Gardens of the World. Botanic Gardens Conservation International, Kew, Richmond, Surrey, United Kingdom.

Wyse Jackson, P. S., and Sutherland, L. A. 2000. International Agenda for Botanic Gardens in Conservation. Botanic Gardens Conservation International, Kew, Richmond, Surrey, United Kingdom.

CHAPTER 13

LAWS AND TREATIES: IS THE CONVENTION ON BIOLOGICAL DIVERSITY PROTECTING PLANT DIVERSITY?

Kerry ten Kate and W. John Kress

DURING THE LAST 20 years, a number of laws and treaties that specifically deal with the protection and conservation of plants and animals have been enacted within nations and between nations. The Endangered Species Act in the United States and the Convention on International Trade of Endangered Species of Wild Fauna and Flora (CITES) are two of the most well known. Perhaps the most far-reaching international agreement on biodiversity conservation is the Convention on Biological Diversity (CBD), which was opened for signature at the 1992 Earth Summit in Rio de Janeiro. As of 2004, 188 countries and the European Union have ratified the CBD. Article 1 establishes the treaty's objectives: the conservation of biological diversity, the sustainable use of its components, and the fair and equitable sharing of the benefits arising from access to genetic resources. The next 18 articles set out the parties' obligations on issues ranging from taxonomy to biotechnology (table 13.1; see the Convention on Biological Diversity Web site).

The CBD is constantly evolving. Since it was opened for signature in 1992, the parties have made various important commitments to implementation of its articles through the Subsidiary Body on Scientific, Technical, and Technological Advice (SBSTTA) and the primary political decision-making body, the biennial Conference of the Parties (COP). As of 2004 there have been nearly 240 COP decisions and SBSTTA recommendations, two further treaties (on biosafety and on agricultural plant genetic resources), and 23 rosters of experts. The parties have adopted the Global Taxonomy Initiative, Guiding Principles on Invasive Alien Species, Global Strategy for Plant Conservation, and Guidelines for the Sustainable Use of Biodiversity, among many other science-based tools. National biodiversity strategies required by the CBD have been undertaken in 150 countries, and the CBD's financial mechanism, the Global Environment Facility, has allocated billions of dollars to developing countries to implement the CBD (ten Kate 2002).

Despite these achievements, the CBD, unlike its climate-change cousin, generally does not contain timetables and targets. Two developments at the sixth COP, which took place in The Hague in April 2002, and reinforced at the seventh COP held in Kuala Lumpur, Malaysia, in 2004, have helped. First, the parties agreed on

Table 13.1 Key provisions of the Convention on Biological Diversity

Article 1	Objectives: conservation of biological diversity, sustainable use of its components, and the fair and equitable sharing of benefits arising from the use of genetic resources
Article 6	General measures (national strategies)
Article 7	Identification and monitoring
Article 8	In situ conservation
Article 9	Ex situ conservation
Article 10	Sustainable use of components of biological diversity
Article 12	Research and training
Article 13	Public education and awareness
Article 14	Impact assessment and minimizing adverse impacts
Article 15	Access to genetic resources
Article 16	Access to and transfer of technology
Article 17	Exchange of information
Article 18	Technical and scientific cooperation (and the clearinghouse mechanism)
Article 19	Handling of biotechnology and distribution of its benefits

an ambitious but achievable strategic plan for the CBD, with the purpose of substantially reducing the rate of loss of biodiversity by 2010. Second, the adoption of the Global Strategy for Plant Conservation, with its 16 targets, established a helpful precedent for quantified targets at the international and national levels (see table 12.1). Furthermore, at the seventh COP, three major new programs of work under the CBD, on technology transfer, protected areas, and mountain biodiversity, were adopted.

If it is to be successful, the CBD's fundamental contribution to science must be to conserve the resource base for life sciences (and life itself): namely, biological diversity. Treaties are agreements between states, but by and large it is organizations, and not governments, that are equipped to do the work. The CBD offers a bargaining chip to universities, research institutes, commercial enterprises, and communities. In return for helping governments achieve their commitments, scientific organizations can participate in national policymaking, raise their own political profiles, derive a fresh mandate and renewed legitimacy for their work, and perhaps use the CBD as a lever to help fund their work. As individual countries implement CBD work programs, apply COP guidelines, and execute national strategies, the influence of the CBD on science is likely to grow. One mechanism will be the allocation of public funding, another the advent of laws and policies that control the direction and methodologies of scientific research.

A significant example of policy that is already influencing science is the regulation of access to genetic resources and benefit sharing. It is a fact that the world's biodiversity is found in roughly inverse proportion to its scientific and technological capacity (Macilwain 1998). Behind the measured text of the CBD lies the hope of a "grand bargain" (Gollin 1993), in which biotechnology-rich countries compensate biodiversity-rich countries for access to their genetic resources, leading to valuable new products, creating sustainable livelihoods for people living near centers of biodiversity, boosting the provider country's gross national product, and paying for conservation along the way. The CBD sets out provisions according to which states should regulate access to genetic resources. In response to the CBD, some 100 countries, largely those that are home to the bulk of the world's biodiversity, have introduced or are now considering laws that regulate access by scientists to genetic resources, biochemicals, and associated traditional knowledge. These laws typically require national and foreign scientists alike to obtain permission for access to biodiversity and to work with partners from the countries providing the genetic resources, in the process sharing benefits such as royalties, technology, joint research, and information.

As defined in the CBD, "genetic resources" are any material of plant, animal, microbial, or other origin containing functional units of heredity of actual or potential value. Access to this significant chunk of life (humans are excluded) is vital for education and research in the life sciences, as well as for research on the conservation and sustainable use of biodiversity. Access also underpins commercial discovery and development. Global sales of products derived from genetic resources (pharmaceuticals, botanical medicines, major crops, horticulture, crop protection products, cosmetics and personal care products, and a broad range of biotechnologies) in 1999 ranged from US$500 billion to $800 billion (ten Kate and Laird 1999).

The CBD seeks to balance the sovereignty and the authority of national governments with the obligation for states to facilitate access to genetic resources for environmentally sound purposes. Access is to be subject to governments' prior informed consent on terms, mutually agreed by the provider and recipient, that promote the fair and equitable sharing of benefits. Similarly, subject to national law, access to the knowledge, innovations, and practices of indigenous and local communities requires the prior approval of the holders of that knowledge.

Scientists, communities, government, and business have watched the development of international policy and national laws in this field with interest to see whether the goals of facilitating science, respecting rights, and ensuring fairness have been achieved. Overall, partnerships are indeed becoming fairer. Biological samples and the rights to use genes and compounds have been exchanged, sometimes under agreements, for decades. In the wake of the CBD, benefit-sharing agreements are increasingly common. Most benefits have flowed to local scientific institutions in the form of training and technology.

The story is not one of unalloyed success, however. First, commercial demand for access is unreliable. Over the past 30 years, interest in accessing biodiversity for pharmaceutical development has been cyclical. In many sectors, research dollars are flowing out of natural products and into synthetic chemistry for rational drug design, combinatorial approaches, and genetics that focus largely on human material. A goal in many national biodiversity strategies is to help alleviate poverty, to support sustainable livelihoods, and to raise living standards. Countries might do well to use the untapped potential for research on genetic resources to meet domestic needs, for example, through low-cost botanical medicines, rather than seeking only to supply fickle international markets. They could also ensure that regulations distinguish between commercialization and the more steady demand for access for vital conservation research in fields such as ecology and systematics.

Second, benefits are not always forthcoming to countries facilitating access to genetic resources. Much genetic material used for research and development is obtained from collections made before the CBD entered into force, for which there are generally no benefit-sharing arrangements. Any benefits that are negotiated rarely "trickle down" to local communities or to conservation. Scientific organizations tend to benefit most, although in a knowledge-based sense rather than an economic one. Countries could require a certain proportion of benefits to be dedicated to conservation, as Costa Rica and Western Australia have done. Countries could also adapt growing experience with trust funds and other mechanisms to ensure that local people benefit and have an incentive to support conservation measures.

Third, evidence is growing that the anticipated bureaucracy, delay, and expense of compliance with the first wave of access laws have deterred foreign and domestic scientists and thus have unwittingly stifled not only commercial research but also essential conservation work. Confusion over which government bodies are authorized to grant access to genetic resources has not helped.

Encouraging ongoing developments, however, provide a growing acknowledgment of the need for a more strategic and flexible approach. In 2002, the parties to the CBD adopted the Bonn Guidelines on Access and Benefit-Sharing (see Convention on Biological Diversity Web site). These recommendations provide guidance to countries in the development of law and policy on access to genetic resources, associated traditional knowledge and benefit sharing, and to stakeholders such as university researchers, companies, and communities in the negotiation of access and benefit-sharing agreements. The guidelines encourage countries to take a strategic and flexible approach. They set out provisions on prior informed consent and mutually agreed terms and list key elements of the roles and responsibilities of countries and organizations as they provide and use genetic resources and associated traditional knowledge. They also contain suggestions on administrative aspects such as a national focal point for each country and potential functions of any competent national authorities established by governments to regulate access.

The guidelines encourage the participation of local end users of biodiversity, which, in the field of health care, might include health research organizations and pharmaceutical companies, as well as traditional healers and indigenous and local communities who provide knowledge associated with genetic resources. Appendixes to the guidelines suggest what to include in the material-transfer agreements commonly used to access and exchange genetic resources and typical monetary and nonmonetary benefits.

An important recent development on access and benefit sharing of plant resources, relevant to crop development but not to pharmaceuticals, is the International Treaty on Plant Genetic Resources for Food and Agriculture (IT). This treaty was finalized in Rome in 2001, in harmony with the CBD, and entered into force on

Table 13.2 Principles on Access to Genetic Resources and Benefit-Sharing endorsed by participating institutions

1. **Convention on Biological Diversity (CBD) and laws related to access to genetic resources and associated traditional knowledge and benefit sharing**
   *Honor the letter and spirit of the CBD, the Convention on International Trade in Endangered Species of Wild Fauna and Flora (CITES), and laws relating to access and benefit sharing, including those relating to traditional knowledge.

2. **Acquisition of genetic resources**
   *In order to obtain prior informed consent, provide a full explanation of how the genetic resources will be acquired and used.
   *When acquiring genetic resources from in situ conditions, obtain prior informed consent from the government of the country of origin and any other relevant stakeholders, according to applicable law and best practice.
   *When acquiring genetic resources from ex situ collections (such as botanical gardens), obtain prior informed consent from the body governing the ex situ collection and any additional consents required by that body.
   *When acquiring genetic resources from ex situ sources, whether from ex situ collections, commercial sources, or individuals, evaluate available documentation and, where necessary, take appropriate steps to ensure that the genetic resources were acquired in accordance with applicable law and best practice.

3. **Use and supply of genetic resources**
   *Use and supply genetic resources and their derivatives on terms and conditions consistent with those under which they were acquired.
   *Prepare a transparent policy on the commercialization (including plant sales) of genetic resources acquired before and since the CBD entered into force and their derivatives, whether by the participating institution or a recipient third party.

4. **Use of written agreements**
   *Acquire genetic resources and supply genetic resources and derivatives using written agreements, where required by applicable law and best practice, setting out the terms and conditions under which the genetic resources may be acquired, used, and supplied and resulting benefits shared.

5. **Benefit sharing**
   *Share fairly and equitably with the country of origin and other stakeholders, the benefits arising from the use of genetic resources and their derivatives including nonmonetary and, in the case of commercialization, monetary benefits.
   *Share benefits arising from the use of genetic resources acquired prior to the entry into force of the CBD, as far as possible, in the same manner as for those acquired thereafter.

Table 13.2 *(continued)*

6. Curation

In order to comply with these principles, maintain records and mechanisms to

*Record the terms and conditions under which genetic resources are acquired.

*Track the use in the participating institution and benefits arising from that use.

*Record supply to third parties, including the terms and conditions of supply.

7. Prepare a policy

*Prepare, adopt, and communicate an institutional policy setting out how the participating institution will implement these principles.

29 June 2004. One of its important elements is a multilateral system for facilitated access to 35 crop genera and 29 forage species in the public domain for "food and agriculture" and associated benefit sharing, through the exchange of information, access to and transfer of technology, capacity building, and a commercial benefit-sharing package. "Food and agriculture" explicitly excludes "chemical, pharmaceutical and/or other nonfood/feed industrial uses." The IT requires seed companies to pay royalties on patented products derived from the genes accessed. In addition, the International Agricultural Research Centres of the Consultative Group on International Agricultural Research are invited to enter into agreements with the governing body of the IT concerning access not only to Annex I materials, but to other materials in their extensive collections (see Commission on Genetic Resources for Food and Agriculture Web site).

Progress has also been achieved by a range of individual companies, professional associations, gene banks such as botanical gardens, and indigenous communities' groups that have developed institutional policies in line with the CBD that provide principles and practical guidance for their employees and associates (Laird 2002). An example is the Principles on Access and Benefit-Sharing for participating institutions, in which 28 botanical gardens and herbaria from 21 countries, led by the Royal Botanic Gardens, Kew, developed common standards on access to genetic resources and benefit sharing (table 13.2; Latorre et al. 2001; Royal Botanic Gardens). As a result of these standards the Limbe Botanic Garden in Cameroon and a range of other institutions worldwide working with indigenous peoples and local communities have endorsed these principles, and then developed in more detail their own guidelines for staff to translate them into action (Laird and Mahop 2001; People and Plants International 2004). Practical tools such as material-transfer agreements, steps for obtaining prior informed consent, guidelines for the preparation of strategies, and illustrative examples of benefit-sharing arrangements are now available (ten Kate and Wells 2001; Convention on Biological Diversity).

The above discussion demonstrates how the principles of the CBD are finding their way into national laws and policies and into the working practices of scientists. Scientists can and do participate in guiding and developing international and national law. However, scientists can have more input to the treaty by lobbying or joining national government delegates, by participating as nongovernmental organizations at meetings, and by serving on expert panels. Scientific organizations can become accredited and attend meetings of the CBD's COP and SBSTTA. Such participation is vital to ensure that the treaty is based on sound science and promotes, rather than hinders, conservation. Scientific organizations can also participate in the implementation and evolution of the Bonn Guidelines on Access and Benefit-Sharing (e.g., by endorsing the principles for participating institutions), so that regulations on access to genetic resources worldwide facilitate science and support fair partnerships. The knowledgeable and rational voices of biologists need to be heard on the broad range of scientific and technical issues covered by the CBD. Finally, scientific organizations should also work with federal, state, and local governments to ensure a coordinated approach for consistent decisions in the broad range of environmental agreements under the auspices of the United Nations, as well as with commercial issues in the World Trade Organization.

In conclusion, the human assault on the earth's natural environments was universally acknowledged by the international representatives at the Rio Earth Summit as one of the gravest issues facing the international community. The CBD marked a turning point in the understanding, conservation, utilization, and sharing of the planet's biological riches. Its entry into force in 1993 was a pivotal event in ensuring that governments and members of civil society take responsibility for the quality of the air we breath, the water we drink, and the conservation of the plants and animals of the earth. The concerted effort of scientists, communities, and corporations, as well as governments, will be needed to achieve the vital goal of the CBD's strategic plan: namely, to halt the loss of biodiversity.

ACKNOWLEDGMENTS

This chapter draws substantially on an earlier article by the first author (*Science* 295 [29 March 2002]: 2371).

LITERATURE CITED

Commission on Genetic Resources for Food and Agriculture. News and Events. http://www .fao.org/ag/cgrfa/News.htm.
Convention on Biological Diversity. Secretariat of the Convention on Biological Diversity. http://www.biodiv.org.
Gollin, M. A. 1993. An intellectual property rights framework for biodiversity prospecting. In Reid, W. V., Laird, S. A., Meyer, C. A., Games, R., Sittenfeld, A., Janzen, D. H., Gollin, M. A., and Juma, C., eds., Biodiversity Prospecting: Using Genetic Resources for Sustainable Development. World Resources Institute, Washington, DC.

ten Kate, K. 2002. Science and the Convention on Biological Diversity. Science 295:2371.

ten Kate, K., and Laird, S. 1999. The Commercial Use of Biodiversity: Access to Genetic Resources and Benefit-Sharing. Commission of the European Communities and Earthscan Publications, London.

ten Kate, K., and Wells, A. 2001. Preparing a national strategy on access to genetic resources and benefit-sharing. A pilot study. Royal Botanic Gardens, Kew, and UNDP/UNEP Biodiversity Planning Support Programme.

Laird, S. A., ed. 2002. Biodiversity and Traditional Knowledge: Equitable Partnerships in Practice. People and Plants Conservation Series. Earthscan Publications, London.

Laird, S. A., and Mahop, T. 2001. The Limbe Botanic and Zoological Gardens Policy on Access to Genetic Resources and Benefit-Sharing, Limbe, South West Province, Cameroon. Limbe Botanical Garden, Limbe, Cameroon.

Latorre García, F., Williams, C., ten Kate, K., and Cheyne, P. (based on contributions from 36 individuals from 28 botanic gardens and herbaria from 21 countries). 2001. Results of the Pilot Project for Botanic Gardens: Principles on Access to Genetic Resources and Benefit-Sharing: Common Policy Guidelines to Assist with Their Implementation and Explanatory Text. Royal Botanic Gardens, Kew. http://www.rbgkew.org.uk/conservation/argbs-policy.html.

Macilwain, C. 1998. When rhetoric hits reality in debate on bioprospecting. Nature 392:535–541.

People and Plants International. 2004. http://www.rbgkew.org.uk/peopleplants/.

Royal Botanic Gardens, Kew. Conservation and Wildlife. http://www.rbgkew.org.uk/conservation.

CHAPTER 14

Conservation activists come from all walks of life. Some are academics and professionals with many years of training in biology and ecology, while others have little formal education in conservation biology. Yet they all share a strong desire to take action on local conservation issues. These activists meet at local gatherings with the purpose of protecting the natural habitats and native plants of their communities. These meetings range from natural history clubs to more visible protests at controversial construction sites. Each activist, like each conservation biologist, has a personal view as to what should be protected and what is the best method of protecting the remaining native vegetation on our planet and in our neighborhoods. This chapter was written from the perspective of one person, a botanist with a long history of involvement in native plant societies, concerning the role that individuals can play in protecting biological diversity at the local level. This essay traces the history of native plant societies, debates the distinction between native versus alien species, and emphasizes the inevitable need for a basic change in lifestyle if the long-term conservation of nature is to succeed.

GRASSROOTS CONSERVATION

Stanwyn G. Shetler

GRASSROOTS CONSERVATION is a broad subject for a single chapter. I can only touch on a few topics, and in treating them I have confined myself mainly to the American scene and my own experience. Accordingly, the latter half in particular is largely a personal essay, drawing from my previous writing (Shetler 1991a, 1991b, 2003) and offering what I hope are useful if at times controversial views, intended as perspectives, not prescriptions. This is not a documented analysis of all sides of the issues raised, which are the subjects of a voluminous literature.

I have lived and worked in the Washington, DC, area for more than 40 years, and during that time I have fought on behalf of the environment in numerous local preservation battles, involving cases in the District of Columbia and both Maryland and Virginia, including all their counties adjacent to DC. I have conducted field inventories, prepared position papers, published articles, briefed officials, testified before various bodies including courts, written letters, even marched. The cases have ranged from defending habitats (from neighborhood greenspace to bogs, swamps, and shale barrens) to protesting tree poaching and testifying against a county weed ordinance in court on behalf of a developer who wanted to leave a bottomland unmowed despite objections from some nearby homeowners.

As a charter member of the Virginia Native Plant Society (VNPS), founded in 1982 (as Virginia Wildflower Preservation Society), and a member of the state board of directors since 1996, I have firsthand knowledge of its development and program (for an overview of its program, see the VNPS Web site). Thus, my section on the role of native plant societies, which is of a general nature, is based on my own experience, augmented by a review of newsletters and Web sites of other state societies and e-mail correspondence with some of the societies (see acknowledgments, and Web sites in Literature Cited). I also served for about a year as the Smithsonian Institution's first representative to the Native Plant Conservation Initiative (now Plant Conservation Alliance).

THE ROLE OF NATIVE PLANT SOCIETIES

The New England Wild Flower Society (NEWFS), founded in 1900 as the Society for the Protection of Native Plants, claims to be the oldest institution in the United States dedicated to the conservation of wild plants, which makes it the oldest native plant society (DeKing et al. 2001). It was soon followed in 1902 by the Wild Flower Preservation Society of America, formed at the New York Botanical Garden, which became the Wild Flower Preservation Society, Inc., in 1925, when its operation moved to Washington, DC, and which continued until about 1974 (Walker 2003).

Today, across North America there are numerous native plant societies, under one name or another, at the regional, state, and local levels. (For addresses and Web sites of the many state and provincial societies, see NANPS 2002; Walker 2002; VNPS 2002b). There is also a continentwide organization, the North American Native Plant Society (NANPS), founded in 1984 as the Canadian Wildflower Society.

The mission of every native plant society might be characterized as a variation of the mission of the NEWFS. This mission, as stated in the masthead of the 100th anniversary issue of *New England Wild Flower*, is "to promote the conservation of temperate North American plants through education, research, horticulture, habitat preservation, and advocacy" (DeKing et al. 2001).

The rise of the modern native plant society in North America is largely a phenomenon of the latter part of the 20th century. The cause that began with the NEWFS more than 100 years ago has come into full bloom only in the native plant movement of the last 35 years or so. Statewide societies have been established in all but a few states. The North Carolina Wild Flower Preservation Society, founded in 1951, was one of the first of these. The largest and probably the best known, the California Native Plant Society, was founded in 1965. Many were formed in the 1980s, and new ones continue to be formed. One of the newest is the Native Plant Society of British Columbia, officially founded in 1997. Of the few states without a statewide society, New York is perhaps the most conspicuous, but it does have a number of native plant organizations within the state. Canada has several provincial native plant societies, which were chapters of the Canadian Wildflower Society before it became the NANPS.

In some respects, native plant societies represent a contemporary expression of the time-honored traditions of academic botanical societies, broadened and popularized to appeal to a wider audience of professionals, amateurs, and lay public. Many botanical societies, which exist to promote serious study of floras of all types near and far and plant biology generally, have always welcomed and attempted to serve the interests of both amateurs and professionals. Examples are the New England Botanical Club, Michigan Botanical Club, and Botanical Society of Washington (DC).

Yet native plant societies also represent something quite different—a movement to champion *all* native species, not just the beautiful or showy ones, and *only* native species, and to enlist in this cause, in addition to natural constituents, the many casual wildflower enthusiasts. Many members of the gardening, horticultural, and landscape communities who are interested in using native plants have been drawn into the movement. The twofold aim is to transform the traditional, popular interest in wildflowers into an ardent concern for the whole native flora in the wild, not just in the garden, and at the same time to draw attention to the naturalized, alien element in the flora at large, which often is pestiferous and invasive.

The change in emphasis from wildflowers to native plants has sometimes been reflected in a name change, as when the Virginia Wildflower Preservation Society, founded in 1982, became the Virginia Native Plant Society in 1989 and took on a broader agenda. This change was not without controversy at the time. Nowadays, however, the lines between botanical, wildflower, and native plant societies are blurred, with the latter two being essentially the same. All stress the need to protect and conserve the native flora. The native plant societies seem to be most effective when, like the more academic botanical societies, they have a strong scientific orientation and motivation.

Already in 1900, when the NEWFS was formed, there was concern for the future of the native wildflowers, because it was the practice of the day for commercial florists to gather and sell flowers from the wild (DeKing et al. 2001). Something comparable to the nascent Audubon movement, which over time became a powerful force for the conservation of birds and other animals, was needed for the plants. Such a force for plants has been very slow in coming, despite the early example of the NEWFS.

Since World War II, but especially in recent years, development has been rampant in many places across North America, causing a growing alarm not only in the scientific community but also among the public at large about the ever accelerating destruction and fragmentation of habitat and the consequent extirpation of the native plant communities and flora. This deep concern has sparked a remarkable conservation movement, a major expression being the native plant societies. In bridging the gap between the professional botanical societies and the garden or horticultural clubs, they have made the conservation of biodiversity an issue in gardening as well as botanical circles.

Native plant societies vary greatly in size, staffing, and budget. Most state societies have fewer than 1,000 members and some fewer than 300. Typically, staffing is all by volunteers, but a few have one or more full-time, part-time, or contract employees. The annual budget is usually no more than about US$50,000 and often much less. Some state societies operate at a much higher budget level, however. The Washington (State) Native Plant Society, for instance, had a budget of US$145,000 in 2002, thanks in part to grants. Most maintain a statewide presence through chapters; the number of chapters varies from none to more than 30 but is usually less than 15. Chapters are societies within a society and, except where there is paid central staff, are largely responsible for carrying out the society's mission at the local level statewide.

The California Native Plant Society, like the NEWFS, is in a league of its own. In 2002, it had 10,000 members, 32 chapters, a budget of US$800,000, and 14 full-time or part-time, paid staff. The NEWFS had 6,000 members in six state chapters, 24 full-time and part-time employees, and an annual budget of US$2.3 million in 2002. Its operation includes the world-famous Garden in the Woods. The relatively young NANPS, with over 1,000 members, is small by these standards, but growing.

CONSERVATION PROGRAMS

If there is a single purpose motivating and uniting all native plant societies, it is conservation. All the other areas of their mission—education, research, horticulture, advocacy, publication, and so forth—ultimately are directed toward advancing the protection and conservation of the native flora. This is accomplished through the direct action of the members and, indirectly, by educating the public at large and thereby arousing general concern and action. What follows is a brief discussion of some of the most common conservation-directed programs, not necessarily in priority order.

Most societies lack the resources to take on a conservancy role, that is, to purchase land outright or accept gifts of land to preserve in perpetuity. Consequently, efforts to save habitat mostly take the form of lobbying and testifying and of educational activities and publications to achieve preservation through the agencies of government, charitable nongovernmental organizations, and philanthropic corporations and individuals. The preservation aims of native plant societies closely parallel those of the Natural Heritage programs in the states, initially established by the Nature Conservancy, and strong links may develop at the state level.

Gifts and easements may be brokered between owners and land trusts, and there are voluntary programs to preserve natural heritage, such as the VNPS's registry of Virginia native-plant sites. In this program, public or private landowners of habitats with regional or state significance for their native plants agree to register their sites. Registry neither preserves nor guarantees access to the sites, but flags a natural heritage feature and thereby conveys a degree of recognition and visibility that, it is hoped, brings an important site a big step closer to preservation. Sites that already are protected at least to some degree may also be registered.

Plant rescue is a very common activity, even sometimes a driving force, of native plant societies. In the usual situation, plants that are in harm's way because of imminent habitat destruction, such as for commercial and residential development or road construction, are rescued and transplanted to a protected location. This might be elsewhere in the wild, a native plant garden operated or sponsored by the society, a public park, or even the gardens of members. Some question the purpose and value of rescues. Often the species rescued are only the showiest species, and they may not be particularly rare or threatened. What to do with the plants after they have been rescued can be controversial, particularly if they are transplanted to wild sites where the species have never occurred naturally. Developers may welcome rescues as a justifiable alternative to in situ protection. The VNPS has deemphasized plant rescue in recent years and is developing a policy and guidelines for rescues.

Few issues have energized the native plant societies in recent years as much as the growing scourge of invasive alien plants in the natural landscape, and attention continues to increase. Many societies have set up special projects to address the issues and have taken steps to educate the nursery and seed trades and agricultural interests in general, as well as the public at large, about known and potential invasive alien plants, and to lobby in various ways for voluntary and legal restraints on the sale and planting of such species. In particular, the planting of invasives along highways, along rights-of-way, and in other reclamations has been a target.

Efforts may be coordinated with government agencies. The cooperative Invasive Alien Plant Project of the VNPS and the Virginia Department of Conservation and Recreation is one such example (see VNPS 2002a). It has produced a list of known invasives and a set of widely used fact sheets, and a representative of the VNPS, using a slide show, has taken the message to many audiences. The VNPS also was effective recently in getting the state to prohibit the sale and distribution of purple loosestrife (*Lythrum salicaria*).

At the local level, "pulling parties" and other eradication actions often are organized, providing park authorities and other government or private agencies with ready volunteer brigades of willing hands. The Internet has greatly multiplied the power of organizers to mobilize the members, even on short notice, for these and such other conservation efforts as plant rescues, letter writing, or testifying at public hearings. Scarcely a day goes by without one or more such appeals appearing in my own e-mail.

Gardening interests strongly influence the activities of the native plant societies. Thanks in large part to the continued growth of these societies, the popularity of gardening and landscaping with native plants is growing, as more and more people learn of the beauty and practicality of native plant gardens.

Native plant gardens are by no means new. Probably the most famous wildflower garden in North America is the Garden in the Woods in Framingham, Massachusetts, started in the 1930s by two dedicated native plant gardeners and deeded to the NEWFS in 1965 (DeKing et al. 2001). Some societies have promoted native specialty gardens (e.g., butterfly gardens). These demonstration gardens are invaluable in educating the public.

Many societies or their local chapters sponsor native plant sales, and these sales are very popular with the members and the public, often providing the most visible face of the society to the public and thereby constituting the prime recruiting opportunity and chief fund-raising activity for the society. Sale events may involve nurseries as well, as in the Pacific Northwest native plant sale, said to be one of the largest of its kind in the region, which is cosponsored by the Native Plant Society of British Columbia (see the Native Plant Society of British Columbia Web site).

The use of native plants for roadside and wayside beautification projects is usually encouraged. An early champion of this among native plant organizations was the Lady Bird Johnson Wildflower Center in Austin, Texas, founded in 1982. The Garden Club of America and the Federal Highway Administration have also promoted the practice in recent years.

Conservation education takes many forms, including field trips, tours, conferences, workshops, classes, school programs, publications, and Web sites. Field trips, essential to native plant education, are a feature common to the programs of all the societies. The standard means of society communication is a newsletter, which carries, in addition to calendars and event announcements, short articles on native plants, relevant topics such as invasive plants, and society programs. Some of the larger societies also publish a journal or magazine with substantial articles on botanical topics, including research results (e.g., *Fremontia* of the California Native Plant Society). Some publish fact sheets on various topics. At least two societies, the Kansas Wildflower Society ("2001 Kansas wildflower" 2001) and the VNPS, highlight a species each year as the "wildflower of the year." The VNPS's selections are featured in leaflets and on its Web site. Many of the societies now have a Web site and are making increasingly sophisticated use of the Web to market themselves and provide serious information content, including photo and other illustration galleries.

The native plant movement has given rise to a number of regional conferences that meet each year. Chief among them is the Cullowhee Native Plant Conference at Western Carolina University in Cullowhee, North Carolina. It has been meeting since the mid-1980s and has spawned other such conferences. It has been especially important in promoting native plants in landscaping. The renowned Wildflower Pilgrimage in the Great Smoky Mountains, in its 54th year in 2004, is one of the oldest and longest-running events held annually to promote native plants.

Advocacy, whether at the local, state, or federal level, is one of the most important activities of native plant societies, and it takes many forms—direct lobbying of officials and legislators, testifying before legislative bodies, writing letters and position papers, joining protests, even legal action. Society officials speak for their organizations and call for members to get involved as well. Members are kept energized for action through timely information and alerts on critical pending issues in newsletters, on Web sites, and by e-mail.

The whole native plant movement has been a call to arms not just to save indigenous plants and their habitats but also, in this struggle for nature, to gain parity for the plants with the animals, both in the public mind and especially in governmental attitudes, regulations, and expenditures. Although there is still a long way to go in this crusade, much progress has been made, thanks in no small measure to the native plant societies. A number of regional and national alliances have been formed to intensify the effort.

In 1991, the NEWFS founded the New England Plant Conservation Program. "More than 65 individuals and organizations were enlisted to form an organized regional plant conservation effort" (Mehrhoff 2001). At the national level, the Plant Conservation Alliance, founded in 1994 as the Native Plant Conservation Initiative, "is a consortium of ten federal government Member agencies and over 145 non-federal Cooperators [individuals, organizations, agencies] . . . within the conservation field" (Plant Conservation Alliance 2004). Recently, the California Native Plant Society and the Center for Biological Diversity initiated the Native Plant Conservation Campaign, "to assemble a national network of native plant societies, botanical gardens, and other plant conservation organizations that will collaborate . . . to create a strong national voice to advocate for native plant conservation" (California Native Plant Society 2002). Many state native plant societies have joined the campaign.

The total impact of the advocacy actions taken by native plant societies over the years is hard to gauge, but certainly it has been enormous. I am sure that every society, not just the largest ones, could cite significant victories. The Native Plant Society of Oregon, for instance, states that without its tireless efforts in the mid-1980s Oregon would not have an endangered species act.

Research is part of the mission of most native plant societies, but the nature and extent of actual research, if any, varies greatly among the societies. Support for research takes many forms, from advice and encouragement to financial support, volunteer participation, leadership, and publication of results. Floristic and ecological surveys are the most common type of research. In at least three states, Arkansas, Oregon, and Virginia, the societies are actively supporting new state flora projects (see Arkansas Native Plant Society 2000; Native Plant Society of Oregon 2002; VNPS Web site).

The Washington Native Plant Society, for example, has supported "numerous" research projects with society funds. Large societies are able to raise sufficient money or obtain grants and contracts to hire staff and manage or even conduct in-house research projects. The California Native Plant Society's program includes both floristic and ecological research. Especially noteworthy are its local floras and its *Inventory of Rare and Endangered Plants of California*, published by the society and in its sixth edition as of August 2001 and available online. The Lady Bird Johnson Wildflower Center has been a leader in conducting research and providing information on landscape restoration with native plants.

NATIVES AND ALIENS (WHEN NATIVES BECOME ALIENS)

What is a "native"? Everything hinges on the meaning of this term, which is not so simple, because a plant can be native on one level and not another (e.g., continent but not region, region but not state, state but not county, county but not locality, locality but not site). Most would subscribe to a definition such as the one used by the Plant Conservation Alliance: a species "that occurs in a particular region, ecosystem, and habitat without direct or indirect human actions" (attributed to J. Kartesz and L. Morse; Alien Plant Working Group 2000). For North America, European settlement is usually regarded as the starting point, and thus a species is native to North America only if it occurred somewhere on the continent in pre-Columbian times. As for its present distribution, a corollary is that it can be native to parts of the continent other than its presettlement or historical range only if in the meantime it has migrated there without any kind of human assistance; and it is *non*native (alien) wherever on the continent it has been *moved* by human agency even, I would add, if *inside* this natural range.

As a matter of process, what is the difference between moving a plant outside its historical range and moving it around inside its historical range? Is a species, even a very rare, localized species, to be regarded as automatically native at any suitable habitat inside its historical or current natural range, regardless how it gets there? Put another way, is any *individual* of a species automatically native anywhere it will grow within the historical or current natural range of the species, even if planted? If so, one can justify various dubious practices, for example, transplanting species from one rare habitat such as a bog to another until all those habitats of a given area have the same flora. It also follows that introducing a native species anywhere within its defined natural range, that is, homogenizing its distribution within its range, would always be a proper thing to do and more justifiable than planting an alien species, regardless of reason.

Much is overlooked, perhaps necessarily for practical purposes, by ignoring the individual and focusing almost exclusively on the species, ultimately an abstraction, in defining the terms and boundaries of nativeness. The only tangible unit of distribution is the individual plant, and thus the only tangible basis for making an absolute judgment rather than a relative one between native and nonnative plants is the distribution of individuals. Only individuals provide the phytogeographer with dots on a map.

As I see it, however, a native can become an alien in its own land. Every plant sowed or transplanted by someone is an introduction and thus, in a sense, becomes an alien or exotic, regardless of its source. Deliberate introduction, by definition, makes aliens of otherwise native plants. It is not the distance from the source that determines what is alien, but the act of planting. The phytogeographer, in plotting and explaining plant distributions, must be able to rely on the authenticity of the individual records and the relative naturalness of the landscape. In its natural setting, each individual plant, population, or habitat has its own history to tell, which, however, is easy to falsify by human introduction or disturbance. The very act of transplanting or sowing in some measure, large or small, falsifies the history of plant migration and establishment in the area. Those who use native species in landscaping should always recognize, therefore, that they are putting together artificial landscapes, simulating but not creating natural ones. Just as a zoo is a zoo no matter what species it contains, a garden is a garden no matter what it contains.

To be sure, human disturbance and anthropogenic effects on plant distribution, past and present, are virtually everywhere in North America. In any original or ideal

302 | *Stanwyn G. Shetler*

sense, therefore, there is precious little truly "natural" or "wild" landscape left. But we have to deal with what we have, which means looking at the remaining fragments of habitat, large or small and even if pale versions of what they once were, as in some sense still natural or wild by contrast with highly developed areas. This is especially true in rapidly urbanizing areas. To do otherwise is to give up on conservation.

Human activity since well before European settlement has been falsifying plant distribution across North America in numerous indirect as well as direct ways, such as destroying native habitats and with them their native plants, changing habitats so as to allow native species to grow where they did not previously occur, blocking natural dispersal routes, and introducing nonnative animals that provide whole new avenues for native plant dispersal. All this said, it does not negate the fact that plant introductions of our own doing, whether of native or alien species, are under our own direct control.

Ironically, from the point of view of the phytogeographer, natives transplanted outside of gardens and other formal landscapes are much more difficult to detect as introductions than obvious exotics that clearly stand out. What appear to be "natural" dispersal events are in fact artificial ones. This is not to defend naturalized species but to point out the relativity of the terms *alien* and *introduction*. While recognizing the particular scourge of exotic invasives and the need to control them, why, one might ask, is it more acceptable to play Johnny Appleseed with native introductions than with exotic introductions? From a process standpoint, a plant that arrives at a local site by natural means of dispersal without direct human introduction, whether a native or an exotic species, could be seen as a more "natural" element in the local landscape than one that is planted. Admittedly, *any* dispersal in the present-day environment will be affected by past or present indirect anthropogenic forces of one kind or another.

THE FUTURE OF NATIVE PLANT SOCIETIES

After more than a century of increasing activism, the native plant societies and the movement they have sparked have had a very large and often crucial impact in the cause of conserving North America's natural habitats and indigenous flora. Their members compose an ever growing army of grassroots conservationists who daily are making a difference at the "frontlines." But will the strategies and solutions of the past be adequate to meet the infinitely greater conservation challenges of the new century? I see our societies at a crossroad of sorts with a need to sharpen our primary aims and avoid letting subordinate activities crowd out the main mission.

Surely the most urgent task is to save native flora *in the wild*, which means wherever native populations of native species are still clinging to native habitats, no matter how compromised they may be. Saving species always comes down to saving the places where they grow, and thus one might say that the three basic rules for the preservation of the native flora are save habitat, save habitat, save habitat! The native plant societies and their various alliances should face this task with singular focus, and their prime energies should be invested in accomplishing it. The best ways to do this are (1) by directly facilitating preservation through purchases, gifts, and activism, and (2) by educating all levels of society, but especially the children and the leaders and officials, about the values of natural areas and native species in situ.

Horticulture, it can be argued, is a subordinate focus that can be very distracting. Many, perhaps a majority, of the members of native plant societies are drawn in through their gardening interests, and native plants become a new outlet for expressing these interests. As a consequence, horticultural perspectives can come to dominate the actions of the society. These perspectives may be at odds with purely botanical perspectives and lead to conflicting objectives and a mixed message to the public.

Gardens of all types, using "garden" in the broadest sense to include the whole realm of agriculture and horticulture, are an essential part of human existence, and in today's world it is appropriate *within the planted or formal landscape* that native plant societies encourage gardening and landscaping with native plants. Native plant gardens, such as the Garden in the Woods in New England, are not only beautiful and satisfying accomplishments but, as already indicated, extremely valuable places for demonstration and teaching.

Gardens are gardens, and planted landscapes are planted landscapes, however. *All* plants in them except the unplanned "weeds," native or alien, are introduced. Native plant gardens are fine, but gardening is gardening. A landscape planted with native species is just as much a cultivated landscape as a typical garden of exotics from the world's flora, even though the former stands out less clearly as planted.

Will we in the native plant movement be remembered more for saving habitats or for adding to the planted landscape? *Surely the primary business of native plant societies should be to save wild places and their native species, not to add to or promote the planted landscape.* Civilization is busily turning natural landscape into planted landscape at an ever faster pace, and native plant societies should be trying to slow down, not fuel, that process.

Native plant societies are contributing to the clamor for planted landscapes by promoting the planting of native species along roadsides and in other sites being reclaimed or restored. Consequently, the landscape is being homogenized and the *natural* landscape thoroughly compromised, as the line between *the natural* and *the artificial* (planted) is being completely blurred. This practice also provides a copout for those developers who want to clear-cut rather than save fragments of wild land, because if they then landscape with native species, they can claim that they are appropriately mitigating the damage, perhaps even enhancing the environment. Of course, mitigation of damage and disturbance with plantings is often necessary, and the use of native species should be given top consideration. There are many situations, however, where the cheaper and ecologically superior option would be to let grow what naturally would grow, but this option is rarely promoted or chosen.

The growing "plant native" movement has spawned a growing market for native plants and a whole industry to supply them, which is rapidly going from a cottage industry to a big-time business among mainstream nurseries. Because there are always some unscrupulous suppliers who sell wild-collected rather than nursery-grown stock, the larger the industry gets the more endangered wild populations become, especially of specialty groups like the cacti and orchids. Helping to stimulate and supply the market are the native plant societies themselves with their own native plant sales and conferences on landscaping with native plants.

Shouldn't our native plant societies be strong advocates for natural process in the revegetation of land, minimizing intervention and letting nature be nature whenever possible? Active reclamation, beautification, and restoration with or without native species are forms of gardening that add to the planted, not the natural, landscape. Artificial roadside wildflower gardens, beautiful and cheery as they may be, nevertheless stand out as plantings and are unlikely to persist without maintenance, whether or not the species are native locally. Compounding seed mixes for these gardens can pose many controversial choices, and any roadside planting is potentially a risk to the ecological and genetic integrity of nearby natural habitats and native plant populations. Most parcels would revegetate quickly and maintain themselves by natural means from local species and thus retain the proper sense of place and harmony with the surroundings. Unwanted aliens such as tree of heaven (*Ailanthus altissima*) may invade whether or not there has been deliberate planting.

Letting nature take its course through the natural stages of colonization and succession whenever possible would save governments and businesses, and ultimately taxpayers and customers, much money, especially now that the canon, from the highest levels of government down, calls for native species in reclamation and landscaping. This requirement is forcing the U.S. federal government into the expensive business of being its own supplier and maintaining regional stocks across the nation. Wild genetic stocks are being compromised by nursery-propagated stocks, and, increasingly, the genetic origins of native flora are being confused.

Eradication can also become a diversion from the main focus. No one can dispute the growing, costly menace of invasive alien plants with the ever-increasing globalization of civilization. Good-faith efforts must be made to stem the tide, particularly through education. The risk is to be unrealistic, however, letting the fight against invasives get out of proportion. After all, only a fraction of the introduced flora is truly invasive, and not all invasives are aliens. The war on invasives tends to lead unintentionally to a phyto-xenophobia that demonizes *all* nonnative plants in the public mind, painting them all in shades of black. This can be counterproductive for the cause by disillusioning the many wildflower lovers who do not understand the distinction between native and nonnative plants and have always cherished such naturalized wildflowers as Queen Anne's lace (*Daucus carota*). It also can stimulate unenlightened efforts that result in eradicating the wrong species (e.g., sumac [*Rhus*] instead of tree of heaven) or in using herbicides unwisely.

The problem of alien invasives is massive and long-standing. Targeted local campaigns within well-delimited areas (e.g., parks, restricted habitats) can be and have been quite successful. The battles have to be chosen carefully, however. Eradication efforts without clearly limited goals are likely to be an exercise in futility. Trying to undo the past or prevent the inevitable and uncontrollable present and future is as hopeless as trying to carry away the beach one sand grain at a time.

LIFESTYLE ISSUES AND PLANT CONSERVATION

Grassroots conservation comes ultimately to the doorstep of every individual as a personal challenge to protect the environment. Our collective lifestyles, therefore, determine the future of nature around us and, through our consumption, in far-flung places. To stem the tide of habitat conversion and urbanization, we need a breakthrough in reducing our consumptive behavior and a new stewardship ethic that places the intrinsic and intangible values of natural habitat and biodiversity above the market value of the land. The simple secret, of course, is to live with less.

Consumerism promotes gross overconsumption and thus becomes a principal, preventable cause of environmental degradation, habitat loss, and, ultimately, species loss. It feeds and is fed by the throwaway mentality of our time, which manifests itself in numerous ways. From our clothing and household furnishings or equipment to our houses, cars, and recreational toys, all too frequently we tire of our possessions prematurely, before they can serve out their useful life, and we move on to something newer and trendier, further fueling the resource/habitat-eating habit. The spoils of indulgence are filling dumps on land and polluting waters at sea at a sobering rate. On a larger scale, it is the same mentality that gives us, in effect, "throwaway" stores and malls, regularly abandoned to uncertain fates when they lose their edge. A single new shopping mall can exact an awful price from the natural landscape, and redevelopment of existing ones certainly should be mandatory.

Consumption is essential to human existence and welfare, of course, but excessive consumption needlessly destroys the natural estate. Consumption, by nature, strikes multiple blows against habitat and the environment generally, from the extraction of the raw materials to the vast mechanisms and facilities of manufacturing, advertising, marketing, and sales, all of which constantly despoil or eat up land and natural resources. The final blows are struck when the wastes of consumption foul the waters or preempt and contaminate the land.

It is easy to fall prey to simple solutions, feel-good lifestyle tokenisms that may only add to the problem. "Green" products, for example, may indeed represent a better way, but they also may beg more questions than they answer. Separating hype from fact is the first problem, but the root problem—consumption—is not addressed. In fact, "green" can be little more than a marketing ploy to increase sales while making the consumer feel virtuous in further consumption. Regardless of what the marketplace might have us believe, there is no ecologically friendly way to use up the environment. For all who would save nature, the first question should always be, "Do I really need this?"

Commercialization of native plant species, in effect placing a price on them, may jeopardize their wild populations, especially when the species are rare, unless growers can propagate and cultivate the species. In some cases, cultivation may compete directly for rare habitats, as in the case of growing blueberries and especially cranberries. The increasing popularity of natural diet supplements and herbal remedies poses an increasing threat, perhaps even the threat of extinction, to such native plant species as ginseng (*Panax quinquefolius*). Likewise, the growing popularity of native plant gardening may be putting a broad range of native plants at increasing risk of extirpation, at least locally, by private and commercial collectors.

From fake turf and plastic greenery and flowers to whole theme parks, ours is an age of synthetic surrogates for nature. Even the bouquets on our graves are plastic, surely the ultimate mockery in perpetual care. The nature that surrounds us is often so cultured and contrived that, as a society, we have come to accept counterfeit biomes as the real thing, seemingly neither noticing the difference nor caring. Indeed, we actually "restore" the wild as a conservation measure, and obviously this may be the best alternative in some cases. While substitution of the artificial for the natural would seem to spare our native habitats and species and thus be a good thing, in fact it may only cheapen and lead to disrespect for what is real and thus weaken the will to save it as is. The sense of what is real becomes thoroughly deadened as the natural is replaced with the artificial.

More and more, ours is an engineered landscape, and our future seems to lie in the managed commons of our lives—parks, gardens, reconstructed ecosystems, roadsides and meadows with consciously planted wildflowers. Even the parks, those last bastions of preservation, are highly compromised. The keepers keep on developing them, because the public demands ever more recreational amenities, regardless of the cost to the natural domain.

It's on the home front at the grassroots level, especially in the suburbs and urbanizing areas, that the battle to save native habitat and species is being lost every day. Too often in the subdivision, we either manicure and sanitize or trash and pillage our common natural areas, and we saturate our lawns with herbicides and our runoff with poisons. Ignorance, prejudice, and apathy are major foes of the natural realm.

Education, particularly of the younger generations, is the key to changing conservation ethics and lifestyles. People first have to have knowledge before they can care and have the wisdom to do the right thing. The public is largely illiterate about the diversity of species and habitats and their ecological interactions in nature. How can anyone who can't tell one flower or tree from another be expected to care about saving native wildflowers and forests? Yet science education today largely lacks real contact with nature. A child can graduate from high school and even college without ever learning to recognize a single plant or animal. There are many exceptions, of course, and, clearly, even a little nature training can make a dramatic difference in a child's understanding and attitude about the natural environment. Education at all levels must include old-fashioned natural history training, with fieldwork, if there is ever to be nature literacy and conservation concern in the general populace. For this to happen, school administrations must take the need for outdoor laboratories as seriously as they take the need for sports facilities. Such thinking is scarce today in the right places.

Fortunately, numerous examples of field-driven natural history programs for children exist in every region across the country. On a national scale, the Roger Tory Peterson Institute of Natural History in Jamestown, New York, for instance, has a model program. In the Washington, DC, area, the Chesapeake Bay Foundation and the Audubon Naturalist Society are among the organizations well-known for their children's programs (see their Web sites).

LOCAL CITIZEN ACTIVISM AND ENVIRONMENTAL CONSERVATION

Conservation battles are fought on all levels. Some, such as the battle over oil drilling in the Arctic National Wildlife Refuge in Alaska, are waged on the national or even international stage and involve high stakes for all parties and high-profile forces of governments and interest groups. Most battles, however, are local and small-scale, involving local citizens, officials, and politicians, and they are often below the radar of the usual conservation agencies of government and the private sector.

Countless land-use decisions are made at the local level every day across North America. In the aggregate, they have untold consequences for the conservation of nature and the environment. It means that the fate of the environment rests to a large degree in local hands and that local citizen-activists bear perhaps the major responsibility for defending and urging the protection of natural habitats and their plants and animals. For an anonymous, numberless army of volunteers, this is an awesome responsibility, and the importance of activism on the local scene, which takes many forms, can hardly be overstated.

People living in urbanized areas often seem to be so preoccupied with their daily lives and so impermanent in their roots that they show little or no apparent concern for, or motivation to protect, what natural environment is still around them. Added to this is a pervasive ignorance, even fear, of nature. Despite this apparent lack of understanding and concern, however, it quickly becomes clear whenever something comes along to upset the status quo in their environment that most denizens of cities and suburbs count on the continuity of their environment as they have known it.

Once change looms and the moment of truth has arrived, many suddenly take notice and reach out desperately for ways to save what is being threatened, commonplace as that might be. They want someone, anyone, to find their ordinary wayside to be nothing short of a patch of Eden. Unfortunately, there are few neighborhood Edens and no magic fixes, and eleventh-hour panic is usually too late. The die has usually been long since cast behind the scenes in zoning and other legally binding decisions, and more often than not there is no longer any remedy available. Suddenly, the cold reality is that freedom to control one's own surroundings, even if just as a viewshed, is not one of the freedoms of life in the subdivision. Vigilant actions to save habitats have to begin long before the development gauntlet has been thrown down.

As a veteran of many local attempts to save habitats, I know how hard it is to make the case for preservation to decision makers and the public. Thousands of ordinary lives and patches of nature are disappearing from our neighborhoods every year. Seldom does an officially "endangered" or "threatened" species come to the rescue. Arguments must be fashioned more broadly, because extinction comes about in increments, beginning at home with local extirpation within the shadow of our daily lives long before species reach the critical stage that puts them on special protection lists. History has shown that general extinction usually is not perceived until it is too late, as a species dies a thousand deaths across its range before disappearing. Species by species, the biotic bleeding will become a great gathering river that drains away the biotic lifeblood of a continent. The descent into darkness is imperceptibly gradual—like turning out the lights of a vast city one at a time.

Local threats to the environment in an area of rapid urbanization and the perceived needs and forces that cause them are seemingly inexhaustible and perennial. There is virtually no such thing as either complete or permanent victory in the conservation struggle. The concerned activist has to remain ever ready to do battle again without warning. Road-building projects are some of the most divisive in any urbanizing area. Like most development projects, they never die, just fade away for a time and wait for a better day.

The tug-of-war between developers and conservationists epitomizes the now-you-win-it, now-you-lose-it nature of environmental battles in general. The road builders reincarnate and come forward in a new guise, perhaps with much bally-hooed environmental tokenisms, or simply bide their time until the political climate is right. It is a perennial debate, with no clear rights and wrongs, because, of course, everyone needs roads. But whether for good or bad reasons, the natural habitat always seems to lose out.

Conventional wisdom in the realm of environmental protection too often subverts the true conservation of habitats and species. Local planners, constrained by the "proffer" system and/or other statutory environmental conventions, may have no choice but to accept half-measures or trade "an orchard for an apple tree" in the name of environmental safeguards. Consequently, the planning process, far from conserving the biotic environment, may actually license its destruction.

The "environment," conventionally, is defined primarily in physical terms, and the biological factors receive little consideration beyond health concerns. Usually, there are local ordinances, for example, to prevent soil erosion and runoff; protect waterways from excessive flooding, stream-bank erosion, and pollution; safeguard the groundwater from contamination; regulate landfills; and prohibit building in wetlands and on shores and steep slopes. In biological terms, however, aside from some measure of protection afforded fish and aquatic life by the stream ordinances and perhaps specimen trees by a tree ordinance, little if any protection is guaranteed by ordinance for the habitats and the biota. Historically, local jurisdictions have not had effective regulations to protect even the special habitats, such as bogs, much less to protect a critical mass of the various habitats and vegetation types and their biodiversity in the areas of advancing development.

Built-in biological protections may consist of little more than perfunctory mitigations, such as preserving narrow buffers of trees or replanting stripped edges between upland developments and retaining only the slightest of riparian borders along streams in bottomlands. These are not viable habitats. Floodplains, therefore, have little protection; in particular, the fig-leaf riparian borders certainly do not constitute the essential stream-corridor habitats that plants and animals need to survive in viable populations or in passing through during migration. Even wetlands, now largely off limits to development under various federal, state, and local regulations, are routinely violated and destroyed, thanks to ingenious ploys, loopholes, and lax enforcement.

Without statutory protections, every battle for a habitat or species becomes a special case to be argued before the local authorities. It must stand on its own merits, which can seem quite esoteric and inconsequential up against the economic forces of development, without the benefit of a larger context and greater biological/ecological literacy and concern among the general public. This is especially true in the case of a habitat such as a bog, where there may be no charismatic headliner species to wave about and which because of its fragility must remain relatively anonymous and off limits.

A front page article in the *Washington Post*, headlined "Density Limits Only Add to Sprawl," pointed out that the 14-county region ringing Washington, DC, has more protected open space than any other urban region in the country of comparable size; but it questioned the down-zoning generally in the surrounding counties, quoting critics from the planning community who say that it only increases sprawl, which adds to air and water pollution and is wasteful of space for the benefit of the wealthy at the expense of affordable housing (Whoriskey 2003). Saving habitat within the growth matrix apparently was not among the environmental concerns on the minds of the critics.

From the growing concern in recent years about urban sprawl, environmental degradation, and other consequences of traditional development patterns, a promising new philosophy of planning known as "smart growth" has emerged (see Smart Growth; Smart Growth Online). Although guided by a set of principles that embody a strong environmental ethic, "smart growth" is often used as a mantra, without clear indication of how the laudable principles apply, and it can mean different things to different people.

If the principles are followed, growth should be much more environmentally responsible in the future. One of the principles speaks directly to this point: "preserve open space, farmland, natural beauty and critical environmental areas." Another is to "strengthen and direct development towards existing communities." These two principles often conflict with each other, however, and to apply them both will always require hard choices or compromises. Those who are concerned about saving habitats, communities, and species cannot rest easy just because the principles of smart growth are being espoused. Although an essential approach in today's planning environment, smart growth is not a panacea, no matter how "smart" the growth, because from the point of view of nature there is in reality no such thing as "safe" growth.

Planned growth itself, though essential in urbanizing areas, is often the enemy of natural habitat. The premises of planning, supposedly self-evident and incontrovertible, are that planned growth is always better than unplanned growth and that new development should be concentrated where development has already occurred. Such planning can slow the march of growth and put off the day of reckoning for the land beyond the present perimeter of advancing development. Consolidating growth, compelling as it may be for a whole variety of reasons, is not necessarily conservation-friendly, however. Such consolidation leads inevitably to the loss not only of the wonderful informality of America's countryside but also of countless pockets of natural habitat and biodiversity.

Planned growth is deadly efficient in using up space and manages to fill all the natural interstices that in unplanned America were too steep, rocky, or wet for exploitation. Every corner of nature is leveled and snuffed out, and in its place comes sculpted scapes, amenity landscaping, and synthetic greenspace. The value of the real vegetation counts for little, as it is bulldozed and artificial, feel-good, restorative formulas are practiced, such as "a tree planted for every tree destroyed," pretending that the natural value of the planted seedling or sapling is instantly equivalent to the natural heritage of the perhaps centuries-old tree removed. A mall developer near my home boasted to the citizens who complained about his clearcutting that the trees he would plant would be much better than the ones he had just bulldozed. He was either thumbing his nose at the citizens, exposing his ecological ignorance, or both.

Even with smart growth, *consolidated* development will, as it expands outward in all directions, continue like a giant amoeba to engulf and absorb the open space into the gathering urban matrix until there is no more land to swallow and only token natural area is left behind. This approach, at the limit, can yield only unrelenting city from ocean to ocean. Attempts to buck this trend and preserve significant open space *within* the built environment are immediately denounced by development interests and critics of urban sprawl, as Loudoun County, Virginia, the nation's fastest growing county in 2004, has learned since revising its comprehensive land-use plan and enabling zoning ordinances.

Plants and animals do not subscribe to the principles of consolidation in the conventional wisdom of development, and we cannot herd them into reservations or keep driving them ahead of us. They live in their own quirky ways all across the land in habitats large and small. Without some effort to save and leave behind a permanent mosaic of sustainable natural habitats within the built environment as development proceeds across the land, nature is ultimately doomed virtually in its entirety; the land will eventually become nothing but continuous city, and the biodiversity of our land will vanish. An occasional big park here and there certainly will not save the wilds. Local conservation activists bear the responsibility for keeping habitat preservation foremost in the planning and zoning process at the local level and must not trust everything to smart growth.

ACKNOWLEDGMENTS

Over the years I have benefited greatly from the views and experiences of my friends in the Virginia Native Plant Society and especially my colleagues on the society's board of directors. I wish to thank the following individuals for kindly providing me with statistical information about their societies: David G. Anderson, Deborah Dale, Verna Gates, Catherine E. Hovanic, David Johnson, Lorna Konsis, Marion Millin, Bruce Newhouse, Barbara Pryor, Eric Sundell, Susan Thompson, Dana Tucker, M. Ross Waddell, and Kim Zarillo. Thanks are due as well to Nicky Staunton, who, as president of the Virginia Native Plant Society, made the society's library of current issues of other native plant society newsletters available to me; and to Deanne Eversmeyer, for information about the Cullowhee and other regional conferences. I am especially indebted to an anonymous reviewer whose criticisms and suggestions have led to a greatly improved chapter. The views expressed here are my own, however, and those who have helped me should not be held responsible for them.

LITERATURE CITED

Alien Plant Working Group. 2000. Weeds Gone Wild. Plant Conservation Alliance. http://www.nps.gov/plants/alien/bkgd.htm.

Arkansas Native Plant Society. 2000. Vascular Flora of Arkansas Project. http://www.anps.org.

Audubon Naturalist Society. 2002. http://www.audubonnaturalist.org.

California Native Plant Society. 2002. Native Plant Conservation Campaign. http://www.cnps.org/NPCC/.

Chesapeake Bay Foundation. 2002. Save the Bay. http://www.cbf.org/site/PageServer.

DeKing, D., Shonbrun, S., and Slavsky, P., eds. 2001. 100 years. New England Wild Flower 5: 1–28.

Lady Bird Johnson Wildflower Center. 2002. http://www.wildflower.org.

Mehrhoff, L. J. 2001. New England Plant Conservation Program. New England Wild Flower 5:26.

NANPS (North American Native Plant Society). 2002. Associations. North American Native Plant Society. http://www.nanps.org/associations/frame.shtml.

Native Plant Society of British Columbia. 2002. Pacific Northwest Plant Sale. http://www.npsbc.org.

Native Plant Society of Oregon. 2002. About Us. http://www.npsoregon.org/abus.htm.

North Carolina Wild Flower Preservation Society. 2002. http://www.ncwildflower.org.

Plant Conservation Alliance. 2004. Frequently Asked Questions (FAQ). http://www.nps.gov/plants/faq.htm.

Roger Tory Peterson Institute of Natural History. 2002. http://www.rtpi.org.

Shetler, S. G. 1991a. Biological diversity: are we asking the right questions? Pp. 37–43 in Dudley, E. C., ed., The Unity of Evolutionary Biology, vol. 1. Proceedings of the Fourth International Congress of Systematic and Evolutionary Biology. Dioscorides Press, Portland, Oregon.

————. 1991b. Three faces of Eden. Pp. 224–247, 267–268 in Viola, H. J., and Margolis, C., eds., Seeds of Change: A Quincentennial Commemoration. Smithsonian Institution Press, Washington, DC.

————. 2003. Native plant societies and grassroots conservation. Wildflower 19:42–44.

Smart Growth. 2002. About Smart Growth. U.S. Environmental Protection Agency. http://www.epa.gov/smartgrowth/about_sg.htm.

Smart Growth Online. 2002. http://www.smartgrowth.org/about/default.asp?res=800.

2001 Kansas wildflower of the year. 2001. Kansas Wildflower Society 23:47.

VNPS (Virginia Native Plant Society). 2002a. Invasive Alien Plant Species in Virginia. http://www.vnps.org/invasive.html.

————. 2002b. Native Plant Societies. http://www.vnps.org/links.htm.

————. 2002c. Virginia Native Plant Site Registry. http://www.vnps.org/registry2.html.

Walker, M. M. 2002. Native Plant Societies of the United States and Canada. New England Wild Flower Society. http://www.newfs.org/nps.htm.

————. 2003. The early Wild Flower: The Wild Flower Preservation Society of America: a short history. Wildflower 19:31–32.

Washington Native Plant Society. 2001. History and Accomplishments. http://wnps.org/history.html.

Whoriskey, P. 2003. Density limits only add to sprawl: large lots eat up area countryside. The Washington Post, March 9, sect. A.

CONCLUSION

DOCUMENTING AND CONSERVING PLANT DIVERSITY IN THE FUTURE

W. John Kress and Gary A. Krupnick

TO UNDERSTAND the future of plant diversity and its conservation, one must understand the history of humanity's view of nature and the accumulation of knowledge about the natural world. The field of conservation biology can be divided into the components of assessing the extent and distribution of biodiversity and the management of biodiversity. The assessment and documentation of plant diversity can be traced back through many ancient cultures, and certainly the use of plants has varied across time and across continents. In Western cultures, museums, botanical gardens, and herbaria have traditionally been the academic centers for the study of plant life. Many of the early botanical institutions began and flourished in 18th- and 19th-century Europe and North America when new plant discoveries were being made as exploratory expeditions returned from faraway lands. The age of discovery of Nature in turn gave way to the development of new schemes of classifying plants and eventually an understanding of the evolutionary principles responsible for the process of speciation. Natural history museums and botanical gardens today continue to be at the center of modern plant research. Yet as the threat to nature increases, these institutions are expanding their efforts in the conservation of biodiversity while continuing the challenge to discover, describe, and inventory the remaining unknown plant species on earth. How will they complete this inventory, what will they find, and how long will it take to complete?

THE GREAT AGE OF PLANT EXPLORATION AND DISCOVERY: THE 18TH AND 19TH CENTURIES

At the start of the great age of exploration by European naturalists in the 18th and 19th centuries, nature seemed mysterious, immense, and infinite. Expeditions to explore uncharted regions of the world were sent out by governments, monarchies, and wealthy patrons to survey and acquire new lands, to bring back new plant products, such as spices and medicines, and to collect natural history specimens for newly established national museums as well as private collections. Most of the preserved and living specimens of plants that were brought back from Africa, South America, and Asia to museums and botanical gardens by such explorers as Alexander von Humboldt, James Cook, Charles Darwin, Ernest Henry Wilson, Frank Kingdon Ward, and Charles Wilkes, to name only a few, were new to science and to horticulture. Discovery and description of biodiversity proceeded at a pace as if the natural world was limitless, enduring, and permanent (fig. C.1). Botanists at botanical gardens at Kew, Edinburgh, Madrid, Berlin, St. Petersburg, Singapore, and Bogor, at natural history museums in Paris, London, Leiden, and later Washington, as well as universities at Uppsala, Oxford, and Harvard, proceeded at a frenzied pace in describing new species of plants, especially from the tropical regions of the world. The great age of exploration starting in the 18th century resulted in an explosion of discovery and documentation of biodiversity in the 19th and early 20th centuries.

The initial period of global exploration was soon followed by colonization by Europeans of these regions in the Americas, Africa, and Asia at an unprecedented pace. In Europe expanding populations, disease epidemics, economic hardships, and religious persecution sent hundreds of thousands of people to these newly opened regions that promised limitless opportunities and riches. The massive, unspoiled landscapes encountered by European settlers in North America, for example, were viewed as an endless, natural garden to be cultivated and exploited, regardless of the native peoples that inhabited these lands (Shabecoff 1995). Yet at the same time this wilderness was forbidding and frightening to these early colonists. In response to both of these perceptions, limitless bounty but frightening wilderness, the woodlands were felled, crops were planted, and towns and cities grew at the great expense of the natural landscapes. The slow but steady threat to the survival of native species in the New World had begun.

THE AGE OF PLANT CLASSIFICATION AND EVOLUTIONARY ANALYSIS: THE 19TH AND 20TH CENTURIES

As the 19th and 20th centuries progressed through unbridled expansion of human populations throughout the world, biologists began to realize that natural habitats, landscapes, and even species were indeed limited, transitory, and ephemeral. The abundant discoveries of species that started in the 18th century led to an intense period of descriptions and taxonomic analysis in the 19th and 20th centuries (fig. C.1). The tremendous influx of new species being described required an overhaul of the earlier classification system of Linnaeus (1753). Major new classification systems of plants were proposed to incorporate the new discoveries, first by the French botanist de Jussieu (1789), followed by the British taxonomists Bentham and Hooker (1862–1883) and later the Germans Engler and Prantl (1887–1915).

After the turn of the 19th century, Darwin's theory of evolution through natural selection and the developing field of genetics preoccupied a different set of biologists in their investigations of the natural world. In the 1940s and 1950s a significant decrease took place in the description of new species of plants, perhaps as a result of the international effects of World War II (fig. C.1). This decrease of new discoveries was coupled with an increase in the reanalysis of the taxonomic hierarchy and relationship of taxa, to reflect new ideas on the nature of species resulting from the

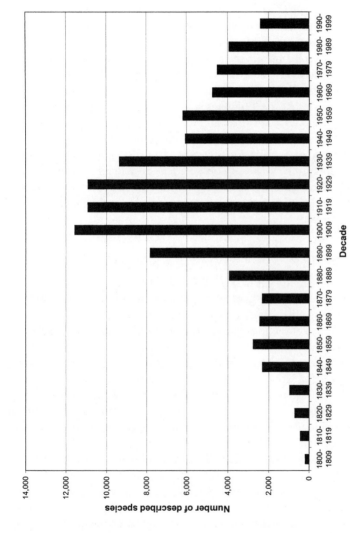

Figure C.1 The discovery and description of new species by decade from 1800 to 1999, based on 93,000 species names in the Type Specimen Register of the U.S. National Herbarium (Smithsonian Institution). Bars indicate total number of species names published per decade. Note that the greatest number of new species represented in the Type Collections at the Smithsonian were described between 1890 and 1939, and that a steady decline has taken place since 1940.

evolutionary synthesis led by biologists Theodosius Dobzhansky, Ernst Mayr, George Gaylord Simpson, and George Ledyard Stebbins Jr. (Mayr and Provine 1980). The intense interest by evolutionary biologists in understanding how species are related to each other, termed phylogenetics, has persisted to the present day. New technological advances using DNA sequence data have revolutionized our concepts on the phylogenetic relationships of plants, and a new classification of flowering plants is gaining wide acceptance (Angiosperm Phylogeny Group 2003).

At the same time that taxonomists and evolutionists were trying to understand the evolution and classification of plants, ecologists and environmentalists were beginning to assess the relationship of people to natural habitats. Aldo Leopold and others in the 1940s were early advocates who clearly saw the threat of unbridled human expansion to natural environments and the species that inhabited them. It was not until the 1970s that most biologists, ecologists, and taxonomists realized that the natural world was under threat and in trouble. In the last three decades of the 20th century, the urgent need to understand and protect the earth's habitats and organisms has resulted in an explosion of new academic programs aimed at studying the environment, professional societies (e.g., the Society for Conservation Biology) and local activist groups (see chapter 14) to unite scientists and citizens in taking action, and even legislation (e.g., the Endangered Species Act) to turn concern for the environment into law.

THE INTERNATIONALIZATION OF BIODIVERSITY

At the turn of the 21st century it has become clear to biologists, conservationists, and a significant segment of the general public that a major extinction of plants, animals, and microorganisms caused by human activities is not only possible but probable unless immediate action is taken. This threat has resulted in the creation of many local, regional, and national governmental and nongovernmental organizations devoted to halting and reversing these activities. One international response to this imminent extinction as a result of increasing degradation of the environment was the Convention on Biological Diversity (CBD) authorized at the Earth Summit in Rio de Janeiro, Brazil, in 1992. This treaty initiated a revolution in the value placed on biodiversity and the intellectual property rights attached to nature. According to the CBD, biodiversity should be conserved and sustainably used, and its benefits shared among all parties. Since the Rio Earth Summit 188 countries have signed the treaty, and a host of additional resolutions, national strategies, and work plans have been developed and implemented (see chapter 13). Although not the initial intent, one result of the CBD was that nature, like other commercial commodities, has now become internationalized. The ownership of nature, whether it be for natural product development through "bioprospecting," establishing logging concessions on indigenous people's land, or collecting plant specimens for scientific study, is now a matter of worldwide concern and law.

THE FINAL AGE OF PLANT EXPLORATION AND DISCOVERY: THE 21ST CENTURY

The globalization of nature and biodiversity coupled with increased species extinction has significantly changed the way that modern-day plant explorers and taxonomists pursue their activities. As habitat destruction accelerates, the pace of discovery, identification, and description of new species of plants has not speeded up. The number of newly described species of plants significantly decreased in the 1980s and 1990s over the previous four decades (fig. C.1). Unlike the predominant perceptions of the 17th and 18th centuries that nature and species were infinite and limitless, we now know this is not true. Estimates of the number of plant species currently present on earth range from 220,000 to over 420,000 (Govaerts 2001; Scotland and Wortley 2003). By extrapolation from what we have already described and what we estimate to be present, it is possible that at least 10% of all vascular-plant species are still to be discovered and described (J. Kress and E. Farr, unpublished data). This number suggests that a considerable amount of work still needs to be done by botanists to find undescribed species.

The recent marriage of biology and advanced technology is leading to the development of novel tools with the potential to transform current methods of plant collecting, such as image-recognition software to be employed in electronic field guides and on-the-spot, rapid DNA sequencing, termed DNA bar-coding (see foreword), for species identification. We envision that 21st-century naturalists will be equipped with palm-top and wearable computers, global positioning system (GPS) receivers, and Web-based satellite communication. These explorers will more swiftly comb the remaining unexplored habitats of the earth, identifying and recording the characters and habitats of plant species not yet known to science. For example, through remote wireless communication, field botanists will be able to immediately compare their newly collected plants with type specimens and reference collections archived and digitized in museums thousands of miles away. The information gathered by these botanists will be sent with the speed of the Internet to their colleagues back in the lab, where the genetic composition and phylogenetic position of each new species will be rapidly determined. The habitat data will be modeled with unparalleled speed and accuracy by ever more powerful computers to determine the place of each species in its respective ecosystem. Moreover, the biochemical constituents of each species will be automatically screened and analyzed for any compounds that may be of benefit to society.

These technological dreams are already being converted into reality by designing new plant explorers of the future. The documentation of the remaining species of plants with the aid of these new tools will provide a solid basis for the precise identification of the species-rich areas of the world for immediate assessment, conservation, and protection. The social, economic, political, and technological changes of the last few decades have ushered in the final age of plant exploration and conservation in the 21st century. We predict that the last new species of plants on the earth will be discovered and described by the year 2040. Natural history museums will continue to play a central role in both the documentation and the conservation of biodiversity.

LITERATURE CITED

Angiosperm Phylogeny Group. 2003. An update of the Angiosperm Phylogeny Group classification for the orders and families of flowering plants: APG II. Botanical Journal of the Linnean Society 141:399–436.

Bentham, G., and Hooker, J. D. 1862–1883. Genera Plantarum. A. Black, and Reeve and Co., London.

Engler, A., and Prantl, K. 1887–1915. Die Naturlichen Pflanzenfamilien. Wilhelm Engelmann, Leipzig.

Govaerts, R. 2001. How many species of seed plants are there? Taxon 50:1085–1090.

de Jussieu, A. L. 1789. Genera Plantarum Secundum Ordines Naturales Disposita. Paris.

Linnaeus, C. 1753. Species Plantarum. Stockholm.

Mayr, E., and Provine, W. B. 1980. The Evolutionary Synthesis: Perspectives on the Unification of Biology. Harvard University Press, Cambridge.

Scotland, R. W., and Wortley, A. H. 2003. How many species of seed plants are there? Taxon 52:101–104.

Shabecoff, P. 1993. A fierce green fire: the American environmental movement. Hill and Wang, New York.

CONTRIBUTORS

Mones S. Abu-Asab
Section of Ultrastructural Pathology
Laboratory of Pathology
National Cancer Institute
Bethesda, MD 20892
mones@box-m.nih.gov

Pedro Acevedo-Rodríguez
Department of Botany, MRC-166
National Museum of Natural History
PO Box 37012
Smithsonian Institution
Washington, DC 20013–7012
acevedop@si.edu

Walter H. Adey
Department of Botany, MRC-166
National Museum of Natural History
PO Box 37012
Smithsonian Institution
Washington, DC 20013–7012
adeyw@si.edu

William S. Alverson
Environmental and Conservation Programs
The Field Museum
1400 South Lakeshore Drive
Chicago, IL 60605–2496
alverson@fieldmuseum.org

Wilhelm Barthlott
Nees-Institut für Biodiversität der Pflanzen
Rheinische Friedrich-Wilhelms-Universität
Meckenheimer Allee 170
53115 Bonn
Germany
barthlott@uni-bonn.de

Richard M. Bateman
Natural History Museum
Botany
Cromwell Road
London SW7 5BD
United Kingdom
r.bateman@nhm.ac.uk

Paul E. Berry
Department of Botany
University of Wisconsin-Madison
132 Birge Hall, 430 Lincoln Drive
Madison, WI 53706–1381
peberry@facstaff.wisc.edu

Stephen Blackmore
Royal Botanic Garden Edinburgh
20a Inverleith Row
Edinburgh EH3 5LR
Scotland
s.blackmore@rbge.org.uk

Emilio M. Bruna
University of Florida–WEC
PO Box 110430
Gainesville, FL 32611-0430
BrunaE@wec.ufl.edu

John L. Clark
Department of Botany, MRC-166
National Museum of Natural History
PO Box 37012
Smithsonian Institution
Washington, DC 20013–7012
clarkjo@si.edu

Paula T. DePriest
Department of Botany, MRC-166
National Museum of Natural History
PO Box 37012
Smithsonian Institution
Washington, DC 20013–7012
depriesp@si.edu

William A. DiMichele
Department of Paleobiology, MRC-121
Smithsonian Institution NMNH
10th and Constitution Avenue NW
Washington, DC 20560
dimichel@si.edu

Laurence J. Dorr
Department of Botany, MRC-166
National Museum of Natural History
PO Box 37012
Smithsonian Institution
Washington, DC 20013–7012
dorrl@si.edu

Robert B. Faden
Department of Botany, MRC-166
National Museum of Natural History
PO Box 37012
Smithsonian Institution
Washington, DC 20013–7012
fadenr@si.edu

Maria A. Faust
Department of Botany
National Museum of Natural History
Smithsonian Institution
4210 Silver Hill Road
Suitland, MD 20746
faustm@si.edu

Vicki A. Funk
Department of Botany, MRC-166
National Museum of Natural History
PO Box 37012
Smithsonian Institution
Washington, DC 20013–7012
funkv@si.edu

Carol Horvitz
Department of Biology
University of Miami
Coral Gables, FL 33124–0421
carolhorvitz@miami.edu

Daniel H. Janzen
University of Pennsylvania
415 South University Avenue
Philadelphia, PA 19104
djanzen@sas.upenn.edu

M. Alejandra Jaramillo
Laboratorio de Biologia Molecular de Plantas
Jardim Botânico do Rio de Janeiro
Rua Pacheco Leão 915
22460-030 Rio de Janeiro, RJ
Brazil
jaramillo@bioqmed.ufrj.br

Gerold Kier
Nees-Institut für Biodiversität der Pflanzen
Rheinische Friedrich-Wilhelms-Universität
Meckenheimer Allee 170
53115 Bonn
Germany
kier@uni-bonn.de

W. John Kress
Department of Botany, MRC-166
National Museum of Natural History
PO Box 37012
Smithsonian Institution
Washington, DC 20013–7012
kressj@si.edu

Gary A. Krupnick
Department of Botany, MRC-166
National Museum of Natural History
PO Box 37012
Smithsonian Institution
Washington, DC 20013–7012
krupnick@si.edu

Diane S. Littler
Department of Botany, MRC-166
National Museum of Natural History
PO Box 37012
Smithsonian Institution
Washington, DC 20013–7012
littlerd@si.edu

Mark M. Littler
Department of Botany, MRC-166
National Museum of Natural History
PO Box 37012
Smithsonian Institution
Washington, DC 20013–7012
littlerm@si.edu

Ai-Zhong Liu
Biological Sciences Department
Vanderbilt University
Station B 351634
2301 Vanderbilt Place
Nashville, TN 37235-1634
aizhong.liu@vanderbilt.edu

Carrie L. McCracken
Laboratories of Analytical Biology
Museum Support Center
Smithsonian Institution

4210 Silver Hill Road
Suitland, MD 20746
cmccrack@lab.si.edu

Denise Mix
Department of Botany, MRC-166
National Museum of Natural History
PO Box 37012
Smithsonian Institution
Washington, DC 20013–7012
mixden@si.edu

Jens Mutke
Nees-Institut für Biodiversität der Pflanzen
Rheinische Friedrich-Wilhelms-Universität
Meckenheimer Allee 170
53115 Bonn
Germany
jens.mutke@uni-bonn.de

S. Miguel Niño
BioCentro, UNELLEZ
Mesa de Cavacas
Edo Portuguesa
Venezuela 3323
smiguel@cantv.net

James N. Norris
Department of Botany, MRC-166
National Museum of Natural History
PO Box 37012
Smithsonian Institution
Washington, DC 20013–7012
norrisj@si.edu

Sylvia S. Orli
Department of Botany, MRC-166
National Museum of Natural History
PO Box 37012
Smithsonian Institution
Washington, DC 20013–7012
stones@si.edu

Paul M. Peterson
Department of Botany, MRC-166
National Museum of Natural History
PO Box 37012
Smithsonian Institution
Washington, DC 20013–7012
peterson@si.edu

Jessica Poulin
Department of Ecology and Evolutionary
 Biology
University of California, Irvine
321 Steinhaus Hall
Irvine, CA 92697
jpoulin@uci.edu

Jonathan Price
Department of Botany, MRC-166
National Museum of Natural History
PO Box 37012
Smithsonian Institution
Washington, DC 20013–7012
pricejo@si.edu

Werner Rauh
Deceased

Harold E. Robinson
Department of Botany, MRC-166
National Museum of Natural History
PO Box 37012
Smithsonian Institution
Washington, DC 20013–7012
robinsoh@si.edu

Ann Sakai
Department of Ecology and Evolutionary
 Biology
University of California, Irvine
321 Steinhaus Hall
Irvine, CA 92697
aksakai@uci.edu

Stanwyn G. Shetler
Department of Botany, MRC-166
National Museum of Natural History
PO Box 37012
Smithsonian Institution
Washington, DC 20013–7012
shetlers@si.edu

Laurence E. Skog
Department of Botany, MRC-166
National Museum of Natural History
PO Box 37012
Smithsonian Institution
Washington, DC 20013–7012
skogl@si.edu.

Bruce A. Stein
NatureServe
1101 Wilson Boulevard, 15th Floor
Arlington, VA 22209
bruce_stein@natureserve.org

Basil Stergios
BioCentro, UNELLEZ
Mesa de Cavacas
Edo Portuguesa
Venezuela 3323
basilven@cantv.net

Kerry ten Kate
Insight Investment
33 Old Broad Street
London EC2N 1HZ
United Kingdom
kerrytenkate@hotmail.com

Warren L. Wagner
Department of Botany, MRC-166
National Museum of Natural History
PO Box 37012
Smithsonian Institution
Washington, DC 20013–7012
wagnerw@si.edu

Dieter C. Wasshausen
Department of Botany, MRC-166
National Museum of Natural History
PO Box 37012
Smithsonian Institution
Washington, DC 20013–7012
wasshaud@si.edu

Stephen Weller
Department of Ecology and Evolutionary
  Biology
321 Steinhaus Hall
University of California, Irvine
Irvine, CA 92697
sgweller@uci.edu

Scott Wing
Department of Paleobiology
National Museum of Natural History
PO Box 37012
Smithsonian Institution
Washington, DC 20013–7012
wings@si.edu

Rebecca Yahr
Duke University
Department of Biology
Biological Sciences Building, Room 139
Science Dr.
Durham, NC 27708
Ry2@duke.edu

Elizabeth A. Zimmer
Department of Botany
National Museum of Natural History
Laboratories of Analytical Biology
Museum Support Center
Smithsonian Institution
4210 Silver Hill Road
Suitland, MD 20746
zimmer@si.edu

INDEX

Note: Page numbers in italics refer to figures and tables.

Cheirodendron, 82
Chelidonium majus, 188
Cherry Blossom Festival, 192
Chesapeake Bay Foundation, 307
chestnut blight (*Endothia parasitica*), 177
Chicago, 247
Chichorium intybus, 188
Chicxulub impact, 52
Chile, 121, 225
China, 8, 16, 18, 20, 22, 50, 86–88, 108, 112, 125, *pl. P.1, pl. 4.4*
Chinese Academy of Sciences, 261
Chionanthus virginicus, 188
Chionopappus, 119
Chiribiquete-Araracuara-Cahuinarí (Colombia), 17
Chirita, 125
Chiropotes, 212
Chloridoideae, 106, *pl. 5.4*
Chlorophyta, 26, 271
chlorophytes, 276. *See also under* algae
Chocó region, 226, 239
Chocó-Costa Rica Center, 17
Chondria, 158
Chondrus crispus, 162, 163
Chonopetalum, 130
Chordaria, 163
Chromolaena odorata, 122
Chrysalidocarpus decipiens, 153
Chrysanthemum, 121
Chrysogonum virginianum, 188
Chrysophyllum, 73
Chrysymenia, 159
Chuquisaca region, 250
Chusquea angustifolia, 73
Chytranthus, 129
Ciceronia, 122
Cichorieae, 119
Cichorioideae, 115, 117–18, *117*
Cichorium, 119
ciliates, 95
Cineraria, 120
cineraria, horticultural (*Pericallis*), 120
Cinna, 74
Circaeaster agrestis, 87
Cirsium, 118
CITES, 247, 286, *290*
Citrus, 250; canker, 176; plantations, 98, 176
Cladonia, 97; *evansii,* 101; *perforate,* 97–102, *pl. 5.2; subtenuis,* 101
Cladoniaceae, *pl. 5.2*
Cladophora, 276; *fascicularis, 157*

Cladoxylales, 6
Clarkia: amoena, 198; franciscana, 198, 199; rubicunda, 198
Clathromorphum circumscriptum, 163
Clavija, 74
Claytonia virginica, 188
cleistogamous flowers, 197
climate change, 8–9, 44, 51, 104, 185
Clusia, 73
Clusiaceae, 73, 153
Cnidaria, 269
cnidarian corals, 263
Coastal Forests of Tanzania/Kenya, 84
coca plantations, 114
Coccolithophoraceae, 26
Cocha Cashu (Peru), *213*
cocklebur (*Xanthium* spp.), 122
coconut cultivation, 80
cod (*Gadus morhua*), 161, 165–68
Codium, 168
Codonanthe crassifolia, 125
Codonopsis convolvulacea var. forrestosa, 87
Coelothrix, 158, 159
coffee (*Coffea arabica*), xii, 74, 75, 114
collections: bias of, 74, 210; in botanic gardens, xv, 258–59, *290;* difficulty in, 94; in herbaria and museums, ix–xii, xv, 88, 97, *212,* 315, 317; historical, 215; incomplete, 210, 214; private, 313; natural history, 209–18, *209;* uses of, 210–18; rare, 119, 129, 130. *See also* specimens
College Park (Maryland), *190, 191*
Colombia, 17, 70, 76, 103, 125, 226, 239
Colombia *Cinchona* missions, xv
colonization, 58, 82–84, 100, 106, 177, 182, 228, 304, 314
Colorado, 52
Colpomenia, 158
Columnea, 125; *sanguinea,* 125
Commelina, 108–11; *benghalensis,* 111; *echinosperma,* 111; *eckloniana,* 111; *latifolia,* 111; *nairobiensis,* 109–11, *pl. 5.5*
Commelinaceae, 108–11, *pl. 5.5*
Commidendron, 120
Committee on Recently Extinct Organisms (CREO), 56
common chickweed (*Stellaria media*), 192
competition, 6, 9, 31, 44, 48, 84, 176, 179–82, 228, 256, 263–69, *264,* 270–75
complementarity, 209, 210, 217
Compositae, 115–24, 245
composites, 10–11, 245

Middle East, 107, 118, 119
migrations, 23, 70, 166, 196, 219, 301, 302, 309
Millepora, 159
Mimulus, 240
mining, 79, 85, 78, 120, 150
minks, 165
Miocene, 32, 148
Missouri, 103
Missouri Botanical Garden, ix
mitigations, 308
Modiolus modiolus, 162
molecular: biology, 27, 84, 198, 200; genetics,
    236–41; phylogenetics, 10, 82, 226, 229, 237
Molecular Ecology, journal, 240
mollusks, 31
Molokai, 56
Monarrhenus, 120
monocotyledons, 245, 271
Monogerion, 122
Monostroma, 163
Monstera, 74
Montanoa, 124
Montserrat, 148
Moquineae, 119
Morona-Santiago (Ecuador), 123
Morondava (Madagascar), 153
mosses, 15–16, 51, 53, 102–4, 242, 245, *pl. 5.3*. See
    also bryophytes
Moyobamba (Peru), *213*
Mozambique, 151–52
Mugil cephalus, 276
Mugilidae, 270
mullet, 270, 273–76
Munnozia luyensis, 123
Murdannia, 108; *clarkeana*, 109–11
museums, xv, 219, 243, 313, 314, 317, 318. See *also*
    specific museums
mussels, 57, *58*
Mustela macrodon, 164
mutation, 180, 194, 200
Mutisieae, 117–18, *117*, 123–24
mutisioid group, 117–18, *117*
Myanmar, 86–88, *pl. P.1*
Myanmar Forest Department, 88
Myopordon, 118
Myrtaceae, 78
Mytilus edulis, 162

Nairobi (Kenya), 109–11, 259, *pl. P.1*
Nairobi National Park (Kenya), 111
Namaqualand, 17, 118, 124
Namib Desert, 20

Namibia, 120
Nannostomus, 212
Napo River (Ecuador), 17
National Botanical Institute (South Africa), 243
National Institute for Amazonian Research, 142
National Museum of Natural History, 88, 211
Native Plant Conservation Campaign, 300
Native Plant Conservation Initiative, 295, 300
native plant societies, 295–304. See *also specific
    native plant societies*
Native Plant Society of British Columbia, 295,
    299
Native Plant Society of Oregon, 300
Natural Heritage programs, 297
natural monuments, 78. See *also* parks; pro-
    tected areas; reserves; nature
natural selection, 195, 314
naturalized species. See invasive species
Nature Conservancy, The, 243, 246, 297
NatureServe, 55–57
Nautilocalyx melittifolius, 125
Navicula, 157
near threatened species, 244
nematodes, 95
Neodypsis decaryi, 155
Neogoniolithon, 159
Neo-Indonesians, 153
Neojeffreya, 120
Neosharpiella, 104
Neotropical Lowlands Research Program, 211
Neotropics, 15, 18, 24, 25, 120, 124, 125, 142, 249
Nephelium: lappaceum, 128; *ramboutan-ake*, 128
Nereocystis, 163
New Caledonia, 84
New England, 168, 303
New England Botanical Club, 296
New England Plant Conservation Program,
    300
New England Wild Flower Society (NEWFS),
    295–98, 300
New Guinea, 17, 18, 20, 82, 221, *pl. 10.5*
New Mexico, 52
New River Triangle (Guyana), 77
New York, 295, 306
New York Botanical Garden, 243, 295
New Zealand, 58, 84, 120
Newfoundland, 163, 165, 166, 168, *pl. 6.4*
niger seed (*Guizotia* spp.), 122
nitrogen oxides, 101
Niue Islands, 80
non-governmental organizations, 151, 259, 261,
    292, 297, 316